FLUID PR   S

Alan Vardy
Professor of Civil Engineering
University of Dundee

McGRAW-HILL BOOK COMPANY

London · New York · St Louis · San Francisco · Auckland · Bogotá
Guatemala · Hamburg · Lisbon · Madrid · Mexico · Montreal
New Delhi · Panama · Paris · San Juan · São Paulo · Singapore
Sydney · Tokyo · Toronto

Published by
McGRAW-HILL Book Company (UK) Limited
Shoppenhangers Road, Maidenhead, Berkshire, SL6 2QL, England
Telephone 0628 23432
Fax 0628 35895

British Library Cataloguing in Publication Data
Vardy, Alan E.,
  Fluid principles.
  1. Fluids. Dynamics
  I. Title
  532′.05
  ISBN 0-07-707205-7

Library of Congress Cataloging-in-Publication Data
Vardy, Alan E., 1945–
  Fluid principles/Alan E. Vardy.
    p. cm.
  Includes bibliographical references (p.   ).
  ISBN 0-07-707205-7
  1. Fluid dynamics. I. Title.
  QA911.V37 1990
  532—dc20
                         89-28557
                            CIP

1234RC 9210

Typeset by Vision Typesetting, Manchester
Printed by Clays Ltd, St Ives plc

To my family

CONTENTS

KEY CONTRIBUTORS TO FLUID DYNAMICS

Archimedes	(*c.* 287–212 BC)	Greek philosopher
Pascal, Blaise	(1623–1662)	French philosopher
Newton, Isaac	(1642–1727)	British mathematician
Bernoulli, Daniel	(1700–1782)	Swiss mathematician
Euler, Leonhard	(1707–1783)	Swiss mathematician
d'Alembert, Jean la Ronde	(1717–1783)	French mathematician
Chézy, Antoine	(1718–1798)	French engineer
Laplace, Pierre	(1749–1827)	French mathematician
Navier, Louis	(1785–1836)	French engineer
Carnot, Nicolas	(1796–1832)	French physicist
Hagen, Gotthilf	(1797–1884)	German engineer
Poiseuille, Jean Louis	(1799–1869)	French physiologist
Magnus, Heinrich	(1802–1870)	German physicist
Darcy, Henri	(1803–1858)	French engineer
Weisbach, Julius	(1806–1871)	German engineer
Froude, William	(1810–1879)	British naval architect
Francis, James	(1815–1892)	American engineer
Manning, Robert	(1816–1897)	Irish engineer
Stokes, George	(1819–1903)	British mathematician
Kelvin, Lord	(1824–1907)	British mathematician
Pelton, Lester	(1829–1908)	American engineer
Mach, Ernst	(1838–1916)	Austrian physicist
Reynolds, Osborne	(1842–1912)	British academic
Boussinesq, Joseph	(1842–1929)	French mathematician
de Laval, Carl	(1845–1913)	Swedish engineer

Joukowski, Nikolai	(1847–1921)	Russian aerodynamicist
Buckingham, Edgar	(1867–1940)	American physicist
Kutta, Wilhelm	(1867–1944)	German mathematician
Weber, Moritz	(1871–1951)	German naval architect
Prandtl, Ludwig	(1875–1953)	German engineer
Moody, Lewis	(1880–1953)	American engineer
von Kármán, Theodore	(1881–1963)	Hungarian engineer
Blasius, Heinrich	(1883–1970)	German academic
Nikuradse, Johann	(1894–1979)	German engineer
White, Cedric	(1898–)	British engineer
Schlichting, Hermann	(1907–)	German academic
Colebrook, Cyril	(1910–)	British engineer

PREFACE

Textbooks inevitably make greater demands on readers than is customary for a good novel. Nevertheless, they can give intense satisfaction if used flexibly and I hope that a few hours with this book will bring enjoyment and challenge. Like anyone else with a burning interest in a particular subject, I derive my true pleasure from sharing my delight—which in this instance is Nature's wonderful world of fluid flows.

This book is first and foremost a teaching text for students in the early stages of higher education reading for diplomas and degrees. However, it is structured to make it accessible to non-specialists and research students alike. The former should read each chapter only as far as their mathematical ability (or interest) permits, and they should find this sufficient to convey the principal ideas. The latter might like to do similarly and then to sweep through the text once again, picking up whatever mathematical derivations they need. All first-time readers are recommended to skip Chapters Two to Five.

Most people should find something of interest in the worked examples, of which there are about a hundred in all. I have tried to include something for everyone, but the selection is inevitably biased towards my own particular interests. In all cases, supplementary comments follow the worked solutions, relating the examples to the real world and to engineering practice. With luck, the examples will stimulate interest, amusement and perhaps even plain disbelief. I hope the latter is never justified, but authors are only human so fallibility should not be ruled out.

If the book turns out to be as free from faults as the original typescript was from typing errors, I shall be well satisfied. I am wholly responsible for the former; Dorothy McCabe is responsible for the absence of the latter. If this preface serves no other useful purpose, it at least affords me the opportunity to express my heartfelt thanks to her and to the surprisingly large number of people who have helped in one way or another with the preparation of this book. Technical contributions range from innocent questions from students to detailed criticisms from reviewers and editors. Practical support has come from typists, colleagues, family and friends.

It is not feasible to mention everyone by name, but I cannot resist acknowledging the huge debt that I owe to my teacher, supervisor and much loved friend, John Fox. His infectious enthusiasm and his exceptional imagination continue to stimulate my curiosity, wonder and fascination in the enchanting behaviour of fluids.

Dundee   Alan Vardy

NOMENCLATURE

a	cross-sectional area of flow
A	cross-sectional area of object
b	breadth
B	Bernoulli sum
c	sonic speed
c_p	specific heat capacity at constant pressure
c_v	specific heat capacity at constant density
C	(i) coefficient
	(ii) couple
	(iii) constant
d	depth
D	diameter
e	specific total energy
E	Young's modulus
E	total energy
f	skin friction coefficient based on mean velocity
F	force
Fr	Froude number
g	gravitational acceleration
h	(i) pressure head
	(ii) boundary layer thickness
h^*	piezometric head
h	specific enthalpy
H	height
H	enthalpy
I	(i) second moment of area
	(ii) moment of inertia

J	angular momentum
k	(i) constant
	(ii) energy loss coefficient
k_s	roughness size
K	bulk modulus of elasticity
l	length
L	length
$[L]$	length dimension
m	mass
M	linear momentum
M	molecular weight
Ma	Mach number
$[M]$	mass dimension
n	natural coordinate (normal to streamlines)
N	(i) rotational speed
	(ii) polytropic exponent
Nu	Nusselt number
p	pressure
p^*	piezometric pressure
Pr	Prandtl number
$\dot{q}$	heat flux per unit length
O	volumetric flow rate
Q	heat
r	radial distance
R	(i) hydraulic radius
	(ii) radius of curvature
Re	Reynolds number
R	gas constant
s	natural coordinate (along a streamline)
s	specific entropy
S	entropy
St	(i) Strouhal number
	(ii) Stanton number
t	time
T	(i) torque
	(ii) time period
$[T]$	time dimension
T	temperature
u	velocity component
u	specific internal energy
U	internal energy
v	velocity component
V	velocity

V volume
w (i) specific weight
      (ii) velocity component
$\dot{w}$ rate of work per unit length
W weight
We Weber number
W work
x, y, z Cartesian coordinates
z elevation
X, Y, Z body forces per unit mass

α (i) slope angle
      (ii) kinetic energy flux coefficient
β momentum flux coefficient
γ shear strain
γ ratio of specific heats
Γ circulation
δ elemental quantity
ε kinematic eddy viscosity
ζ vorticity component about z-axis
η (i) vorticity component about y-axis
      (ii) constant defined in Eq. (5–71)
θ angle
$[\Theta]$ temperature dimension
λ skin friction coefficient
λ thermal conductivity
μ dynamic (absolute) viscosity
ν kinematic viscosity
ξ vorticity component about x-axis
π ratio of circumference: diameter of circle
Π dimensionless group
ρ mass density
σ (i) normal stress
      (ii) surface tension coefficient
τ shear stress
φ angle
ψ stream function
Ω rate of rotation

Suffices (with different meanings from above)

AT   atmospheric
D   drag

l   laminar
t   turbulent
w   wall

Superscripts

$^{-}$ mean value
$\cdot$ rate ($\partial/\partial t$)
$'$ fluctuating component

ONE

LAWS OF NATURE

1-1 THE JOY AND THE CHALLENGE

Fluid motions hold a fascination for many of us. Our ideas about scenic beauty lead us to fill our picture calendars with views of waterfalls, rivers, lakes and seas. We stand at the seashore and spend hours watching waves breaking on the rocks. We decorate our city centres with fountains, our gardens with fishponds, and our homes with liquid-filled lamps. Children delight in playing with soap bubbles, with toy yachts, and with kites and paper aeroplanes. They build dams in country streams and moats around seaside sandcastles. Adults share these pleasures, though often on a grander scale, using sea-going yachts and hang-gliders for example. Swimming is an international pastime, and many sports—golf, baseball, cricket, tennis, etc.—involve fluid mechanics phenomena causing balls to swerve in flight, though not always intentionally.

Many engineers are also fascinated by fluid motions, but they have additional reasons for their interest. It is not sufficient to stand in wonder of the sea and to marvel at the beauty of natural watercourses. Instead, ways must be found to adapt these phenomena for the good of the community. Winds, tides and waves are potential sources of power. Water deposited on hillsides as rain and snow is vital to our domestic needs. Rivers and oceans are used for transportation.

Sometimes the engineers' task is to pit their wits *against* nature. Walls and harbours are needed to protect coastlines and vessels from the destructive power of the sea. Bridges and tall buildings must withstand gale-force winds. Rivers require frequent dredging to prevent the closing of navigable channels by the deposition of silt. Vehicles must be shaped to move through air with minimum drag. Lubrication must be provided to reduce friction and wear in metal surfaces moving past each other at very high speeds.

A great deal of ingenuity is required to achieve these aims in an economical manner. For example, reservoirs must be sufficiently large to store the quantities of water needed for water supply or for a hydroelectric power station, but the associated dams must not be of excessive size because of their enormous capital cost. Aerogenerators must be slender to extract power efficiently from both gentle breezes and strong winds, but they must also be structurally safe in a gale.

Not all engineering is carried out on a large scale. While some of us make detailed studies of wave forces on offshore structures and of lift generating airflows around aircraft wings, others are concerned with the lubrication of bearings, with the mixing of fuel and air streams in combustion chambers and with blood flows in arteries and capillaries.

A wide range of speeds can be encountered, supersonic speeds being commonly developed in gas flows whereas liquid velocities rarely reach even 50 m/s and are usually of an order of magnitude smaller than this. Enormous ranges of temperature and pressure are also found. Superconducting magnets contain liquid helium at about $-269\,°C$ while temperatures in combustion chambers may be as high as $3000\,°C$. The pressures at the inlet to a steam turbine may be about 150 atmospheres whereas the pressure fluctuations involved in ordinary speech are more than a million times smaller.

Philosophers faced a formidable task when they first attempted to find a sense of order in this seemingly complex environment. As long ago as the third century BC, Archimedes understood the behaviour of stationary liquids almost as well as we do today. However, nearly two thousand years were to pass before significant advances were made in the study of moving fluids. Even today, theoretical developments are never accepted until they have been confirmed by experimental observations. Nevertheless, knowledge is now sufficiently far advanced for many engineering designs to be accomplished routinely using established theoretical procedures. It is to these procedures that attention is now directed, and we begin with a discussion of the various types of flow that can be encountered.

1-2 CLASSIFICATION OF FLUID FLOWS

Fluid flows may be classified in many ways. The importance of each classification depends upon the particular application and it is not necessary to list all possibilities. Instead, some of the most commonly encountered terms are introduced.

Steady and unsteady It is often useful to distinguish between *steady* and *unsteady* flows, the former being characterized by conditions that do not change with time at any position.

The overwhelming majority of real flows are unsteady, but we frequently neglect the unsteadiness and concentrate on the steady contribution to the flow when this is easier to analyse. For example, the effects of random gusts of wind might be ignored in a preliminary evaluation of lift forces on an aircraft. Similarly, the pulsatile nature of blood flows is usually neglected when studying flows through capillaries.

When unsteadiness cannot sensibly be ignored, we often go to the opposite extreme and try to neglect complications arising from background steady flows. Common sense tells us that the propagation of sound waves in a room is not significantly influenced by convection currents induced by central heating systems, for instance. Likewise, the behaviour of pressure waves induced by slamming a valve in a pipe network or by vibrating the pipe supports is not much influenced by steady flows along the pipes.

Uniform and non-uniform In pipes and channels, the fluid motion is constrained by geometrical boundaries to be predominantly parallel to the sides. When the conditions at all successive cross-sections are identical at any instant, the flow is termed *uniform*. Otherwise it is described as *non-uniform.*

Notice that this classification is complementary to the steady–unsteady classification which deals with temporal changes. For example, the flow along a conical diffuser might be steady but non-uniform whereas an oscillating water column in a U-tube is unsteady but uniform (Fig. 1-1).

Compressible and incompressible All *fluids* are compressible to a greater or lesser extent. However, fluid *flows* are conventionally described as compressible only when the associated pressure or temperature changes are sufficiently large to cause significant changes in density. For example, isothermal flows in long pipelines connecting offshore gas fields to onshore users are described as compressible. So are airflows past supersonic aircraft with all the associated shock waves. In contrast, steady water flows in distribution systems are regarded as incompressible, as is seepage under a dam or airflow past a building.

In most unsteady flows, compressibility has an important role. It limits the magnitude of the pressure disturbances that can be caused by slamming a valve for instance. However, gently accelerating liquid flows can usually be treated as incompressible.

Internal and external In some flows, such as those in pipes and channels, the fluid is contained within well-defined boundaries. These so-called *internal* flows contrast sharply with *external* flows where the fluid of interest is wholly

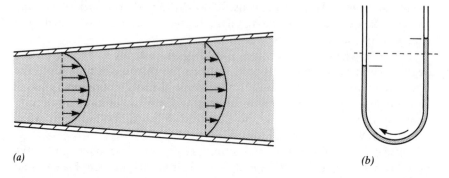

Figure 1-1 Examples of steady, non-uniform flow and non-steady, uniform flow. (a) Diffuser: steady, non-uniform. (b) U-tube: unsteady, uniform.

outside the restraining boundaries. Water motions induced by submarines or ships are typical examples of external flows.

It is not always possible to make a clear distinction between these categories. For instance, the flow around a bridge pier may be regarded as a local external flow within the larger-scale internal flow of the river itself.

Zones of flow Most fluid flows of engineering interest are constrained by solid boundaries. This simple observation is of considerable importance because there is usually no relative motion between the solid and immediately adjacent fluid particles. The movement of any nearby particles is also restrained because the fluid resists the shearing action associated with particles sliding past one another. The restraining force is usually greatest at the boundary and reduces with distance from the boundary. Sufficiently far away, its influence is very small and may be neglected for all practical purposes. The region in which it may not be neglected is aptly termed the *boundary layer*.

In fully developed *internal* flows the influence of the boundaries extends throughout the whole flow-field. It is not usual to employ the term 'boundary layer' to describe such flows because its counterpart, the outer flow-field, sometimes called the *external* flow, is non-existent.

Laminar and turbulent One of the most important classifications from an engineering point of view concerns the detailed nature of the flow structure in a boundary layer or in a fully developed internal flow. When all particles follow clearly defined paths like well-drilled soldiers on parade, the flow is termed *laminar* or *viscous*. In contrast, when they behave more like the members of a defeated army fleeing from a battlefield in disarray, with individuals stumbling about in many directions even though the overall direction of motion is obvious, the flow is described as *turbulent*.

Not surprisingly, laminar flows are more amenable to mathematical analysis than are turbulent flows. This is rather unfortunate because the vast majority of flows encountered in engineering are turbulent. Exact analytical solutions have been found for a few laminar flows, and many more have been obtained numerically with the aid of computers. In contrast, we are totally unable to describe turbulence exactly—although great strides are being made by contemporary theories of chaos. Also, quite accurate numerical simulations are often possible with the aid of empirically based representations of turbulent behaviour.

Because of the great differences between these types of flow, it is important to know which to expect in any given circumstance. Experience shows that the likelihood of a flow being turbulent increases in proportion to (1) the geometrical size of the objects through or past which the fluid is flowing, (2) the fluid density and (3) the speed of the flow. It decreases proportionately with increasing fluid viscosity. Denoting these parameters by the symbols d, ρ, V and μ respectively, we can see that the likelihood of a flow being turbulent will increase with the *Reynolds number Re* defined by

$$Re \equiv \frac{\rho d V}{\mu} \qquad (1\text{-}1)$$

The numerical value of the Reynolds number depends upon the particular choices made for the characteristic parameters. Nevertheless, it is always the case that turbulent flows correspond to relatively high values while laminar flows are associated with relative low values.

1-2-1 Flow Visualization

Figure 1-2 depicts the positions of several motor vehicles at a particular instant on a busy highway. For identification purposes, the vehicles are labelled A, B, C or D to indicate the lane they were in when they passed under the bridge. In Fig. 1-2a the thick broken line is a *pathline* that marks the path (route) that is imagined to have been followed by the particular vehicle B1 between the bridge and its present position. Each vehicle is free to choose its own route independently and many different pathlines are implied by the figure.

Figure 1-2b shows the same vehicles at the same instant. In this case, however, the thick broken line is a *streakline*, namely the locus of different vehicles that have previously passed through a particular point—all cars labelled B in this instance.

Figure 1-2c shows the same vehicles at the same instant and also at a slightly later time. Because two positions are known for each vehicle, the instantaneous directions of travel can be deduced and are depicted by arrows. The thick broken line in the figure is a *streamline*, namely a line which is everywhere tangential to the local velocity vector.

A certain amount of artistic licence is needed to position the lines shown in

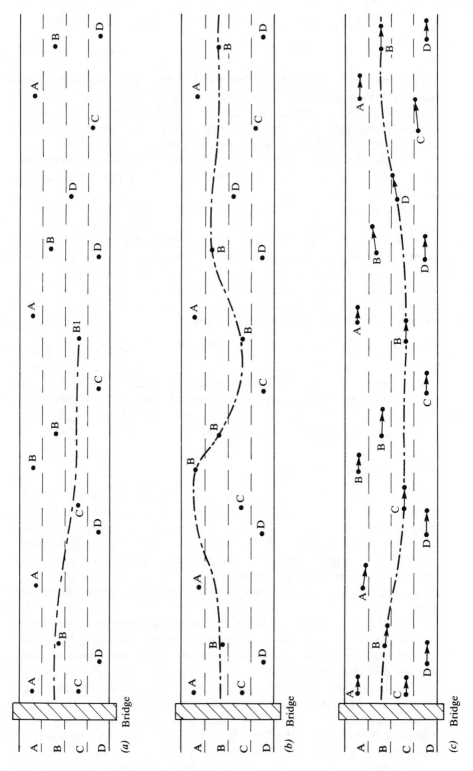

(a) Bridge

(b) Bridge

(c) Bridge

Fig. 1-2 because the individual vehicles are widely distributed. In a fluid, however, there are no gaps between the individual fluid particles and so no ambiguity is possible in the positioning of the lines. Pathlines and streaklines are more easily visualized than streamlines, but the latter are shown in Chapter Nine to be of greater value in mathematical analyses.

In the example shown in the figure, pathlines, streaklines and streamlines are all different. This is because the traffic 'flow' is unsteady. If all vehicles remained in their own lane, however, the three lines would be coincident. This is the case in all *steady* flows.

1-3 NATURAL LAWS

It might be expected that a great many natural laws would together govern the behaviour of fluids. In practice, however, this is not so. Remarkably few independent laws of nature have been found and of these only four have an important influence on most flows of engineering interest. These are the law of conservation of mass, Newton's second law of motion and the first and second laws of thermodynamics.

The first two of these laws place such restrictions on the possible behaviour of fluids that equations derived from them are sufficient to determine the motion of most *liquid* flows and many low speed *gas* flows. However, account must be taken of the first law of thermodynamics when dealing with high-speed gas flows or with the effects of heat and work. The second law of thermodynamics is rarely used explicitly, largely because its requirements are usually satisfied intuitively in the way in which we visualize natural events.

In addition to the natural laws, we need an accurate knowledge of the characteristics of the fluid medium under consideration. These are usually specified in terms of fluid *properties* such as density, pressure, viscosity and compressibility. The relationships between the various properties are the distinguishing features of any particular fluid.

Most of the remaining chapters in this book deal with the natural laws and with fluid properties. Firstly, however, these are all discussed briefly in this chapter so that the role played by each in relation to the others can be clearly seen. It is important to realize that it is not possible to *prove* the truth of any of the natural laws. They are generally *believed* to be true because none of us and none of our predecessors has ever found any evidence to the contrary except in extreme conditions when relativistic or nuclear effects are significant.

Figure 1-2 Road vehicle analogy illustrating differences between pathlines, streaklines and streamlines. (a) Pathline: route followed by vehicle B1. (b) Streakline: vehicles which were in lane B at the bridge. (c) Streamline: parallel to instantaneous directions of motion.

1-3-1 Law of Conservation of Mass

The law of conservation of mass requires that the mass of any particular object such as a fluid particle remains permanently unchanged, no matter how it moves or how it is heated, etc.

In fluid mechanics, it is usual to consider the consequences of this simple statement from a slightly different viewpoint. Instead of focusing attention on particular fluid particles, it is informative to consider how the total mass of fluid inside a given container varies with time. Suppose that the total mass of fluid entering a container—e.g. a reservoir, a surge tank, a combustion chamber, a turbine or a human heart—during a certain time interval is m_{IN}, and that the total mass leaving in the same interval is m_{OUT}. Clearly, the net increase of mass within the container is $m_{IN} - m_{OUT}$. On dividing by the time interval, we find that, for the container contents,

$$
\begin{array}{c}
\text{The rate of} \\
\text{increase of mass}
\end{array}
=
\begin{array}{c}
\text{the rate of} \\
\text{mass inflow}
\end{array}
-
\begin{array}{c}
\text{the rate of} \\
\text{mass outflow}
\end{array}
\qquad (1\text{-}2)
$$

Equation (1-2) is known as the *continuity equation*. It is derived solely from the law of conservation of mass, but the information that it conveys is somewhat more useful than the original statement that the masses of individual fluid particles never change.

Equation (1-2) is particularly useful when an analytical expression can be found for the rate of mass flow $\dot{m}$ (otherwise known as the *mass flux*). It is shown in Chapter Six that a typical expression is

$$\dot{m} = \rho a V \qquad (1\text{-}3)$$

in which ρ denotes the fluid density, a is the cross-sectional area of the flow and V is the velocity of flow. The product aV is a measure of the volumetric quantity of fluid entering (or leaving) the container in unit time. It is often denoted by the symbol Q and is known as the *volumetric flux* or *discharge*.

Example 1-1 Figure 1-3 depicts a compensating reservoir which receives water at a *constant* rate Q_I, but delivers to a distribution main at a variable rate Q_O. Explain how suitable dimensions may be chosen for the reservoir when the variation of Q_O is known.

SOLUTION During the daytime, the rate of outflow will exceed the rate of inflow and so the surface level in the reservoir will fall. At night, the converse will apply. A rise δH in the liquid level implies an increase of $A \, \delta H$ in the volume stored, where A denotes the cross-sectional plan area of the reservoir. If this increase occurs in a time interval δt then the *rate* of increase of volume stored is $A \, \delta H / \delta t$. In the limit, therefore, the continuity equation (1-2) requires that

$$A\frac{dH}{dt} = Q_I - Q_O \qquad (1\text{-}4)$$

which, if A is constant, may be integrated to give

$$A(H_2 - H_1) = \int_{t_1}^{t_2} (Q_I - Q_O)\,dt \qquad (1\text{-}5)$$

On integrating the expression on the right-hand side of Eq. (1-5) between the time t_1 when the influx first exceeds the efflux and the time t_2 when the converse event occurs, we obtain the minimum possible capacity for the reservoir.

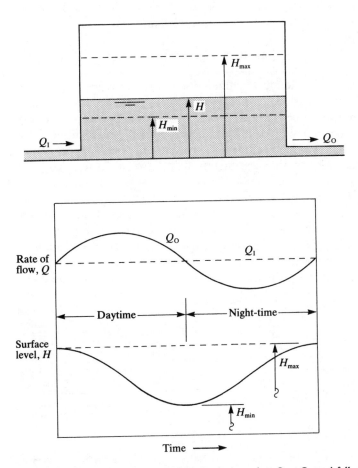

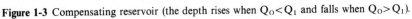

Figure 1-3 Compensating reservoir (the depth rises when $Q_O < Q_I$ and falls when $Q_O > Q_I$).

Comments on Example 1-1

1. The primary function of a compensating reservoir is to smooth out the fluctuations in demand so that the water treatment works further upstream can operate at a steady rate.
2. The above analysis yields the minimum possible capacity of the reservoir. In practice a somewhat larger size would be chosen to ensure an adequate safety margin in the event of malfunctions either upstream or downstream and to avoid draining the reservoir each evening and filling it each morning.

1-3-2 Newton's Second Law of Motion

In its most familiar (but restricted) form, Newton's second law of motion equates *force* with the product of *mass* and *acceleration*. However, this equation is valid only when all parts of the system under consideration accelerate at the same rate. With fluids, this proviso is rarely satisfied and it is usually necessary to use the more fundamental expression

$$\boxed{\text{Force} \quad = \quad \text{rate of change of momentum}} \tag{1-6}$$

In fluid mechanics, the most common forces are those due to pressure, shear and gravity. Occasionally, however, electrical, magnetic and surface tension forces must also be considered. It is the *net* force that appears in Eq. (1-6), that is the algebraic sum of all forces acting on the system. The net force causes a rate of change of momentum of the system *in the direction of the force*.

In many examples of practical interest, fluid flows are both steady and uniform. In this case, there is no rate of change of momentum and it follows from Eq. (1-6) that there will be no net force. For example, when water flows along a river, the friction forces on the bed and the banks of the river may exactly balance the streamwise component of the gravitational force—Fig. 1-4a. Similarly, in a steady uniform flow through a horizontal pipe—Fig. 1-4b— the friction forces are balanced by pressure forces. These examples illustrate the importance of the contribution of shear stresses to fluid motions. The

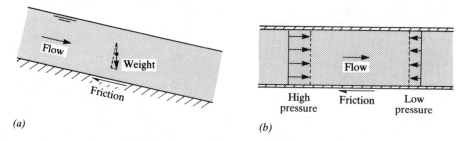

Figure 1-4 Equilibrium flows in an inclined channel and a horizontal pipe. (a) Inclined channel (gravity-induced flow). (b) Horizontal pipe (pressure-induced flow).

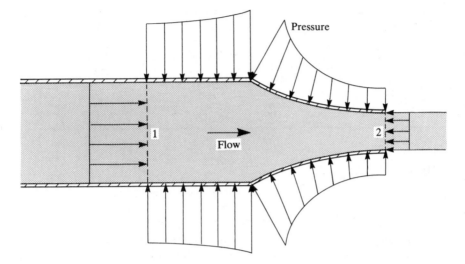

Figure 1-5 Steady, non-uniform flow (pressure acts normal to the surface).

gravitational and pressure forces required to maintain the steady flows are used solely to overcome the frictional resistance of shear stresses on the solid boundaries.

In non-uniform or unsteady flows, the fluid particles undergo accelerations and so the net applied force is not zero. Figure 1-5 illustrates a steady, non-uniform flow through a nozzle. Fluid particles accelerate as they pass from section 1 to section 2 under the action of pressure forces which in this example are much more important than friction forces. The force exerted on the fluid by the nozzle in the axial direction can be related to the rate of change of momentum of the fluid by means of Eq. (1-5). Although the necessary algebra is not presented until Chapter Seven, it should be intuitively obvious that this method of determining the force will be much simpler than the alternative approach of integrating the elemental pressure forces over the whole of the nozzle surface.

Example 1-2 During a period of steady acceleration, the free surface liquid level in a road tanker slopes at an angle of 3° to the horizontal (Fig. 1.6). At what rate is the vehicle accelerating?

SOLUTION Consider the particular liquid sample identified in the figure. Since the pressure at a depth d below the free surface in a stationary liquid of density ρ is $\rho g d$, the pressure on the left-hand face of the sample is $\rho g d_L$ and the corresponding force is $F_L = \rho g d_L A$. Similarly, the force on the right-hand face is $F_R = \rho g d_R A$ and so the net force in the direction of travel is $F = \rho g A (d_L - d_R)$.

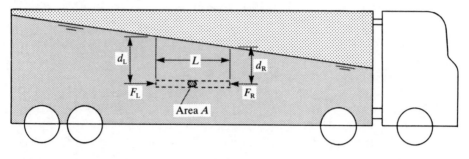

Figure 1-6 Effect of acceleration on liquid in a tank.

This force must equal the rate of change of momentum of the chosen sample and, in this instance, we can equate this to mass × acceleration. Since the mass of the sample is ρAL, it follows that the acceleration dV/dt is

$$\frac{dV}{dt} = \frac{F}{\rho AL} = \frac{g(d_L - d_R)}{L} = g \tan 3° \qquad (1\text{-}7)$$

Thus the fluid sample (and hence the whole vehicle) is accelerating at a rate of $0.514\,\text{m/s}^2$.

Comments on Example 1-2
1. The sloshing of liquids in tanks is a major problem for designers of seagoing tankers as well as road tankers. In practice, the tanks are compartmentalized by the provision of transverse and longitudinal bulkheads and baffles. Otherwise the effects of the free-moving cargo could lead to instabilities during sudden manoeuvres or roll.
2. It is permissible to use the hydrostatic relationship for the pressure at a depth d even though the vehicle is moving relative to the road. There are no vertical accelerations.

1-3-3 First Law of Thermodynamics

The principle of conservation of energy requires that the total energy of any isolated system remains permanently constant. In this context, the adjective *isolated* implies that the system undergoes no heat or work interactions with its surroundings. The observable forms of the energy of the system might change, say from gravitational potential energy to kinetic energy or from electrical potential energy to internal energy, but the *total* energy in all its forms is invariable.

To change the total energy of a system, it is necessary to expose it to heat and/or work. In this case, the first law of thermodynamics shows that the increase in the total energy, ΔE, is equal to the algebraic difference between the net heat Q supplied to the system and the net work W done by it, that is

$$Q - W = \Delta E \qquad (1\text{-}8)$$

Heat and *work* are mechanisms by which energy transfers can take place and, as such, they exist only while the transfers are taking place. When a heat or work interaction takes place between a system and its surroundings, one of them gains energy and the other loses energy.

Energy can exist in various forms of which kinetic, potential and internal energies are the most important in common fluid mechanics applications. The *internal* energy of a system is sometimes alternatively known as the *thermal* energy. It may be thought of as a measure of the randomly directed kinetic energy of the molecules composing the system. It should not be confused with heat.

Example 1-3 A droplet of water of mass m falls from the end of a domestic tap. Describe how its main forms of energy will vary during its fall.

SOLUTION During the early stages of its fall, air resistance will be small and so, after falling through a height h, the loss of gravitational potential energy mgh will be accompanied by a nearly equal increase in kinetic energy $\frac{1}{2}mV^2$. Its downwards velocity will therefore be approximately $\sqrt{2gh}$. As it continues to fall, air resistance will become more important, and because of the work done by the droplet in overcoming this resistance the total increase in its kinetic energy will be less than the reduction in its potential energy. In partial compensation, there will be a small rise in the droplet's internal energy (and hence in its temperature).

Comments on Example 1-3
1. In a rigorous investigation, this simple picture might be complicated by such factors as distortion or breakup of the droplet, evaporation or condensation, instability of the linear motion, and heat transfers with the air. These effects are often of little consequence, but they can greatly influence phenomena such as blade erosion in a steam turbine.
2. In section 11-1-3 it is pointed out that the concept of gravitational potential energy can be discarded if one prefers to evaluate the work done on (or by) the gravitational field explicitly. Kinetic energy cannot be discarded, but notice that its value depends upon the velocity of the reference axes chosen by the observer. Internal energy is independent of the choice of axes.

Example 1-4 Describe the influence of heat and work interactions on the energy of the fluid mixture in the cylinder of a four-stroke internal combustion engine (Fig. 1-7) during the power stroke.

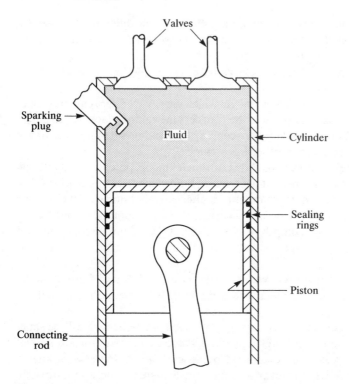

Figure 1-7 Schematic illustration of a combustion chamber in a four-stroke internal combustion engine.

SOLUTION The primary purpose of the engine is to use the chemical energy of petrol to produce useful work. A downward pass of the piston draws in a mixture of air and petrol through the inlet valve. This is then compressed by an upward pass, and the mixture is ignited by an electrically induced spark. The resulting explosion may be thought of as a conversion of part of the chemical energy of the mixture into internal energy. Its effect is to increase the pressure and temperature of the mixture, and the high-pressure fluid pushes the piston downwards. The work done by the fluid in moving the piston through an elemental distance δx is the product of the pressure force on the piston surface and the distance moved, that is

$$\delta W = pA \; \delta x \qquad (1\text{-}9)$$

in which p denotes the average pressure *on the surface* of the piston of cross-sectional area A. In accordance with Eq. (1-7), the work is done at the expense of an equal decrease in the energy of the mixture, heat transfers being small. In practice, a small heat flow to the cylinder will cause an additional fall in the energy of the mixture.

Comments on Example 1-4

1. The total work done during the power stroke can be estimated by integrating Eq. (1-9), making use of a knowledge of the mixture properties to describe the pressure decrease as the exploded mixture expands. Strictly, account should also be taken of spatial variations in pressure within the cylinder.
2. Not all of this work appears as useful shaft work. Even if minor losses are ignored, allowance must be made for work interactions during the exhaust, inlet and compression strokes which are necessary before the next power stroke can occur.

1-3-4 Second Law of Thermodynamics

Everyday experience shows us that real processes always have a time-arrow associated with them. Seeds germinate; men grow old; smoke diffuses in the atmosphere; water settles in a bowl; eggs harden in boiling water; waves steepen and break as they approach a beach. The effects of the time-arrow are such a natural part of our everyday lives that we would immediately recognize an error if, say, a cine-film of breaking waves was shown in reverse.

None of the three natural laws discussed so far accounts for this behaviour and so we must look elsewhere for its cause. As a first step in our search, it is useful to imagine a system undergoing a *cyclic* process, that is a process in which a given set of conditions is exactly reproduced every once in a while. Since our hardened egg cannot get back to its previous 'soft' condition, we expect that the relevant natural law will place restrictions on the types of cyclic process that are physically possible. Indeed, this is precisely so.

The second law of thermodynamics can be stated (in a somewhat abstract form) as follows:

> Anything operating repetitively in constant-temperature surroundings must be receiving work and emitting heat (1-10)

At first sight, one might suppose that the consequences of this law would apply only to repetitive (cyclic) processes. However, we shall see in Chapter Twelve that it also restricts all other processes because these can always be imagined to form a part of a hypothetical cyclic process. Similarly, the law also influences the behaviour of systems undergoing heat interactions at more than one temperature—or none at all. For example, it is not possible for the internal energy of an *isolated* system to be converted into kinetic energy.

From an engineering point of view, one of the most useful consequences of the law is that the existence of a thermodynamic property known as the *entropy* of a system can be established. Any change in entropy can be shown to be related to the heat interactions undergone by the system, and so the property can be used to determine whether or not particular processes are possible. The best-known relationship applies to systems that undergo no heat interactions. For these, the entropy change ΔS during *any* process must be either positive or zero. A negative value would imply a reversal of the time-arrow. The special processes for which $\Delta S = 0$ are termed *isentropic*.

Example 1-5 It can be shown from the second law of thermodynamics that the efficiency of a thermal power plant operating at a maximum absolute temperature T_1 in an environment of temperature T_2 cannot exceed $(1 - T_2/T_1)$. How does this influence the generation of power for domestic and industrial use?

SOLUTION Nearly all power plants make use of steam turbines or gas turbines for which the maximum sustained temperature is about 600 °C, that is about 875 K. Since the environmental temperature on the surface of our planet is usually within about 10 per cent of 280 K, a typical upper limit to the thermal efficiency of a power station will be $(1 - 280/875) = 0.68$. After allowing for the efficiency of the turbines and the combustion processes necessary to generate the high-temperature gases, it is easy to see why practical overall efficiencies are always below 50 per cent.

Comment on Example 1-5 Hydroelectric power stations are far more efficient than thermal power stations and we don't have to destroy the water to extract the power. However, the amount of water available at high elevations is limited by the hydrological cycle and is insufficient for the needs of a developed country. Moreover, the best sites tend to be remote from the main regions of demand, and this leads to transmission inefficiencies.

1-4 FLUID PROPERTIES

Before applying the four natural laws to processes involving fluids, we must first understand the nature of fluids themselves. The basic characteristic of any fluid is its tendency to *flow*. Given sufficient time, a fluid will adapt itself to fit the shape of any container in which it is placed.

Thus air fills irregularly shaped rooms and water covers irregularly shaped river beds. Both liquids and gases continually distort as they flow through the most tortuous of routes, changing their shapes to match the available passages.

The ability of fluids to flow may alternatively be described as an inability not to flow when shear forces are applied. Unlike solids, which distort by a finite amount under the action of shear force, fluids distort continuously until they attain a condition in which shear forces are totally absent. Such a condition is invariably stationary when viewed from a suitable set of axes.

Although all fluids exhibit this common feature, the manner of their various responses to changes in the environment can be far from common. For instance, the rates at which distortions occur in response to applied shear stresses are very different for water and treacle. Likewise, the compressibilities of orange juice and air and the volatilities of oil and acetone are in sharp contrast. It is the particular values of the many properties and the inter-relationships between them that give each fluid its individual physical identity.

Engineering works usually involve the use of fluids that can be regarded as *pure substances*, that is whose chemical composition is both homogeneous and invariable. In practice this simply implies that we rarely need to deal with substances undergoing chemical reactions or with unevenly mixed combinations of non-reacting substances. An advantage of restricting our attention to pure substances is that, no matter how many properties they may have, only *two* can be varied independently when the influences of motion, gravity, magnetism, electricity and capillarity are discounted. For example, if we know, say, the pressure and temperature of a sample of oxygen gas, then we implicitly also know everything else about it, for example its density, its specific heat capacities and its viscosity. These values can be found by reference to data published by previous workers who have carried out the necessary measurements.

In practice, it would be most inconvenient to have to refer to published empirical data every time we want to know the value of some property corresponding to a new pair of values of the pressure and temperature. Happily, this is not usually necessary. Instead we can often regard a property as a constant for the purposes of any particular analysis. For instance, it is usually reasonable to neglect variations in the density of a liquid or in the specific heat of a gas.

Even when we cannot assume that the value of a particular property is independent of the environment, it is often possible to find an analytical expression to relate it to other properties such as the pressure and the temperature. For example, the viscosities of most liquids and gases can be approximated by functions of the temperature alone over a wide range of pressures.

One of the best-known algebraic relationships is the *equation of state* of a perfect gas,

$$p = \rho R T \qquad (1\text{-}11)$$

in which p, ρ and T are the absolute pressure, the density and the absolute temperature respectively. The parameter R is a property known as the *gas constant*. It is related to the universal gas constant R_0 by the expression $R_0 = MR$ in which M is the so-called molecular weight of the gas. Equation (1-11) closely approximates to the behaviour of real gases over a wide range of temperatures and pressures. It holds true at high pressures provided that the temperature is also sufficiently high and at low pressures regardless of the temperature. For example, air may be regarded as a perfect gas for all pressures up to about 10 MPa provided that its temperature exceeds about 250 K, i.e. about $-20\,^{\circ}\text{C}$.

1-5 PHYSICAL DIMENSIONS

Although a great deal is known about the laws that govern natural processes and about the properties of the fluids to which the laws are applied, it is only rarely possible to obtain exact solutions of the resulting equations. In some cases, this is a consequence of the complex geometry of the boundaries enclosing the fluid. In others, it is because the fluid motions themselves are too involved, e.g. when the flow is turbulent. Often approximate solutions can be obtained by making simplifying assumptions, but the ensuing results are not wholly reliable, and care is needed in their interpretation. Particular difficulty is experienced when the boundaries themselves respond to the fluid motions— when turbine blades flutter or when silt is carried along a river bed for example.

Because of the inadequacy of theoretical techniques, expensive projects are rarely undertaken without first obtaining experimental confirmation of the design. Fortunately, the required evidence can often be deduced from physical *models* which can be much smaller than the prototype. Provided that certain rules are obeyed by the experimenter, the fluid behaviour in a model will accurately simulate the full-scale behaviour. It is common practice to use physical models to investigate topics such as the effects of bridge piers on the regime in a river, the suitability of a breakwater at a harbour entrance, the aerodynamic drag on a motor vehicle or aeroplane, and the characteristics of a turbine equipped with blades designed according to some new theoretical ideas.

The rules that must be followed in the design of the models are deduced by a procedure discussed in Chapter Five and known as *dimensional analysis*. Notionally, any person with a knowledge of this procedure could design a suitable model even with very little prior knowledge of fluid mechanics. In practice, however, it is found that experience is invaluable in the selection of the parameters that must be modelled and in the interpretation of the results.

A closely allied topic of great importance concerns the dimensional homogeneity of physical equations. It is obvious to us all that $6 + 8 = 14$, that $6\,\text{kg} + 8\,\text{kg} = 14\,\text{kg}$, and that $6\,\text{s} + 8\,\text{s} = 14\,\text{s}$, but what does $6\,\text{s} + 8\,\text{kg}$ equal? The correct response of course is that the question is meaningless; seconds cannot be added to kilograms. This trivial example is given because it illustrates the need for equations representing physical phenomena to deal with quantities that are all of the same type. With this in mind, let us consider a question to which the answer is less obvious.

Example 1-6 One of the most useful relationships in fluid mechanics is the Bernoulli equation. For steady, incompressible flows, this may be written as

$$p + \tfrac{1}{2}\rho V^2 + \rho g z = \text{constant} \tag{1-12}$$

in which p, ρ, V, g and z denote the pressure, density, velocity, gravitational acceleration and vertical elevation respectively.

(a) In what SI units should the constant be expressed?

(b) For what types of flow could the following equation be valid?

$$p + \tfrac{1}{2}\rho V^3 + g z^2 = \text{constant}$$

SOLUTION

(a) The *dimensions* of the terms p, $\tfrac{1}{2}\rho V^2$ and $\rho g z$ are all $[ML^{-1}T^{-2}]$ in which $[M]$, $[L]$ and $[T]$ denote the fundamental dimensions of mass, length and time respectively. For Eq. (1-12) to be dimensionally homogeneous, it follows that the 'constant' must also have the dimensions $[ML^{-1}T^{-2}]$. Its units could be written as kg/m s^2 or alternatively as N/m^2 (otherwise known as pascals, Pa).

(b) The three terms on the left-hand side of the second equation have the dimensions $[ML^{-1}T^{-2}]$, $[MT^{-3}]$ and $[MT^{-2}]$ respectively. By inspection, therefore, the terms are of three different types and so the equation can have no physical meaning.

Comments on Example 1-6

1. Dimensional arguments such as this enable us to determine which forms of an equation are *possible* and which are not. They do not enable us to take the further step of deciding which of a group of notionally possible equations is *correct*.

2. In this context, a *dimension* may be regarded as a generalization of a *unit*. For example, the distance between two churches on the same street could in principle be expressed in terms of metres, feet, wavelengths of light or even 'blocks'. Similarly, the mass of a rocket could be described in kilograms or pounds. These are all units whose purpose is to quantify the fundamental dimensions of length and mass.

1-6 USE OF THIS TEXTBOOK

First-time readers are recommended to skip the next four chapters and to read as many of Chapters Six to Twelve as they see fit before returning to Chapters Two to Five. This will enable them to see how the natural laws can be utilized to solve quite complex problems in fluid mechanics without having to worry about finer points of detail. Most people already have a working understanding of the main concepts discussed in Chapters Two to Five.

All readers may prefer to skip the sections at the ends of chapters dealing with differential forms of the various equations.

FURTHER READING

Japan Society of Mechanical Engineers (1988) *Visualized Flow*, Pergamon.

Kermode, A.C. (1987) *Flight Without Formula*, 4th edn, Pitman.

Merzkirch, W. (1987) *Flow Visualization*, 2nd edn, Academic Press.

Rouse, H. and Ince, S. (1954/56) *A History of Hydraulics*, 12 part supplement to *La Houille Blanche*.

Vallentine, H.R. (1967) *Water in the Service of Man*, Pelican.

Van Dyke, M. (1982) *An Album of Fluid Motion*, Parabolic Press.

TWO

FLUIDS AND THEIR PROPERTIES

2-1 THE THREE PHASES OF A PURE SUBSTANCE

In our everyday experience we become aware that water can exist as a solid, a liquid or a gas. All other pure substances behave similarly, their particular state depending upon the environmental conditions. For example, steel is a liquid at the high temperatures required for its manufacture and oxygen is a liquid at the low temperatures encountered in an air separation plant.

The three phases of a pure substance can be identified in pressure–temperature diagrams such as those shown in Fig. 2-1 for (a) substances like water that expand on freezing and (b) most other substances that contract on freezing. At most combinations of pressure and temperature, the whole of the substance is in the same phase, but it is also possible for some of it to be in one phase and the remainder to be in another. For example, liquid water can coexist with either ice or steam. Similarly, ice can coexist with steam, as it does in winter when a hoar frost vaporizes into the atmosphere. It is also possible for all three phases to exist simultaneously, but this happens at only one value of pressure and one value of temperature for any particular substance. This condition is commonly known as the *triple point* (TP).

The fusion and sublimation lines extend as far as measurements have been made for all pure substances, but the vaporization line terminates at the *critical point* (CP). At higher pressures and temperatures, it is not possible to classify the fluid as either liquid or gas in a rigorous manner. Fortunately, this phenomenon presents no difficulties for the purposes of this book.

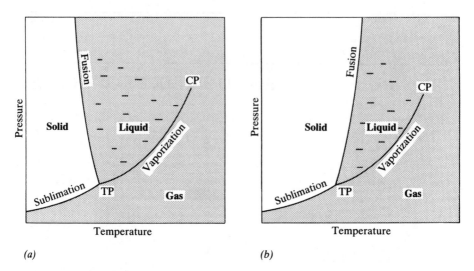

(a)   *(b)*

Figure 2-1 Phase diagrams for pure substances (not to scale). (a) Expands on freezing. (b) Contracts on freezing.

2-2 FLUIDS AND CONTINUA

When a tank of liquid is left to stand in a vibration-free environment, all perceptible motion eventually ceases. For practical purposes, it is satisfactory to say that every particle of the liquid is at rest. Nevertheless, such a statement is in conflict with the fundamental nature of fluids, which are made up of vast numbers of molecules that are continually in motion. Therefore, the statement must be qualified by the proviso that the liquid particles under consideration are sufficiently large for the influence of the motion of individual molecules to be negligible. Fortunately, this does not imply that the particles must be of any great size. For example, a 1 μm diameter sphere of water at atmospheric pressure and 15 °C contains almost 1.75×10^{10} molecules, and the same volume of air in the same conditions contains about 1.33×10^{7} molecules. From an engineering point of view, it is therefore perfectly satisfactory to define the *velocity* of a particle as the average velocity of all the molecules within it.

Before we can use differential calculus in the study of fluid motions it is necessary to give meaning to such terms as the velocity and acceleration *at a point* in a fluid. With a real fluid, these will have widely fluctuating values according to whether or not there happens to be a molecule at the point at any instant. This will not therefore serve as a useful basis for a definition as far as analyses dealing with macroscopic motions are concerned. Instead, a hypothetical fluid is imagined in which the macroscopic behaviour of the fluid is reproduced exactly, but the microscopic behaviour is not. The velocity *at a point* is regarded as the velocity of a very small particle which has its centre of mass at that point. The particular particle chosen is the smallest one possible, consistent with the notion that no individual molecule shall have a significant influence on the result. From the values quoted in the preceding paragraph, it is apparent that this condition will be satisfied with particles that are much less than one micron (one micrometre) in diameter. The hypothetical fluid in which velocities (and accelerations etc.) are defined in this manner is an example of a *continuum*, so called because it does not exhibit the discontinuous fine structure of a real fluid.

By definition, the behaviour of the hypothetical fluid differs from that of a real fluid only when it is studied in such fine detail that individual molecules can have a significant effect on the average motion of a particle. It follows that for nearly all practical purposes engineers may ignore the distinction. Nevertheless, it is prudent to be aware that there are limits to the range of validity of the continuum approach. It should not be used when extreme detail is required—in high-vacuum technology for instance.

The continuum concept is also used extensively in the study of solids. Indeed, even materials such as concrete are frequently regarded as homogeneous continua. The overall deflected shape of a structural beam or column can be predicted with good accuracy in this manner, but local effects such as the stress distribution around a piece of aggregate require more careful attention.

It is worth noting in passing that the above description is slightly misleading because it implies that the value obtained by the averaging process is independent of time. In practice, the averaging process is designed to take account of temporal as well as spatial variations in the distribution of molecules.

2-3 TEMPERATURE

The concepts of 'hot' and 'cold' are familiar ones to us all. They are qualitative terms which can be used to indicate the relative *temperatures* of different objects. When two objects initially at different temperatures are allowed to communicate, a *heat* interaction occurs. If the communication is allowed to continue for a sufficiently long time, the temperatures become equal and the heat interaction ceases. It is important to realize that although heat interactions occur solely as a consequence of temperature differences, the converse is not true. The temperature of an object can be altered by either heat or work interactions. For example, when air is compressed in a bicycle pump, its temperature increases. The 'hot' air then heats the walls of the pump so that the latter also becomes hot.

For temperature to be a meaningful property, it is necessary to define a temperature scale. The *absolute temperature scale* is used throughout this book, but its precise definition is deferred until Chapter Twelve. Occasionally, it is convenient to use the *Celsius* temperature. This is defined to be numerically smaller than the absolute temperature in kelvins by 273.15 degrees. Thus, for example, $20\,°C = 293.15\,K$.

Any reader who feels uncomfortable with the use of the property temperature before it has been defined might like to use the old Centigrade scale. This can be defined in terms of the length of a mercury column in a glass thermometer, with $0\,°C$ and $100\,°C$ points marked at the freezing and boiling points of water at standard pressure. For most practical purposes, the differences between the Centigrade and Celsius scales may be ignored.

2-4 PRESSURE

In a *solid* the relative positions of the molecules do not change. Attractive forces between them give the material its cohesive nature. Similar forces exist in *fluids*, but they are much less important because the molecules are in a continual state of motion that has the collective appearance of randomness. Collisions between the molecules give rise to an internal compressive stress in the fluid that we observe as its pressure (sometimes called its *absolute pressure*).

At an interface between a fluid and solid, the movement of the fluid molecules results in a bombardment of the solid surface, tending to push it away from the fluid. It is this force, for example, that must be resisted by a dam containing the water in a reservoir. The pressure *at a point* on the surface is defined at the centroid of an element of the surface. It is found by dividing the force normal to the element by the area of the element. The pressure at a point within the fluid is defined similarly by considering the force exerted on an imaginary elemental surface within it. In the usual case when the chosen imaginary surface is at rest relative to the fluid, the pressure is called the *static* pressure.

In the SI system the basic unit of pressure is the *pascal*, Pa, defined as $1\,N/m^2$. Unfortunately, one pascal represents a very small pressure and so engineers commonly use kPa or MPa. For some purposes, an intermediate unit called a *bar* is used. One bar is equal to $100\,kPa$ and so it is nearly equal to a standard atmosphere, $101.325\,kPa$.

Gauge pressure In engineering practice, it is commonplace to use *gauge* pressures, namely the excess of the absolute pressure over the local atmospheric pressure. There are two main reasons for this practice. Firstly, many of the instruments designed to measure pressures (section 3-6) actually measure the *difference* between the fluid pressure and the pressure in the environment— hence the term *gauge*. Secondly, engineers are frequently concerned with the ability of a fluid to do work. It is an important limitation in the real world that the gauge pressure is usually more reliable than the absolute pressure as a measure of this ability. That is, fluids are normally discharged into the atmosphere, not into a vacuum.

It is necessary to note in passing that the term *gauge pressure* has occasionally been used for a different purpose, namely to describe the difference between the pressure at a point in a moving fluid and the pressure at the same point when the fluid is at rest. This enables us to distinguish clearly between pressure differences associated with fluid motion and those associated with elevation. (*NB:* Even in a stationary fluid, the pressure varies with depth —as pilots and submariners well know.) This parameter is not used in this book. Elsewhere, it is more correctly known as a *non-gravitational pressure*.

Pressure head Fluid pressure is often specified as a *head of liquid*. For example, a pressure of one atmosphere might be described as a head of 760 mm mercury or 10.33 m water. This practice is acceptable because pressure varies linearly with depth in a stationary liquid (section 3-3). Doctors measure blood pressures in millimetres of mercury and fan manufacturers often quote air pressures in millimetres water gauge. Water engineers commonly talk simply of metres, the words 'of water' being implied but omitted.

2-4-1 Saturated Vapour Pressure and Cavitation

When the temperature of water is increased at constant pressure—in a kettle, say—its volume increases slightly and it eventually begins to boil. This well-known behaviour corresponds to the broken line AB drawn parallel to the temperature axis in Fig. 2-2 from the typical point A to the vaporization line.

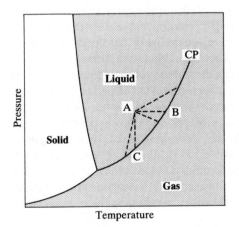

Figure 2-2 Phase diagram for water A→B=heating; A→C=expanding.

This process is by no means the only one that can lead to the evaporation of the liquid. Indeed, by inspection of Fig. 2-2 it is easily seen that an infinity of possible routes exists from the point A to the phase-change line. The most important alternative route corresponds to pressure reduction at constant temperature. Evaporation due to this cause is known as *vaporous cavitation* and the pressure at which it takes place is called the *saturated vapour pressure*. Thus the original water sample A is said to have a boiling point temperature T_B and a saturated vapour pressure p_c.

Water in everyday use behaves in a more complex manner than would be expected from the preceding discussion because it is never free from contaminants, especially dissolved air. When the pressure is reduced by even a small amount, air may be released from solution in a process that closely resembles evaporation and is known as *gaseous* cavitation. At atmospheric pressure, ordinary water contains about 2 per cent air by volume. That is, if all the dissolved air could be removed from solution and held in a container at atmospheric pressure, its volume would be about 2 per cent of the water volume. At a pressure of half an atmosphere, the *same* contamination represents 4 per cent air by volume. The lower the pressure, the greater the mass of air released from solution and the greater the volume occupied by each element of released air. It is this phenomenon that causes pumps to lose their prime and that renders siphons useless when their invert is more than about 7 m above an atmospheric water surface.

A further complication is that the *rate* at which air can be released depends upon factors such as the pressure at which the solution was originally in equilibrium and the amount and nature of other contaminants. Moreover, the rate at which it will redissolve if the pressure is subsequently increased is yet another variable. These complications are important because the rates of change of pressure in an unsteady flow can greatly exceed those at which equilibrium conditions may be assumed. This topic has received considerable attention from research workers, especially since the early 1970s.

When air bubbles released during cavitation move downstream, their subsequent collapse in a high-pressure region can cause considerable damage. The collapse mechanism of bubbles in contact with a solid surface can involve a very thin jet of liquid which may strike the surface at very high speed—values of hundreds of metres per second have been suggested. Whether or not these values are typical, there is no doubt that the collapse can be responsible for the severe erosion of dam spillways and of blades in pumps and turbines. It is sensible design practice to avoid the development of sub-atmospheric pressures in liquids whenever possible. This particular consequence of cavitation is so important that some authors choose to include it within the definition of the term 'cavitation'. However such a practice is not helpful because it leads to confusion when dealing with cavitating flows in which no damage is caused.

2-4-2 Partial Pressures

The pressure in a homogeneous mixture of two or more gases may be regarded as the sum of individual contributions from each gas. As an example, consider 1 kg air consisting of 0.232 kg oxygen and 0.768 kg 'atmospheric' nitrogen. When 0.232 kg oxygen occupies a volume of, say, 0.8 m^3 at a temperature of 5°C, its pressure is 21.0 kPa. Similarly, when 0.768 kg atmospheric nitrogen occupies 0.8 m^3 at a temperature of 5 °C, its pressure is 78.8 kPa.

Experiments such as those first conducted by Dalton in 1801 show that when both gases share the same volume of 0.8 m^3 at 5 °C the resulting mixture (i.e. 1 kg air) has a pressure of 99.8 kPa. It is therefore natural to regard the total pressure as the sum of the so-called *partial pressures* of each gas.

2-5 DENSITY AND SPECIFIC WEIGHT

The *mass density* of a particle is the ratio of its mass and its volume. The mass density *at a point* in a fluid continuum (see section 2-2) is defined at the centre of mass of a real element of mass δm and volume δV as

$$\rho \equiv \frac{\delta m}{\delta V} \tag{2-1}$$

In this book the mass density is usually abbreviated to simply the *density*.

It is often reasonable to treat the density of a fluid as a constant. When this is not acceptable, the density must be determined from published data or from a relationship such as Eq. (1-11). Typical values of the density of water and air are given in Appendix A for temperatures and pressures commonly encountered in practice. Their variations over a much wider range are illustrated in Fig. 2-3. In this and in subsequent figures, pressure and temperature are used as the reference properties because these are usually the easiest to measure in practice. Continuous lines represent the variation of a property with pressure at constant temperature; the relevant pressure axis is at the top of each figure. Broken lines denote variations with temperature at constant pressure; for these, the temperature axis is drawn along the bottom of each figure.

Specific weight Hydraulic engineers commonly use the *specific weight, w,* instead of the density. It is defined as

$$w \equiv \rho g \tag{2-2}$$

in which g denotes the gravitational acceleration. This practice has gained wide acceptance because the product ρg often occurs in equations in common use and because engineers are usually more concerned with force than mass.

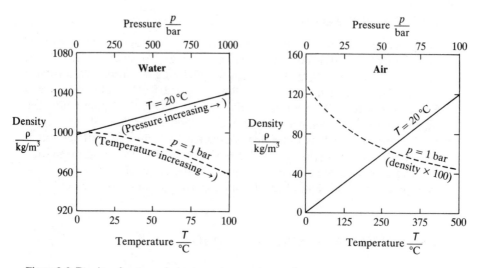

Figure 2-3 Density of water and air at constant pressure and at constant temperature.

Nevertheless, the natural laws relate to mass, not weight, and so it is appropriate to use the density when developing their analytical forms. In this book, the density is normally preferred, but the specific weight is used in certain applications where engineering practice has deemed it to be more appropriate.

Relative density A third method of describing the density of a substance is by means of its relative density (otherwise known as its specific gravity). This is the ratio of the respective densities of the substance and pure water. It is rarely used explicitly in engineering fluid mechanics except in the study of flows such as sediment transport which involve both fluids and solids.

2-6 COMPRESSIBILITY

All substances are compressible. When the pressure of a sample of fluid is increased, its volume decreases and its density increases. Since its mass m remains constant, differentiation of the expression $m = \rho V$ yields $\rho \, dV + V \, d\rho = 0$ and so $d\rho/\rho = -dV/V$, that is the densimetric strain is of equal magnitude but opposite sign to the volumetric strain. The *bulk modulus of elasticity* K of a fluid may be defined as the ratio of an increase in pressure and the increase in densimetric strain that it causes. Thus

$$\frac{dp}{d\rho} = \frac{K}{\rho} \quad \text{and} \quad \frac{dp}{dV} = -\frac{K}{V} \qquad (2\text{-}3,\ 2\text{-}4)$$

Notice that fluids with a high value of K are only slightly compressible and vice versa. The reciprocal of K, known as the *compressibility* of a substance, is rarely used in fluid mechanics.

Because the pressure and density of a fluid can vary independently, Eq. (2-3) cannot be used to determine a value for K unless it is accompanied by additional information describing the particular process. For example, the *isothermal* bulk modulus K_T is obtained when the process takes place at constant temperature and the *isentropic* bulk modulus K_S applies when the entropy remains constant (section 2-9).

Liquids Differences between the isothermal and isentropic bulk moduli of a liquid are usually small. Typical values for water are shown in Fig. 2-4 and in Appendix A. The influence of pressure at a constant temperature of 20 °C is small and almost linear. The influence of temperature at a constant pressure of 1 bar is also small and it is anomalous; no other substance has been found to exhibit a maximum such as that found for water at approximately 46 °C. Water is also unique in that its density exhibits a maximum at about 4 °C— hence 'ice cold' water floats on top of slightly warmer water. Both of these characteristics disappear at pressures in excess of about 300 bar.

Gases It is not usual to use bulk moduli explicitly when dealing with gases because more useful alternatives exist. For example, the isothermal bulk

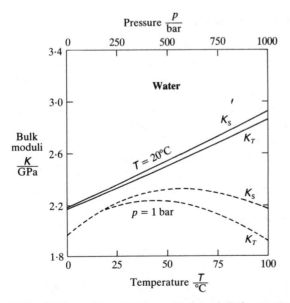

Figure 2-4 Isentropic and isothermal bulk moduli of water at constant pressure and at constant temperature.

modulus of a perfect gas is equal to its pressure. The isentropic bulk modulus is γp in which γ denotes the ratio of the specific heats of the gas (section 2-9).

2-6-1 Transmission of Pressure Waves

The compressibility of fluids is especially important in some classes of *unsteady* flows because it plays a crucial role in the transmission of pressure waves. These can be shown to travel through a fluid at a speed c, where

$$c \geqslant \sqrt{K_s/\rho} \tag{2-5}$$

For very small pressure waves, the equality sign applies and so $\sqrt{K_s/\rho}$ is the sonic speed. An identical expression with identical meaning describes the speed of sound in solids.

Unsteady flows may be classified into two categories, namely those involving small rates of change and those involving rapid changes. In the former category, analogous to the deceleration of a motor car by a gentle application of the brakes, compressibility is unimportant. In the latter, analogous to deceleration due to collision with a wall, say, compressibility is very important. The latter type of unsteadiness is typically generated in a pipeline by the rapid closure of a valve or by a pump trip due to electrical failure. The *magnitudes* of pressure waves generated in this manner are found to be proportional to the wavespeed c and so are strongly dependent upon compressibility.

2-6-2 Compressible and Incompressible Flows

Notwithstanding the importance of compressibility in highly unsteady flows, it is conventional to reserve the term *compressible flow* for those flows in which compressibility plays a significant role in determining the *steady* component. In practice the term is used only for certain gas flows.

Steady, low-speed gas flows involving no external heat or work may be analysed using incompressible-flow formulae. Significant errors (> 1 per cent) do not result from this approximation until the fluid particle velocities exceed about 30 per cent of the local sonic velocity. Thereafter, the errors become increasingly more important until, at supersonic velocities, incompressible analyses can be so inaccurate as to fail to predict correctly whether a fluid will accelerate or decelerate in given conditions.

The *Mach number, Ma*, defined by

$$Ma \equiv \frac{V}{c} \tag{2-6}$$

in which V is the local particle velocity, is frequently used to characterize a flow. Whenever Ma is sufficiently small, flows involving no external heat or work may be regarded as incompressible. At higher values—$Ma > 0.3$, say—account must be taken of the compressibility of the fluid.

2-7 VISCOSITY AND THE NO-SLIP CONDITION

Figure 2-5a depicts a steady laminar flow along a circular pipe. The broken line 1 indicates the position at which a fine line of dye is imagined to have been released into the fluid along the pipe diameter without causing any disturbance to the flow. The lines 2, 3 and 4 depict successive positions of the dye as the fluid flows along the pipe. It can be seen that particles at different radii move at different velocities, these being at a maximum on the pipe axis and zero at the wall. This behaviour contrasts sharply with that which would be expected of a close-fitting solid cylinder moving along the pipe. The fluid does not slide; it flows.

The radial variation of the axial velocity is depicted in Fig. 2-5b. Except on the axis, the *velocity gradient* du/dy is everywhere positive. Experiment shows that shear stresses must be set up in the fluid to resist this velocity gradient, and it is usually possible to relate these parameters at constant temperature and pressure by an expression of the form

$$\tau = f\left(\frac{du}{dy}\right) \tag{2-7}$$

For the special case of *Newtonian* fluids such as water and air, the shear stress τ at constant temperature and pressure is directly proportional to the velocity gradient, that is

$$\tau = \mu \frac{du}{dy} \tag{2-8}$$

in which the coefficient μ is known as the *dynamic viscosity* (or *absolute viscosity*). A generalized form of this expression suitable for use in one-, two- or three-dimensional flows is presented in section 4-6.

The dynamic viscosity of a typical fluid varies only slightly with pressure but quite strongly with temperature. Normally, it *decreases* with temperature for liquids but *increases* with temperature for gases, this being a predictable consequence of the very different molecular densities in the two phases. Typical values are presented for water and air in Fig. 2-6 and in Appendix A.

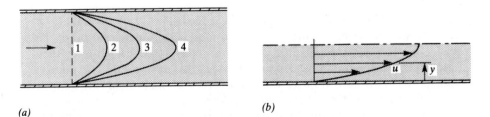

(a)   *(b)*

Figure 2-5 Velocity distribution in laminar flow along a constant-diameter pipe.

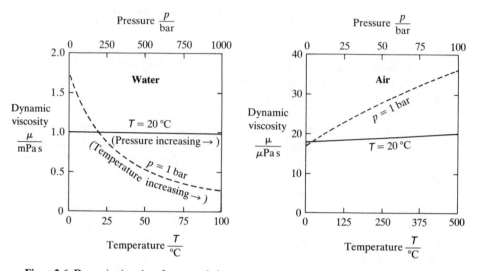

Figure 2-6 Dynamic viscosity of water and air at constant pressure and at constant temperature.

Turbulent flow The conditions depicted in Fig. 2-5 imply a very orderly sort of behaviour in which individual particles follow well-defined paths parallel to the pipe walls. This is a good approximation to observed behaviour at sufficiently low speeds, but it is a poor approximation at higher speeds when the flow becomes turbulent. In these circumstances, the behaviour of individual particles appears more or less random at first sight although their average behaviour is broadly similar to that depicted in Fig. 2-5. Viewed in sufficiently fine detail, the flow within individual turbulent eddies is more orderly and has the characteristics of a laminar flow. In practice, however, the necessary degree of detail is far too fine for this approach to be possible with present-day computers. Instead, we redefine local average velocities (local in time as well as space) to average out the influence of individual eddies, and then attempt to use continuum concepts as before. As a result, turbulent viscosity models have to account for eddy motions as well as molecular motions and they are accordingly less precise. We shall not use them explicitly in this book.

No-slip condition In Fig. 2-5, the velocity at the wall of the pipe is zero. To understand the reason for this, it is useful to consider the interface between the fluid and the solid in detail. Since all solid surfaces appear rough when they are greatly magnified, Fig. 2-7 may be regarded as typical. The fluid is imagined to be flowing from left to right past the stationary solid, and we wish to explain why the fluid at the surface is stationary.

Because of the highly irregular geometry, fluid particles close to the

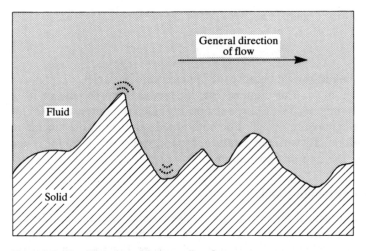

Figure 2-7 Magnified view of a pipe wall surface.

surface are constrained to move along strongly curved paths involving extremely small radii of curvature. Particles slightly further away are similarly constrained, but the various radii of curvature are very different. Now, for reasons explained in Chapter Ten, the existence of significant proportional differences in the radii of curvature of adjacent streamlines implies the existence of significant proportional differences in velocity. This in turn would imply the existence of large shear stresses unless the absolute velocities were very small (see Eq. 2-8). In practice however experiment shows the shear stress to be very small indeed and we therefore conclude that velocities close to the surface must be negligible.

Stokes offered an alternative explanation for the no-slip condition. In effect he argued that since layers of fluid cannot 'slip' past one another it is implausible to imagine that they can slip past solids. In other words if the molecular attractions between adjacent fluid molecules can resist fluid 'slip' then molecular attraction between adjacent fluid and solid molecules should be able to resist interfacial slip.

Inviscid fluids Although all real fluids exhibit viscosity, it is sometimes useful to consider the behaviour of an imaginary *inviscid* fluid in which viscosity is absent. In such cases, we always also discard the no-slip condition and imagine that the fluid can flow freely over solid surfaces.

2-7-1 Kinematic Viscosity

In many analyses of fluid flows, the dynamic viscosity appears as a ratio with the density—see Eq. 1-1 for example. The ratio has come to be known as the *kinematic* viscosity v, defined as

$$v \equiv \frac{\mu}{\rho} \qquad (2\text{-}9)$$

The term 'kinematic' is used because the dimensions of v, namely $[L^2/T]$, are independent of mass. For liquids, the kinematic viscosity shows little dependence on pressure. For gases, a strong dependence on pressure is found except when the pressure is high. For both liquids and gases, the kinematic viscosity varies rapidly with temperature. Typical values are given for water and air in Fig. 2-8 and in Appendix A.

2-7-2 Non-Newtonian Fluids

Many fluids of practical interest satisfy the Newtonian relationship (2-8) at constant temperature and pressure. There are, however, some notable exceptions such as blood, milk, bitumen and slurries. For these, the general expression (2-7) normally takes the form

$$\tau = \tau_0 + \mu' \left(\frac{du}{dy}\right)^N \qquad (2\text{-}10)$$

in which μ' and N are not necessarily constants. Some examples of non-Newtonian behaviour are illustrated in Fig. 2-9. It can be seen that plastics behave like solids until some threshold shear stress τ_0 is exceeded; at higher stresses they behave as fluids. This leads to intriguing possibilities such as flow along a pipe with the outer layer in fluid form surrounding a solid core moving at the maximum velocity.

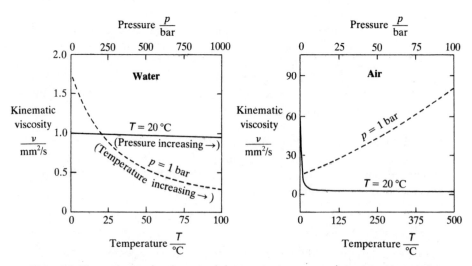

Figure 2-8 Kinematic viscosity of water and air at constant pressure and at constant temperature.

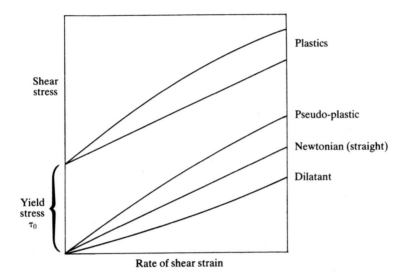

Figure 2-9 Rheological chart.

The parameters μ' and N in Eq. (2-10) are sometimes found to be dependent upon the history of the fluid particles as well as upon the instantaneous conditions. In some cases, this may be a consequence of using the expression to describe motions in a non-homogeneous substance in which flexible, solid particles are suspended in a fluid. For example, whole blood is often regarded as a non-Newtonian fluid composed of corpuscles suspended in a plasma that by itself is approximately Newtonian. The flow of non-Newtonian fluids is not dealt with explicitly in this book. Nevertheless, much of the work presented is applicable to any fluid that can be regarded as a pure substance. Obvious exceptions occur whenever shear stresses are considered either explicitly, as in laminar flows, or implicitly, as in the derivation of empirical relationships for pipe friction.

2-8 SURFACE TENSION

Engineers normally deal with fluids in such large quantities that *surface tension* effects may be neglected. However, important exceptions occur whenever the flow has a free liquid surface (liquid–gas or liquid–liquid) with a short boundary. Examples include the growth and collapse of bubbles in a cavitating liquid, and model ships built to simulate the behaviour of prototypes. Another common consequence of surface tension is the phenomenon of *capillarity*, namely the elevation of a liquid surface in a capillary tube or in the pores between soil particles.

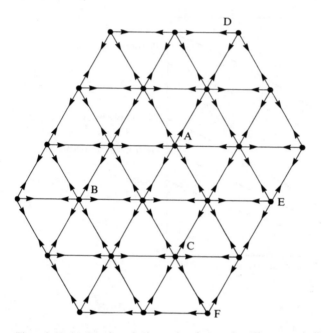

Figure 2-10 Analogy for cohesion and surface tension. The arrows indicate forces experienced by the nails (molecules).

The origin of surface tension is interesting because it is simply the surface manifestation of a phenomenon that exists throughout the liquid, namely molecular attraction. Each molecule is attracted by all other molecules in its neighbourhood, making the substance *cohesive* and causing it to resist disintegration. On average, an internal molecule is attracted equally in all directions and so experiences no *net* force. In contrast, a molecule in or near the surface experiences a net inwards force, and it is this that gives rise to the so-called 'surface' tension. Figure 2-10 depicts a two-dimensional analogy in which a group of nails has been hammered into a flat piece of wood. Their heads have been joined by pieces of elastic. If the tension is the same in each piece of elastic, then internal nails such as A, B and C will experience no *net* force. However, nails such as D, E and F around the perimeter will experience a net force directed inwards towards the main body of nails. The tension in the elastic around the perimeter is closely analogous to the surface tension in a liquid.

In a liquid, the molecules (nails) are not fixed in position, but are free to move in response to the cohesive forces. Since all surface molecules are attracted towards the interior, the surface tends to contract until its area is the minimum possible for which equilibrium conditions can be maintained— hence the spherical shape of soap bubbles and liquid droplets in the absence of external forces. The proof of this result is straightforward and is based on the minimization of the *free-surface energy* which can be shown to exist as a consequence of the surface tension (section 11-1-3).

The nail–elastic analogy gives a molecular view of the phenomenon. On a larger scale, the effect is similar to that of a very thin elastic sheet stretched over the whole of the surface of the liquid. To pursue this analogy, it is necessary to imagine that the tension in the membrane is the same at all points and in all directions, and that it is independent of the size of the surface. The strain energy stored in the surface is proportional to the area of the surface.

When the interface between the fluids is flat, the pressure is the same on both sides of it. However, when it is curved, there is a pressure difference Δp which satisfies

$$\Delta p = \sigma \left(\frac{1}{R_1} + \frac{1}{R_2} \right) \tag{2.11}$$

in which σ is the coefficient of surface tension (for which the SI unit is N/m) and R_1 and R_2 are radii of curvature of the surface in two mutually perpendicular directions. For a spherical surface R_1 and R_2 are equal and the higher pressure is on the inside of the surface. For other shapes, such as saddles, account must be taken of the signs of R_1 and R_2. Appropriate values of the coefficient of surface tension for a clean water–air interface at atmospheric pressure are given in Fig. 2-11. The figure may also be used to find the surface tension between water and steam at saturation. For this case, the curve could be extended to the critical point. The value of the coefficient varies almost linearly from a maximum at the triple point temperature (0.01 °C) to zero at the critical point temperature (374.15 °C).

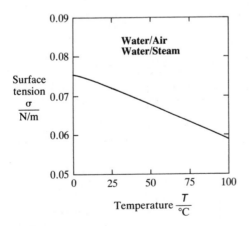

Figure 2-11 Surface tension of a clean water/air or water/steam interface (pressure and temperature both vary).

2-9 THERMODYNAMIC PROPERTIES

Internal energy The *internal energy* of a fluid substance is energy that is stored mainly in the random movement of its molecules. It is the only form of energy that can be influenced directly by heat transfers and for this reason it was historically regarded as a sort of 'stored heat'. This practice has long been discredited however because it is also possible to increase or decrease the internal energy by means of work—by compression for instance. In such cases, it would be more logical to regard internal energy as a sort of 'stored work'. Nevertheless, it is well known that 'stored heat' can be used to do work and that 'stored work' can be used for heating. In reality, the stored energy is the *same* in both cases; it is sensibly called the internal energy.

For pure substances, it is usual to define a *specific* internal energy u, namely the internal energy *per unit* mass. Typical variations of this property with pressure and temperature are shown for water and air in Fig. 2-12 and in Appendix A. It is usually acceptable to regard the specific internal energy of a liquid or a gas as a linear function of the temperature alone.

Enthalpy In many analyses of fluid flows, the specific internal energy u occurs as a sum with the ratio p/ρ. The sum has come to be known as the *specific enthalpy h*, defined as

$$h \equiv u + \frac{p}{\rho} \tag{2-12}$$

Typical values of the specific enthalpy of water and air are given in Fig. 2-13

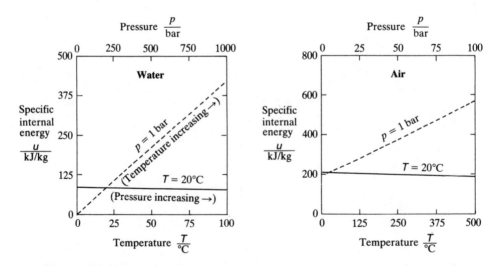

Figure 2-12 Specific internal energy of water and air at constant pressure and at constant temperature.

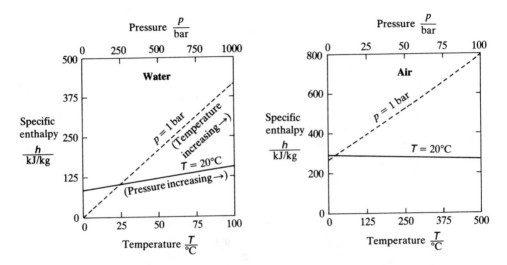

Figure 2-13 Specific enthalpy of water and air at constant pressure and at constant temperature.

and in Appendix A. For gases (but not for liquids) it is usually acceptable to regard the specific enthalpy as a function of the temperature alone.

Specific heat capacities The specific heat capacities c_v and c_p of a pure substance are defined as the rates of change of the specific internal energy and the specific enthalpy with temperature at constant density and pressure respectively. That is,

$$c_v \equiv \left(\frac{\partial u}{\partial T}\right)_\rho \qquad (2\text{-}13)$$

and

$$c_p \equiv \left(\frac{\partial h}{\partial T}\right)_p \qquad (2\text{-}14)$$

When u and h are functions of T alone, it is not necessary to stipulate that the derivatives must be obtained at constant density or pressure. When the dependence is also linear, as it is for perfect gases, the specific heat capacities are constants and Eqs (2-13) and (2-14) can be integrated to give

$$(u_2 - u_1) = c_v(T_2 - T_1) \qquad (2\text{-}15)$$

and

$$(h_2 - h_1) = c_p(T_2 - T_1) \qquad (2\text{-}16)$$

Ratio of specific heat capacities The ratio of the specific heat capacities γ, defined by

$$\gamma \equiv \frac{c_p}{c_v} \qquad (2.17)$$

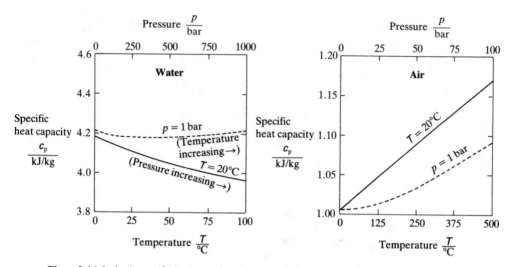

Figure 2-14 Isobaric specific heat capacity of water and air at constant pressure and at constant temperature.

is a useful property. For liquids, it is often nearly equal to unity but this is not so for gases, typical values for air and steam being about 1.4 and 1.3 respectively at low pressures and moderate temperatures. The dependence of the specific heat capacities of water and air on pressure and temperature is illustrated in Figs 2-14 to 2-16 and in Appendix A. For perfect gases, it can readily be shown that the gas constant R is related to the specific heats by the simple relationship

$$R = c_p - c_v \qquad (2\text{-}18)$$

so that

$$R = c_v(\gamma - 1) \qquad (2\text{-}19)$$

and

$$R = c_p(1 - 1/\gamma) \qquad (2.20)$$

Entropy The *entropy* of a system is discussed in section 12-4. It is a measure of the extent to which the energy of the system is unavailable for doing work. For a pure *fluid* substance, elemental changes in the *specific* entropy s (entropy per unit mass) in the absence of magnetism, electricity and capillarity satisfy

$$T\, \delta s = \delta u + p\, \delta\left(\frac{1}{\rho}\right) \qquad (2\text{-}21)$$

in which T denotes the *absolute* temperature. Typical values of the specific entropy are given for water and air in Fig. 2-17 and in Appendix A. The above expression is especially useful when dealing with *isentropic* processes ($ds = 0$).

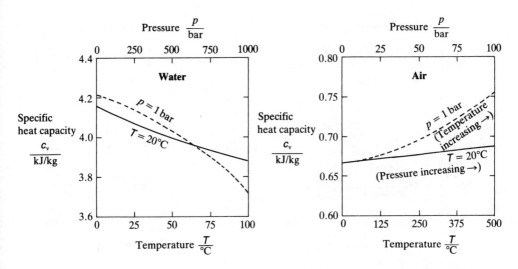

Figure 2-15 Isochoric specific heat capacity of water and air at constant pressure and at constant temperature.

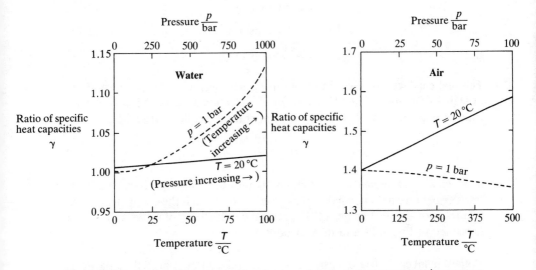

Figure 2-16 Ratio of specific heat capacities of water and air at constant pressure and at constant temperature.

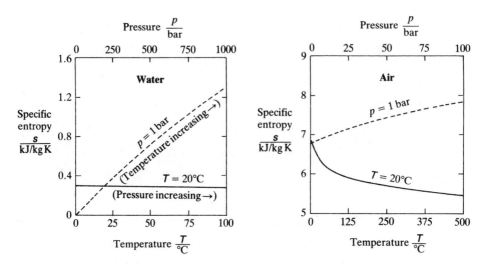

Figure 2-17 Specific entropy of water and air at constant pressure and at constant temperature.

For an isentropic process with a *perfect gas*, Example 12-2 gives

$$p/\rho^\gamma = \text{constant} \tag{2-22}$$

$$T/\rho^{\gamma-1} = \text{constant} \tag{2-23}$$

and

$$T/p^{1-1/\gamma} = \text{constant} \tag{2-24}$$

Thermal conductivity Heat interactions occur solely as a consequence of the existence of temperature differences. In a *conduction* process, the *rate* of 'heat' transfer is proportional to the local temperature gradient. It satisfies

$$\dot{q} = \lambda \frac{\partial T}{\partial x} \tag{2-25}$$

in which λ is the *thermal conductivity* of the substance and $\dot{q}$ is the rate of heat transfer per unit area normal to the direction x in which the energy transfer is taking place. Typical values of the thermal conductivity of water and air are presented in Fig. 2-18 and in Appendix A.

Prandtl number In heat transfers involving convection, 'warm' fluid particles mix with 'cool' particles and, because of their temperature differences, local heat conductions occur. Although the energy transfers occur solely by conduction, the mixing of the warm and cool particles permits a much more rapid thermal diffusion within the fluid than could occur by conduction alone. The same mixing process necessarily also involves momentum transfers, and the relative importance of the two phenomena is of interest. The *Prandtl number*, *Pr*, is a measure of the relative ability of the fluid to allow momentum

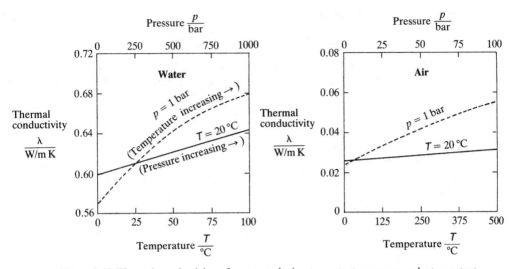

Figure 2-18 Thermal conductivity of water and air at constant pressure and at constant temperature.

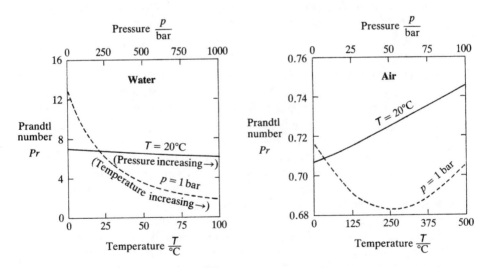

Figure 2-19 Prandtl number of water and air at constant pressure and at constant temperature.

diffusion and thermal diffusion. It is defined by

$$Pr \equiv \frac{\mu c_{\mathrm{p}}}{\lambda} \tag{2-26}$$

in which μ is the dynamic viscosity. Typical values of the dimensionless Prandtl number are presented for water and air in Fig. 2-19 and in Appendix A.

2-10 USE OF FLUID PROPERTIES

Example 2-1: Speed of sound Estimate the speed of sound in water at a temperature of 20 °C and a pressure of (a) 1 bar and (b) 500 bar. Also estimate the speed of sound in air at 20 °C and at pressures of (a) 1 bar and (b) 50 bar assuming that its gas constant $R = 287$ J/kg K.

SOLUTION At this temperature, Fig. 2-4 gives the isentropic bulk modulus of water as approximately $K_S = 2.19$ GPa and 2.55 GPa at pressures of 1 bar and 500 bar respectively. The corresponding densities in Fig. 2-3 are $\rho = 998$ kg/m^3 and 1091 kg/m^3. Using Eq. (2-4), we find the speeds of sound are

$$c = \sqrt{K_S/\rho} \tag{2-27}$$

namely 1481 m/s and 1582 m/s.

For air, the isentropic bulk modulus is equal to γp. Figure 2-16 gives $\gamma \simeq 1.40$ at 1 bar and $\gamma \simeq 1.495$ at 50 bar and Eq. (1-11) gives the corresponding densities as $\rho \simeq 1.19$ kg/m^3 and 59.4 kg/m^3 respectively. Using Eq. (2-5), the two speeds of sound are approximately 343 m/s and 355 m/s.

Comments on Example 2-1
1. Experimentally, it is very difficult to measure values of the bulk modulus of a fluid with high accuracy, especially at extreme conditions of temperature and pressure. In practice, it is usual to measure the speed of sound in the fluid and to deduce the bulk modulus using Eq. (2-5).
2. The speed of sound in a mixture of water and air can be much smaller than the corresponding speed in either fluid alone. When free air bubbles exist in water, the compressibility of the mixture is governed largely by the air whereas the density is governed by the water. If, say, 5 per cent of the mixture (by volume) is in the form of air bubbles and 95 per cent is water, then the overall bulk modulus will be approximately 20 times that of air while the density will be 95 per cent that of water. At 20°C and 1 bar, these values yield a sonic speed of approximately

$$c \simeq \sqrt{2.8\,\text{MPa}/(948\ \text{kg/m}^3)} \simeq 54\ \text{m/s}$$

Example 2-2: Skin friction Figure 2-20 depicts a laminar flow between two parallel plates, the flow being induced by sliding the upper plate over the lower at 5 m/s. Estimate the shear force per unit area on the underside of the upper plate if the viscosity of the fluid is 0.001 Pa s (Pascal seconds). The gap between the plates is 0.2 mm.

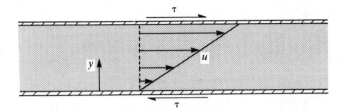

Figure 2-20 Skin friction in a shear flow between parallel plates.

SOLUTION The depicted velocity gradient is constant and equal to $5\,\text{m/s} \div 0.2\,\text{mm}$, namely $25\,000\,\text{s}^{-1}$. Using Eq. (2-8), we find the shear stress is

$$\tau = \mu\frac{du}{dy} = 25\,\text{Pa} \tag{2-28}$$

and the shear force on an area $A = 1\,\text{m}^2$ of the plate is $\tau A = 25\,\text{N}$.

Comments on Example 2-2
1. Suppose that the upper 'plate' is the underside of a smooth shoe and that the lower plate is an ice-covered pavement. If we assume a shoe contact area of $0.01\,\text{m}^2$, the resistance to sliding will be $0.25\,\text{N}$, which is certainly too little for someone trying to walk safely. Viewed more positively, this explains why children can slide so easily over ice, bearing in mind that the pressure induced by their weight causes the surface of the ice to melt.
2. This example could be used as a starting point for an introduction to the principles of lubrication. As shown in Example 7-9 (p. 198), however, it is necessary to extend the discussion to include the effects of a non-parallel gap between the two surfaces. Lubrication is successful only if pressure forces induced by the sliding action tend to hold the solid surfaces apart.

Example 2-3: Surface tension Deduce the excess pressure inside (*a*) a spherical water droplet of 2 mm diameter falling steadily through air and (*b*) a 2 mm diameter air bubble rising steadily through water. Assume the coefficient of surface tension to be $0.073\,\text{N/m}$ in both cases.

SOLUTION Figure 2-21*a* depicts the pressures and surface tension acting in the *y*-direction on an imaginary rectangular box containing one half of a spherical particle. By regarding the figure as a plan view and choosing axes moving with the particle in the *z*-direction, we can avoid complications due to gravity and assume that the depicted forces are in equilibrium. For a particle of radius R we obtain

$$(p + \Delta p)\pi R^2 - p\pi R^2 - \sigma 2\pi R = 0 \tag{2-29}$$

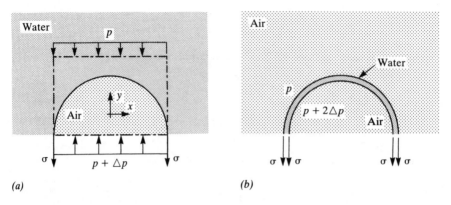

Figure 2-21 Pressure differences across curved interfaces ($\Delta p = 2\sigma/R$). (a) Air bubble in water. (b) 'Soap' bubble.

which leads to

$$\Delta p = \frac{2\sigma}{R} \qquad (2\text{-}30)$$

This expression is valid irrespective of whether the particle is a water droplet in air or an air bubble in water. In both cases, the internal pressure exceeds the external pressure by $2\sigma/R$.

Comments on Example 2-3

1. It was pointed out in section 2-4-1 that the rate at which dissolved air is released from solution in water is different from the rate at which it will subsequently redissolve. This is partly explained by Eq. (2-30) which shows that the excess pressure inside a bubble is inversely proportional to its radius. It follows that bubbles of infinitesimal radius cannot exist stably.
2. The excess pressure inside a 'soap' bubble is double the value given by Eq. (2-30). This is because there are two water–air interfaces, each with its own surface tension. In effect, the soap bubble is an air bubble inside a water droplet of only marginally greater diameter. Figure 2-21b illustrates this interpretation.

Example 2-4: Isentropic compression The pressure and temperature of the air in a bicycle pump are initially 1 bar and 20 °C respectively. After compression, the volume of each element of air has reduced to a quarter of its initial value. By assuming that the compression takes place isentropically and that $\gamma = 1.4$, estimate the pressure and temperature of the air available for delivery to the bicycle tube.

SOLUTION With the initial and final conditions denoted by the suffices i and f respectively, Eq. (2-22) yields

$$p_\text{f}/p_\text{i} = (\rho_\text{f}/\rho_\text{i})^\gamma \qquad (2\text{-}31)$$

Since the density ratio ρ_f/ρ_i is 4 and the initial pressure is 1 bar, we obtain $p_f = 6.96$ bar. The final temperature may be obtained from either of Eqs (2-23) and (2-24). Using the former, we find

$$\frac{T_f}{T_i} = \left(\frac{\rho_f}{\rho_i}\right)^{\gamma-1} \tag{2-32}$$

Since $T_i = 20\,°C \simeq 293\,K$, we obtain $T_f \simeq 475\,K \simeq 202\,°C$.

Comment on Example 2-4   Many people are surprised to discover that such high pressures and temperatures occur in this commonplace device. The need to generate such a high pressure restricts the acceptable diameter to the pump because the force exerted by the operator is approximately equal to the product of the area of the piston and the pressure difference across it. With a 20 mm bore pump the force need to maintain the *gauge* pressure of 5.96 bar would be 187 N.

PROBLEMS

1 The temperature of water at its triple point is defined to be 273.16 K. Express this temperature in degrees Celsius and explain why it differs from the value quoted for the temperature of freezing water in everyday discussions.

[0.01 °C]

2 If the atmospheric pressure on a particular day is 98 kPa, what is the absolute pressure 45 m below the surface of the sea ($\rho = 1029$ kg/m^3)? Also what gauge pressure is implied by a blood pressure recorded as 135 mmHg? The gravitational acceleration is 9.81 m/s^2 and the density of mercury is 13.6 Mg/m^3.

[522 kPa, 18.0 kPa]

3 A 4 m^3 vessel contains a mixture of air ($R = 287$ J/kg K) and natural gas ($R = 520$ J/kg K) at an absolute pressure of 140 kPa and a temperature of 30 °C. If the total mass of the contents is 6 kg, determine the mass of air.

[5.46 kg]

4 A 4 m layer of oil with a relative density of 0.86 floats on sea water ($\rho = 1026$ kg/m^3) in an open storage vessel. Estimate the gauge pressure at a depth of 3 m below the oil–water interface. Take $g = 9.81$ m/s^2 and $\rho_w = 998$ kg/m^3.

[38.6 kPa]

5 The bulk modulus K of water may be assumed to vary linearly from 2.193 GPa at a pressure of 1 bar to 2.933 GPa at a pressure of 1000 bar. If the density at 1 bar is 998 kg/m^3, determine (*a*) the density at 1000 bar and (*b*) the percentage error in the estimated density change arising from assuming a constant value of K equal to the mean of the above values.

[1038 kg/m^3, 0.70 per cent]

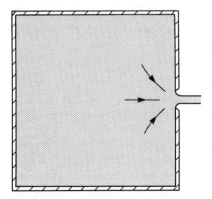

Figure 2-22 Steady discharge from a high-pressure vessel.

6 The isentropic bulk moduli of air, water and steel at an atmospheric pressure of 100 kPa are 140 kPa, 2.193 GPa and 172.5 GPa respectively and the corresponding densities are 1.2 kg/m³, 998 kg/m³ and 7850 kg/m³. Estimate the speed of sound in the three substances.

[342 m/s, 1482 m/s, 4688 m/s]

7 Figure 2-22 depicts a fluid discharging steadily to the atmosphere through an orifice in the side of a pressurized tank. The velocity of flow just downstream of the orifice is 250 m/s. By comparing the velocity of flow with the speed of sound evaluated in Problem 6, determine whether the compressibility of the fluid has a significant influence on the flow. Assume that the fluid is (a) air and (b) water.

[yes, no]

8 A 10 mm cylindrical rod is pulled axially through a 10.02 mm diameter cylinder, the width of the gap being uniform around the circumference (Fig. 2-23). If the cylinder is 125 mm long and the viscosity of the oil that fills the gap is 0.01 Pa s, what force F must be applied to the rod to maintain a constant speed of 20 mm/s?

[0.118 N]

9 The absolute viscosities of air ($R = 287$ J/kg K) and water ($\rho = 998$ kg/m³) at a pressure of 100 kPa and a temperature of 20 °C are 18.07 µPa s and

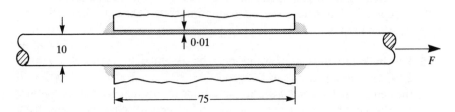

Figure 2-23 Axial movement of a cylindrical rod (all dimensions in millimetres).

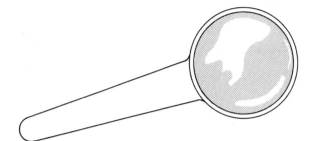

Figure 2-24 Soap film in a circular hoop.

1.002 mPa s respectively. Show that the kinematic viscosity of the air is about 15 times greater than that of the water.

10 The pressures inside two spherical oil droplets exceed the pressure in the surrounding water by 10 Pa and 1 kPa respectively. Estimate the diameters of the droplets if the surface tension coefficient of the oil–water interface is 0.021 N/m.

[8.4 mm, 0.084 mm]

11 A flat soap film ($\sigma = 0.067$ N/m) is contained in a circular hoop as illustrated in Fig. 2-24. Estimate the induced axial force around the 75 mm diameter hoop.

[5.02 mN]

12 Steam at an initial pressure of 1 bar and a temperature of 100 °C is heated at constant pressure to a temperature of 500 °C. Estimate the increase in its specific entropy, approximating the substance to a perfect gas for which $R = 458$ J/kg K and $c_p = 2030$ J/kg K.

[1480 J/kg K]

13 Air at an initial pressure of 1 bar and a temperature of 100 °C is heated and compressed isentropically to a temperature of 500 °C. Estimate the resulting pressure, treating the air as a perfect gas for which $R = 287$ J/kg K and $\gamma = 1.40$.

[1.28 MPa]

FURTHER READING

Engineering Sciences Data Unit (1978) *Physical Data, Mechanical Engineering.*

Haywood, R.W. (1972) *Thermodynamic Tables in SI Units*, 2nd edn, Cambridge University Press.

Kennan, J.H. and Kaye, J. (1987) *Gas Tables* 2nd edn, Wiley.

Reid, R.C., Prausnitz, J.M. and Sherwood, T.K. (1987) *The Properties of Gases and Liquids*, 4th edn, McGraw-Hill.

Vargaftik, N.B. (1983) *Handbook of Physical Properties of Liquids and Gases*, 2nd edn, Hemisphere.

Walton, A.J. (1983) *Three Phases of Matter*, 2nd edn, McGraw-Hill.

THREE

FLUID PRESSURE

3-1 CLASSIFICATION OF FORCES

The science of fluid mechanics deals principally with relationships between fluid motions and the forces that cause them. Pressure forces exerted by jet engines enable huge aircraft to take off and fly. Shear forces on the walls of a pipe make it necessary for water and blood to be pumped through their distribution networks. Hydrostatic pressure forces attempt to cause dams to overturn. Gravitational forces cause stones to sink, ships to float and balloons to soar.

In this chapter and the next, attention is focused on fluid forces rather than on the motions they cause. Examples cited in the development deal exclusively with cases where the fluid may be regarded as being in equilibrium, and the simplest example is undoubtedly a stationary fluid. Nevertheless, most of the derivations apply equally to fluids in which forces are not in equilibrium, as is the case in much of Chapters Six to Twelve.

It is convenient to classify fluid forces in two groups, namely *body* forces and *surface* forces. The former result from field effects such as gravitation while the latter result mainly from interactions between the fluid and its surroundings.

Body forces Figure 3-1 depicts water flowing along a channel and passing over a weir. Arbitrarily chosen elements A, B and C are shown at an instant when A is within the fluid stream, B is at the free surface and C is touching a solid boundary. The earth's gravitational field exerts a force $\rho g V$ on each of

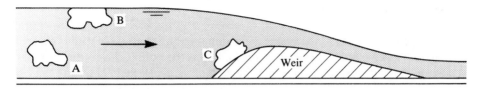

Figure 3-1 Free surface flow over a weir.

these elements, where ρ is the density, g is the local gravitational acceleration and V is the volume of A, B or C. In more general cases where the density of the fluid varies, the weight force is $\int \rho g \, dV$.

The gravitational force acts directly on each bit of the fluid. It is experienced by particles simply as a result of their presence in a gravitational field and it is independent of whatever they happen to be doing. Other forces of this sort can be caused by magnetic or electric fields if the fluid is responsive to them. There may also be inertial forces when analyses are carried out relative to rotating or linearly accelerating axes.

All of these are examples of *body* forces, that is forces that act directly on the body of the substance. In this book, gravity is the only source of body forces unless otherwise stated.

Surface forces The fluid elements A, B and C also experience normal and tangential forces on their *surfaces*. These are imposed by the surrounding water and also by the atmosphere on element B and by the weir on element C. Surface tension forces act on the element B where its boundary meets the free water surface.

The normal surface forces acting on element A arise from the fluid pressure. By definition, they act normal to the chosen surface and in practice, they act inwards—that is, they tend to compress the element. Particles within the element push outwards with exactly equal strength. Similar forces exist on the surfaces of particles B and C, but the external forces are applied in part by the atmosphere and the weir.

The tangential surface forces on the elements arise from the fluid motion which causes the particles to distort as they move downstream. The distortion involves shear strains which (with the possible exceptions of some super-conducting fluids) are resisted by shear stresses. In practice, however, shear stresses in fluids are often very small and a considerable body of literature exists on the study of (hypothetical) ideal fluids in which shear stresses are absent.

In this chapter, we deal with normal stresses, that is with fluid pressure. Chapter Four deals with tangential stresses, namely shear.

3-2 NORMAL STRESS

The pressure *at a point* in a fluid is defined in section 2-4 at the centroid of an imaginary elemental surface within the fluid. It is the ratio of the normal force acting on the elemental surface and the area of that surface. In choosing this definition, it is implicitly assumed that the result obtained will be independent of the orientation of the imaginary surface within the fluid, that is we assume that the pressure acts in all directions simultaneously. This is exactly true in a stationary fluid and approximately true in a moving fluid. It is a consequence of the apparently random movement of the fluid molecules and it accounts for the ability of a balloon to maintain its shape for instance. We shall now prove the validity of this assumption, which is sometimes called *Pascal's law*.

Consider the fluid element depicted in Fig. 3-2 in which the z axis is chosen vertically upwards. The normal stresses at the centres of the faces parallel to the x, y and z axes are σ_y, σ_z and σ_θ and shear forces are absent. The normal *force* on each face is the product of the stress and the area of the face, and the only other force is assumed to be the body force due to gravity. The length of the sloping side is equal to both $\delta y \sec \theta$ and $\delta z \csc \theta$, and so the normal force on the inclined face may be written as either $\sigma_\theta \, \delta x \, \delta y \sec \theta$ or $\sigma_\theta \, \delta x \, \delta z \csc \theta$.

For vertical equilibrium of the element, the difference between the force on the lower face and the vertical component of the force on the inclined face must balance the weight of the element whose volume is $\frac{1}{2} \delta x \, \delta y \, \delta z$. Therefore

$$\sigma_z \, \delta x \, \delta y - (\sigma_\theta \, \delta x \, \delta y \sec \theta) \cos \theta = \tfrac{1}{2} \rho g \, \delta x \, \delta y \, \delta z \tag{3-1}$$

and so
$$\sigma_z - \sigma_\theta = \tfrac{1}{2} \rho g \, \delta z \tag{3-2}$$

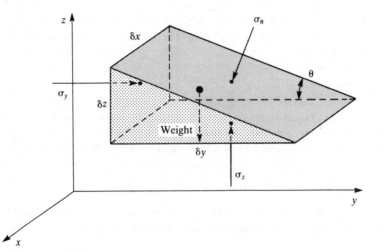

Figure 3-2 Normal stresses (pressure) on a fluid element.

Eq. (3-2) may be interpreted as a statement that the difference between σ_z and σ_θ is negligible when δz is sufficiently small. Therefore, in the limit

$$\sigma_z = \sigma_\theta \qquad (3\text{-}3)$$

Similarly, for horizontal equilibrium of the element, the normal force on the vertical face must balance the horizontal component of the force on the inclined face. Thus

$$\sigma_y \; \delta x \; \delta z = (\sigma_\theta \; \delta x \; \delta z \; \text{cosec} \; \theta) \sin \theta \qquad (3\text{-}4)$$

and so $$\sigma_y = \sigma_\theta \qquad (3\text{-}5)$$

Eqs (3-2) and (3-5) together show that the magnitude of the normal stress in the arbitrary direction θ in the $(y\text{-}z)$ plane is equal to both σ_y and σ_z. Since the orientation of the element in the horizontal plane is also arbitrary, it follows that the magnitude of the normal stress at a point is the same in all directions provided that shear stresses are absent. It is therefore appropriate to define the pressure as

$$p \equiv \sigma_x = \sigma_y = \sigma_z \qquad (3\text{-}6)$$

This proof is valid for any stationary fluid and for the steady flow of an ideal (inviscid) fluid. It could be extended to include *unsteady* inviscid flows by introducing inertia terms that would vanish along with the gravitational term. However, it is not valid for the flow of a viscous fluid because account must be taken of shear stresses and these are not necessarily negligible. Nevertheless, the differences between σ_x, σ_y and σ_z are still found to be small and they are ignored for most practical purposes. In general, the fluid pressure is defined as the average value of the three components of normal stress, i.e.

$$p = \tfrac{1}{3}(\sigma_x + \sigma_y + \sigma_z) \qquad (3\text{-}7)$$

3-3 HYDROSTATIC PRESSURE DISTRIBUTION

Although the fluid pressure is the same in all directions at any particular point, it is not usually the same at different points. In a stationary fluid, for instance, pressure variations can result from gravitational or other field effects. In a moving fluid they can also occur when accelerations or shear stresses are induced.

In this section the effects of acceleration and shear stress are assumed to be absent and it may be helpful to imagine that the whole fluid is at rest. Nevertheless, some of the results can be applied in a moving fluid, perpendicular to the direction of flow—as we shall see in subsequent chapters.

For the special case of a stationary fluid experiencing no body forces, the pressure is the same at all points. When the only body force is due to gravity acting vertically downwards, we find that

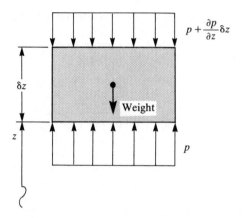

Figure 3-3 Hydrostatic equilibrium of an elemental column.

1. there is no horizontal variation in pressure, and
2. there is a vertical pressure distribution which counterbalances the fluid weight.

A pressure distribution which satisfies these conditions is termed *hydrostatic* (*hydro* as in water, *static* as in stationary).

Horizontal distribution The absence of horizontal pressure variations is usually taken for granted. Nevertheless, it is a most important result. Without it, there would be no means of, say, evaluating pressure forces on the undersides of objects such as boats.

Vertical distribution Consider the column of fluid of height δz and cross-sectional area A shown in Fig. 3-3. For clarity, pressure forces acting on the vertical sides of the element have been omitted because they do not influence vertical equilibrium. The pressure force acting vertically upwards on the lower face is pA and the force acting downwards on the upper face is $(p + \partial p/\partial z\ \delta z)A$. For vertical equilibrium, the difference between the two forces must balance the gravitational force on the element. Therefore

$$-\frac{\partial p}{\partial z}\ \delta z\ A = \rho g A\ \delta z \tag{3-8}$$

and so

$$\boxed{\frac{\partial p}{\partial z} = -\rho g} \tag{3-9}$$

is the rate of change of pressure with height. Notice that the minus sign indicates that the pressure in a stationary fluid always decreases with height.

Liquids When the density is assumed constant, Eq. (3-9) may be integrated to give the linear pressure distribution

$$p - p_0 = -\rho g z \tag{3-10}$$

in which p_0 denotes the pressure at the arbitrary datum $z = 0$. When dealing with liquids with a free surface, it is usual to express the pressure as a function of the depth d below the surface (Fig. 3-4). If the surface is at atmospheric pressure,

$$p - p_{AT} = \rho g d \tag{3-11}$$

or, using the specific weight $w = \rho g$,

$$p - p_{AT} = w d \tag{3-12}$$

It is verified in Example 3-1 (p. 66) that the small compressibility of real liquids may be neglected when determining hydrostatic pressure variations.

Gases For many engineering purposes, Eq. (3-10) may also be used for gases because the rate of change of density with height is usually very small. Indeed, reductions of pressure with height can often be disregarded without introducing serious errors because the rates of change of pressure are also usually small. However, it is sometimes necessary to integrate Eq. (3-9) for a compressible fluid—when evaluating the pressure distribution in the atmosphere for example. The procedure is illustrated by considering the *isothermal* hydrostatic pressure distribution in a perfect gas. By combining Eq. (3-9) with the equation of state for a perfect gas, (1-11), we obtain

$$\frac{dp}{dz} = -\frac{pg}{RT} \tag{3-13}$$

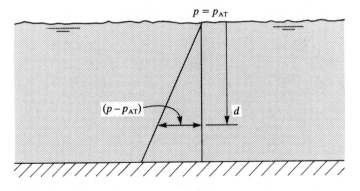

Figure 3-4 Hydrostatic pressure distribution in a liquid, $p - p_{AT} = \rho g d$.

or

$$\frac{dp}{p} = -\frac{g\,dz}{RT} \qquad (3\text{-}14)$$

Integrating:

$$\ln\left(\frac{p}{p_0}\right) = -\frac{gz}{RT} \qquad (3\text{-}15)$$

in which p_0 denotes the pressure at the arbitrary datum $z = 0$. This relationship is approximately valid in the earth's stratosphere. An equivalent expression applicable in the troposphere is derived in Example 3-2 (p. 67).

3-4 HYDROSTATIC FORCES ON SUBMERGED SURFACES

It is commonly necessary to determine the pressure *force* acting on the surface of an object that is partially or totally submerged in a fluid (e.g. the triangular object shown in Fig. 3-5). The total force is the vector sum of all the elemental forces $p\,\delta A$ acting on elemental areas of the surface. In practice, we usually wish to determine the component of the force in a particular direction—the horizontal, say—and this is the sum of all the elemental components $p\,\delta A \cos\theta$ in that direction, i.e.

$$F_x = \int_A p \cos\theta\,dA \qquad (3\text{-}16)$$

in which θ is the angle between the chosen direction and the local surface normal.

The evaluation of this integral occasionally involves complex algebraic

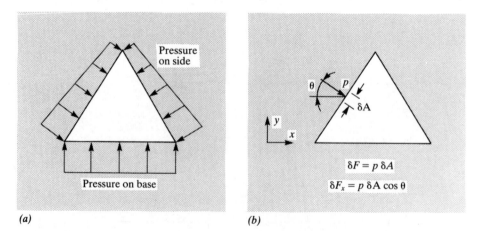

(a)   (b)

Figure 3-5 Pressure on the surface of a submerged object. (a) Pressure distribution. (b) Local pressure force.

manipulations, but it is usually straightforward. With flat surfaces, for instance, $\cos \theta$ is constant and so F_θ is the product of $\cos \theta$ and the total force normal to the surface $\int p \, dA$. When the surface is horizontal, the hydrostatic pressure acting on it is constant and the required force is simply $pA \cos \theta$. Methods of simplifying more general cases are given in sections 3-4-1 and 3-4-2.

Centre of pressure As well as evaluating the magnitude and direction of the resultant force on the surface of an object, it is often necessary to determine its effective line of action. When this is known, it is possible to investigate the overall equilibrium of the object by the usual principles of statics, that is by taking moments and by resolving the forces into, say, horizontal and vertical components. The point of intersection of the line of action of the resultant force with the surface on which it acts is called the *centre of pressure*.

3-4-1 Submerged Flat Surfaces

General expressions are now developed for the force and the position of the centre of pressure on a *flat submerged* surface. Consider the arbitrarily shaped flat surface shown in side and end elevations in Fig. 3-6. The *gauge* pressure at a depth d is $\rho g d$, and so the gauge pressure force acting normal to an elemental strip of area δA is $\rho g d \, \delta A$. Since $d = y \sin \theta$, the elemental force is $\rho g y \sin \theta \, \delta A$ and the total force is

$$F = \rho g \sin \theta \int_A y \, dA \qquad (3\text{-}17)$$

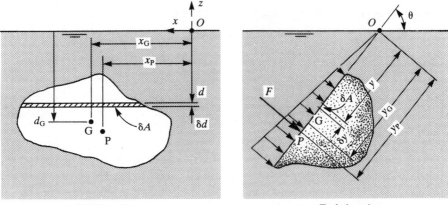

Side elevation   End elevation

Figure 3-6 Pressure distribution on a submerged flat surface. (G = centroid; P = centre of pressure).

The depth of the centre of pressure is found by taking moments about the x axis which is seen in the end elevation as the point O in the liquid surface. The moment of an elemental force $\rho g y \sin \theta \, \delta A$ is $\rho g y^2 \sin \theta \, \delta A$ and the total moment may be equated to the moment of the resultant force to give

$$F y_P = \rho g \sin \theta \int_A y^2 \, dA \qquad (3\text{-}18)$$

By inspection, the integrals in (3-17) and (3-18) are respectively the first and second moments of area of the surface about the x-axis. They may be written as

$$\int_A y \, dA = A y_G \qquad (3\text{-}19)$$

and

$$\int_A y^2 \, dA = I_x \qquad (3\text{-}20)$$

Resultant force By combining Eqs (3-17) and (3-19), we obtain

$$F = \rho g \sin \theta \cdot A y_G \qquad (3\text{-}21)$$

Now $y_G \sin \theta$ is the depth of the centroid and $\rho g y_G \sin \theta$ is the corresponding gauge pressure. This relationship may therefore be expressed as

| The force normal to a flat submerged surface | = | the area of the surface | × | the pressure at its centroid | (3-22) |

which is a general result. It applies even when the liquid does not have a free surface. Also, it applies to absolute pressures as well as to the gauge pressures used in the above derivation.

Centre of pressure By combining Eqs (3-18) and (3-20) we obtain

$$F y_P = \rho g \sin \theta \cdot I_x \qquad (3\text{-}23)$$

By comparing this with (3-21) we find that the position of the centre of pressure satisfies

$$y_P = \frac{I_x}{A y_G} \qquad (3\text{-}24)$$

which is the ratio of the two moments of area about the x-axis in the liquid surface. In practice, it is unlikely that I_x will be known explicitly, but I_{Gx}—the second moment of area about a parallel axis through the centroid—may be known. In this case, I_x may be obtained from the parallel axes theorem:

$$I_x = I_{Gx} + A y_G^2 \qquad (3\text{-}25)$$

The expression (3-24) is less general than (3-22) because the position of the x-axis cannot be chosen arbitrarily. When using *gauge* pressures, the axis must be chosen at the elevation corresponding to atmospheric pressure. This is usually, but not always, at the free surface. When using *absolute* pressures, the axis must be chosen at the elevation notionally corresponding to zero pressure (e.g. approximately 10 m above a free water surface in typical atmospheric conditions).

When necessary, the *lateral* position of the centre of pressure P can be found by taking moments about the z-axis. Usually, however, this need not be done because the position is known from symmetry or because the object extends so far in the x-direction that only a unit length of it need be considered.

3-4-2 Submerged Curved Surfaces

The extension of these ideas to *curved* submerged surfaces is surprisingly straightforward. To demonstrate this, it is convenient to consider the horizontal and vertical force components separately.

Horizontal component The horizontal component is independent of the shape of the surface. Figure 3-7 shows cross-sections through two tanks with one vertical side and one curved side. The pressure force on an elemental strip of a curved surface of width b normal to the page is $pb \, \delta l$ and its horizontal component is $pb \, \delta l \cos \theta$. Now $\delta l \cos \theta$ is simply the elemental depth δd and so the horizontal force on the strip is $pb \, \delta d$, which is the same as the force on an elemental vertical strip of height δd.

The need for the horizontal component of the force to be independent of the shape of the surface is intuitively obvious. If the forces were different, tanks such as those in Fig. 3-7 would accelerate sideways if placed on wheels!

Vertical component The vertical component of the gauge pressure force acting on the curved surface CD in either figure is simply the weight of the liquid in the region CDE. Example 3-7 (p. 74) illustrates a case where the vertical component acts upwards on a surface.

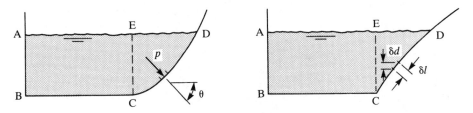

Figure 3-7 Pressure forces on submerged curved surfaces.

3-5 BUOYANCY

When a three-dimensional body is partially or totally submerged in a fluid, it experiences a net pressure force known as a *buoyancy force*. The magnitude of this force in a stationary fluid is well known to be equal to the 'weight of fluid displaced'. This is the famous *Archimedes' principle* which will now be verified in a descriptive but none the less rigorous manner.

Consider the arbitrarily shaped object shown in Fig. 3-8. By inspection, the pressure on its underside is greater than that on its upper surface and so the buoyancy force acts upwards. Unless this force exactly balances the weight of the object, there must be an additional support of some type, but this is omitted for clarity.

Since the fluid surrounding the object is at rest, the pressure distribution is hydrostatic. In particular, the pressure distribution around the surface of the object is exactly the same as it would be if the object was replaced by an equal volume of the fluid. Now the pressure force acting on the replaced fluid would exactly balance its weight, and so the pressure force acting on the original object must also be equal to this weight.

Similar arguments can be applied in the case of an object 'floating' in the free surface between two fluids. Once again, the buoyancy force is equal to the weight of fluid which would be needed in place of the object to maintain the external equilibrium. The part of the object below the free surface would be replaced by the more dense fluid and the part above the free surface would be replaced by the less dense fluid. This is approximately true even when surface tension forces cause local variations in the surface level around the object, but the 'free surface' must be interpreted as the mean level remote from the object. Archimedes' principle fails, however, when the mass of fluid displaced by surface tension is significant in comparison with the mass of the object.

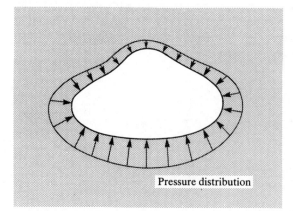

Pressure distribution

Figure 3-8 Pressure distribution around a submerged object.

3-5-1 Stability of Floating Bodies

Engineers dealing with floating bodies are often concerned with their dynamic characteristics, that is with their behaviour when disturbances such as waves cause them to deviate from their stable position. Consider the typical cross-section through a prismatic vessel shown in Fig. 3-9a. The weight of the vessel acts vertically downwards through its centre of mass and the equal buoyancy force W acts upwards through the centre of buoyancy B_1.

Suppose that an external disturbance causes a small rotation of the vessel to the new position shown in Fig. 3-9b. The weight of the vessel will still act through the centre of mass, but the position of the buoyancy force will move to the new centre of buoyancy B_2 at the centroid of the trapezium-like submerged cross-section. The two forces W together form a couple that tends to cause the vessel to rotate back towards its original position (or to displace further if the original condition was unstable).

The subsequent movement of the vessel closely approximates to simple harmonic motion. The vessel rolls back and forth about a horizontal axis close to its centre of mass. When the angular displacement is clockwise, the restoring couple acts counterclockwise, and vice versa. When the displacement is a maximum, so is the restoring couple and so is the angular acceleration.

The magnitude and the frequency of the oscillation are both of considerable interest. Generally speaking, if the restoring couple is large, the oscillation will be small but of uncomfortably high frequency. Conversely, a small restoring couple leads to low frequencies, but to comparatively large magnitudes of roll. The challenge is to find a satisfactory compromise.

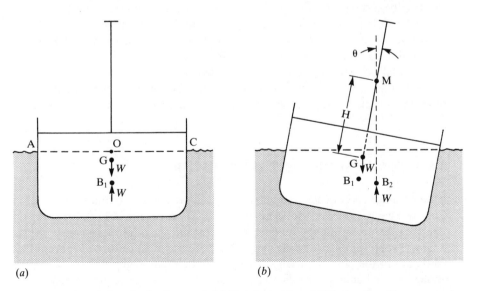

Figure 3-9 Stability of a floating vessel. (a) Upright position. (b) Displaced position.

It is possible to develop a simple analysis of the phenomenon that gives a good description of its main features. In Fig. 3-9b, the lines B_1G and B_2G can be projected to meet at a point M known as the *metacentre*. It is the height H of this point above the centre of mass G (i.e. the *metacentric height*) that determines the magnitude and frequency of the roll. The horizontal distance between the lines of action of the two forces W is $H \sin \theta$ and so the magnitude of the couple is $WH \sin \theta$. At any instant, the angular acceleration is $d^2\theta/dt^2$ and so Newton's second law of motion requires that

$$WH \sin \theta = -I_R \frac{d^2\theta}{dt^2} \qquad (3\text{-}26)$$

in which I_R denotes the moment of inertia of the vessel about the axis of roll. The minus sign indicates that the acceleration and the restoring couple are always of opposite sign. The analysis is applicable only to small angles of roll and so we may use the approximation $\sin \theta \approx \theta$. In this case, Eq. (3-26) describes simple harmonic motion and the period of roll is

$$T = 2\pi \sqrt{\frac{I_R}{WH}} \qquad (3\text{-}27)$$

Metacentric height To determine the position of the metacentre, it is necessary to find the positions of B_1 and B_2 in Fig. 3-9. These are at the centroids of the submerged cross-sections and so can be found geometrically. The position of the metacentre is deduced from the geometrical construction depicted in Fig. 3-9b.

For the special case of a vessel with sides that are vertical at the water-line, it can be shown that the distance B_1M is equal to I_{WL}/V where V is the submerged volume and I_{WL} is the second moment of area of the *water-line section* about a longitudinal axis through its centroid. The water-line section is a horizontal plane through the vessel, bounded by the water-line (Fig. 3-10). In Fig. 3-9a, the water-line section is seen as a line AOC and the longitudinal axis through its centroid appears as the point O. The procedure is illustrated in Example 3-8 (p. 75).

3-6 THE MEASUREMENT OF PRESSURE

If a vertical tube is connected to a hole in the walls of a pipe containing liquid at a positive gauge pressure, the liquid will rise up the tube (Fig. 3-11a). Provided that the pressure inside the pipe remains constant, the liquid level inside the tube will soon stabilize at some height H above the pipe. In this case, even if the liquid in the pipe is moving, the hydrostatic relationship (3-12) may be applied to the liquid *in the tube* to demonstrate that the gauge pressure at the

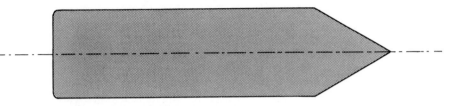

Figure 3-10 Vessel cross-section at the water surface.

wall is wH where $w = \rho g$ is the specific weight of the liquid. This is therefore a simple way of measuring the pressure in the pipe, and the tube is called a *piezometer tube*. However, the method is rarely used in practice because it is not suitable for measuring negative gauge pressures, large gauge pressures (except with a very long tube) or gas pressures.

Barometer The mercury barometer (Fig. 3-11b) is essentially a piezometer tube that is closed and evacuated at the top, the height of the mercury column being a direct measure of atmospheric pressure. Strictly, the space above the liquid cannot be a true vacuum. It contains mercury vapour at its saturated vapour pressure, but this is extremely small at room temperatures (e.g. 0.173 Pa and 20 °C).

U-tube manometer The difficulties associated with piezometer tubes are overcome by a U-tube manometer. In Fig. 3-11c, the top of one arm of a U-tube is connected to the pipe so that the region between A and B contains fluid from the pipe. The region BD contains the manometric liquid—usually mercury if the pipe contains water, but paraffin or water if the pipe contains air. The second arm of the U-tube is open to the atmosphere and so the gauge pressure at C is $w_m z_m$, in which the suffix m denotes the manometric fluid. Since there can be no horizontal variation of pressure in a stationary fluid, the gauge pressure at B is also $w_m z_m$. The pressure at A is $w z_A$ less than that at B, and so

$$p_A - p_{AT} = w_m z_m - w z_A \qquad (3\text{-}28)$$

This expression is also valid for negative gauge pressures, but z_m must be regarded as negative when the elevation of D is below that of B.

In manometry it is usual to neglect vertical variations in atmospheric pressure over the small distances involved. Also, when the working fluid is a gas, Eq. (3-28) is usually simplified by neglecting $w z_A$ in comparison with $w_m z_m$. If these approximations are not considered to be acceptable, p_{AT} must be regarded as the pressure at the point D.

Differential U-tube manometer It is often necessary to establish the *difference* between the pressures at two positions in a flow. For this purpose, the results

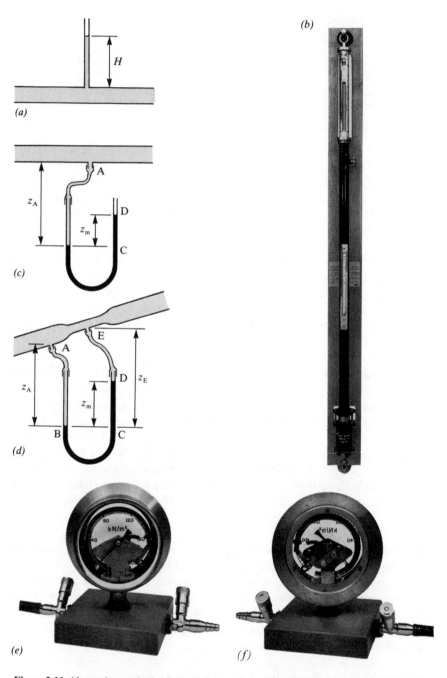

Figure 3-11 Alternative methods of measuring pressures. (a) Piezometer tube. (b) Barometer. (c) U-tube manometer. (d) Differential U-tube manometer. (e) Bourdon gauge (courtesy of Armfield Ltd). (f) Bourdon gauge (rear view).

from two manometers could be compared, but it is more accurate to use a single U-tube as shown in Fig. 3-11*d*. Equation (3-28) may be interpreted as

$$p_A - p_D = w_m z_m - w z_A \tag{3-29}$$

Since the pressure difference $(p_D - p_E)$ is equal to $w(z_E - z_m)$, we obtain by addition

$$p_A - p_E = (w_m - w)z_m + w(z_E - z_A) \tag{3-30}$$

In the special case where the tapping points A and E are at the same elevation, $z_E = z_A$, and so

$$p_A - p_E = (w_m - w)z_m \tag{3-31}$$

When the tapping points are not at the same elevation, Eq. (3-30) is conveniently rearranged as

$$(p_A + w z_A) - (p_E + w z_E) = (w_m - w)z_m \tag{3-32}$$

which shows that the manometer reading z_m is a measure of the amount by which the sum $(p + wz)$ has changed in the pipe between the points A and E. This sum commonly occurs in the study of incompressible fluid flows. It is known as the *piezometric pressure p**, defined as

$$p^* \equiv p + wz = p + \rho g z \tag{3-33}$$

Pressures can be described as an equivalent *head* of liquid (see section 2-4). The piezometric pressure can be specified as a piezometric (or *manometric*) head *h**, defined as

$$h^* \equiv \frac{p}{w} + z = \frac{p}{\rho g} + z \tag{3-34}$$

In this case, the equation for the differential U-tube manometer (3-32) is

$$h^*_A - h^*_E = \left(\frac{w_m}{w} - 1\right) z_m \tag{3-35}$$

Other U-tube manometers Only the simplest forms of the U-tube manometer have been described. Many ingenious modifications have been employed in manometers in common use. For example, one arm is often made of a much larger cross-section than the other. The changes in the liquid level in the larger arm are very much smaller than those in the other arm and, in the limit, may be ignored. For greater sensitivity, the narrow arm can be set at an angle to the horizontal so that small changes in elevation will cause large movements of the meniscus (see Problem 3 at the end of this chapter). When the working fluid is a liquid and the pressure differences are small, the U-tube is sometimes inverted so that air can be used as the manometric fluid instead of mercury. The resulting differences in level are thereby greatly increased.

3-6-1 Pressure Gauges

Any instrument that responds in a definite manner to applied pressure changes can be used as a pressure gauge. Manometers are a special case in which the response can be calculated using hydrostatic formulae. Many other gauges exist, the most common in everyday use being the Bourdon gauge (Fig. 3-11e). The principal component of this gauge is a curved hollow tube of roughly elliptical cross-section. When this tube is pressurized, its cross-section tends to become more circular, and the whole tube tends to become more straight. To make use of this effect, one end of the tube is tightly clamped and the movement of the other end is used to drive a pointer around a scale.

In industrial or research circumstances where automatic controls respond to pressure sensors or where a permanent record of the fluid pressure is required, electronic pressure transducers are used. These gauges contain a small diaphragm—typically a few millimetres in diameter—which is directly exposed to the fluid pressure. Movements of the diaphragm cause the gauge to modulate tiny electronic signals which are then amplified as the user requires. A wide variety of pressure transducers exists for the measurement of both small and large pressures. Some are capable of measuring only slowly varying pressures while others can respond accurately to pressure changes occurring in time intervals of less than a microsecond.

3-7 APPLICATIONS OF FLUID PRESSURE

Example 3-1: Pressure distribution in deep water Derive an expression describing the vertical variation of pressure in a slightly compressible liquid. Hence estimate the percentage errors associated with the use of the incompressible relationship (3-10) in water depths of 100 m, 1000 m and 10 000 m. Take the density and bulk modulus of water to be 1000 kg/m^3 and 2.3 GN/m^2 respectively.

SOLUTION By combining Eq. (3-9) with the compressibility relationship $\rho \, dp = K \, d\rho$ given in section 2-6, we obtain

$$\frac{d\rho}{\rho^2} = -\frac{g \, dz}{K} \tag{3-36}$$

in which K is the bulk modulus of elasticity of the fluid. Over sufficiently small ranges of pressure, it is reasonable to regard the bulk modulus as a constant (see Fig. 2-4). In this case, Eq. (3-36) may be integrated to give

$$\frac{1}{\rho} - \frac{1}{\rho_0} = \frac{gz}{K} \tag{3-37}$$

in which ρ_0 is the density at the arbitrary datum $z = 0$. After rearranging

Table 3-1 Values of $p - p_0$ in deep water

Depth, m	100	1000	10000
Eq. (3-39), MPa	0.98	9.83	100.3
Eq. (3-10), MPa	0.98	9.81	98.1
Difference, %	0.02	0.2	2.2

this equation to give an expression for the density which may be substituted into Eq. (3-9), we obtain

$$\frac{dp}{dz} = -\frac{\rho_0 g K}{K + \rho_0 g z} \tag{3-38}$$

On integrating and choosing $p = p_0$ at $z = 0$, the expression for the pressure distribution is

$$p - p_0 = -K \ln\left(1 + \frac{\rho_0 g z}{K}\right) \tag{3-39}$$

Numerical values of $p - p_0$ obtained for *elevations* of -100 m, -1000 m, and -10000 m are listed in the Table 3-1, together with the corresponding values obtained from the incompressible relationship (3-10).

Comment on Example 3-1 The above analysis can be improved by taking account of the dependence of the bulk modulus on pressure (section 2-6) or by using more accurate average values of the bulk modulus for the different pressure ranges. However, these refinements do not influence the obvious conclusion that Eq. (3-10) may be used instead of more complex relationships for all depth differences likely to be encountered in practice.

Example 3-2 The Troposphere Derive an expression describing the vertical pressure variation in the troposphere (the lower 10 km of the earth's atmosphere), assuming that the temperature decreases linearly with height.

SOLUTION The linear temperature distribution may be written as

$$T = T_0 - kz \tag{3-40}$$

in which k is a positive constant and T_0 is the temperature at an arbitrary datum $z = 0$. By combining this expression with the equation of state for air $p = \rho RT$, we obtain

$$p = \rho R(T_0 - kz) \tag{3-41}$$

Equations (3-9) and (3-41) may both be regarded as expressions for the

Table 3-2 Values of $p_0 - p$ in the troposphere

Elevation, m	100	1000	10 000
Eq. (3-45), kPa	1.18	11.3	73.6
Eq. (3-10), kPa	1.19	11.9	118.6
Difference, %	0.5	4.8	61.1

density; that is

$$\rho = -\frac{1}{g}\frac{dp}{dz} \qquad \text{and} \qquad \rho = \frac{p}{R(T_0 - kz)} \qquad (3\text{-}42a, b)$$

After eliminating ρ and rearranging the terms, we obtain

$$\frac{dp}{p} = -\frac{g\,dz}{R(T_0 - kz)} \qquad (3\text{-}43)$$

Integrating, we find

$$\ln\left(\frac{p}{p_0}\right) = \frac{g}{kR}\ln\left(\frac{T_0 - kz}{T_0}\right) \qquad (3\text{-}44)$$

in which p_0 denotes the pressure at the datum $z = 0$. An alternative form of Eq. (3-44) is

$$\frac{p}{p_0} = \left(1 - \frac{kz}{T_0}\right)^{g/kR} \qquad (3\text{-}45)$$

which is the required pressure distribution.

Comments on Example 3-2

1. Numerical values of the pressure difference $p_0 - p$ at various elevations above ground level are listed in Table 3-2 together with the corresponding values obtained from the incompressible relationship (3-10). In deriving these values, the conditions at ground level have been taken as $p_0 = 100$ kPa and $T_0 = 15\,°C$. The gas constant is $R = 287$ J/kg K and the rate of decrease of temperature with height is $0.006\,°C/m$.

   It can be inferred that the incompressible relationship is acceptable for differences in height of up to about 250 m, but that consideration should be given to the use of a more accurate expression such as (3-45) when dealing with greater changes in elevation. With the above data, the incompressible expression yields negative pressures when the elevation exceeds about 8.4 km!

2. Equation (3-45) is nominally valid for all values of k, the rate of change of temperature with height. However, the conditions become unstable when k is large and so the results obtained are not physically possible (anology:

pencils do not stand freely on their points). In an inversion, the temperature *increases* with height in the lower part of the troposphere. For this case, which corresponds to negative values of k, the cloud base is usually clearly defined.

Example 3-3 Hydraulic jack Two circular cylinders of radius R and $5R$ respectively contain oil and are connected by a short pipe (Fig. 3-12). When a certain load rests on a close-fitting piston in the larger cylinder, a force W must be applied to a piston in the smaller cylinder to maintain equilibrium. Neglecting other forces, estimate the weight of the load.

SOLUTION The cross-sectional area of the smaller piston is πR^2. The pressure in the oil just beneath the piston is therefore $W/\pi R^2$. Since there can be no horizontal variation in pressure in a stationary fluid, the pressure just beneath the larger piston must also be $W/\pi R^2$. The cross-sectional area of this piston is $25\pi R^2$ and so the weight which can be supported is $25W$.

Comment on Example 3-3 This result—that a small load can support a much larger one—surprises many people. It should not. Similar results can be obtained with other mechanical systems involving, for instance, levers or pulleys. The arrangement shown in Fig. 3-12 includes the essential features of a hydraulic jack.

Example 3-4 Coffer dam To enable workmen to operate in dry conditions when building a bridge pier in a river, a coffer dam is formed by driving interlocking sheet steel piles vertically into the river bed. The plan dimensions of the rectangular dam are 4 m × 8 m and the maximum water depth is $d_0 = 5$ m. Estimate the net hydrostatic force on one of the longer sides of the dam and the depth of the centre of pressure.

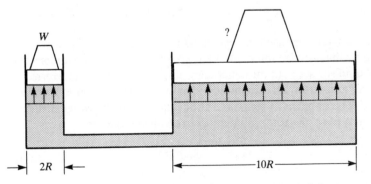

Figure 3-12 Principle of hydraulic transmission (pressures are equal; forces are unequal).

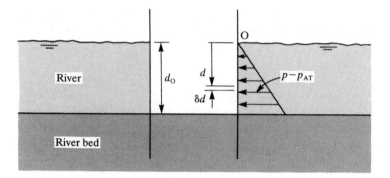

Figure 3-13 Lateral hydrostatic force on a coffer dam.

SOLUTION The general arrangement is shown schematically in Fig. 3-13. The hydrostatic gauge pressure distribution is shown on one side of the dam by arrows which indicate that the pressure force is experienced as a horizontal thrust. The *absolute* pressure p at a depth d below the water surface is found from Eq. (3-11) to be $p_{AT} + \rho g d$ and so the horizontal pressure force acting on an elemental strip of height δd and width b normal to the page is $(p_{AT} + \rho g d)b\ \delta d$. The total horizontal force imposed by the water on an outer face of the dam is therefore

$$F = \int_0^{d_0} (p_{AT} + \rho g d)b\ dd \tag{3-46}$$

i.e.
$$F = bd_0 p_{AT} + \tfrac{1}{2}\rho g b d_0^{\,2} \tag{3-47}$$

On the *inside* of the coffer dam, the air below the level of the water surface exerts a horizontal force $bd_0 p_{AT}$ in the opposite direction. The *net* hydrostatic force on the dam side is therefore $F_N = F - bd_0 p_{AT}$, that is

$$F_N = \tfrac{1}{2}\rho g b d_0^{\,2} \tag{3-48}$$

Using $\rho = 1000\ \text{kg/m}^3$, $g = 9.81\ \text{m/s}^2$, $b = 8\ \text{m}$ and $d_0 = 5\ \text{m}$, the net force is $F_N = 981\ \text{kN}$.

To find the depth of the centre of pressure, consider the moment exerted by the pressure forces about an axis O formed by the intersection of the liquid surface with the coffer dam. The *net* pressure force exerted by the water and air on an elemental strip of the coffer dam at a depth d is $\rho g d b\ \delta d$. The moment of this force about the axis at the water surface is $\rho g b d^2\ \delta d$ and so the total moment exerted on the piles is

$$M = \int_0^{d_0} \rho g b d^2\ dd \tag{3-49}$$

i.e.
$$M = \tfrac{1}{3}\rho g b d_0^{\,3} \tag{3-50}$$

The net force F_N would cause the same moment if it acted at a depth $\bar{d}$ below the water surface, where

$$M = F_N \bar{d} \qquad (3\text{-}51)$$

Using Eqs (3-48) and (3.50), the depth of the centre of pressure is

$$\bar{d} = \tfrac{2}{3} d_0 \qquad (3\text{-}52)$$

The overall effect of the pressure forces on the face of the coffer dam is therefore equivalent to a horizontal force of 981 kN acting at a depth of 3.33 m on the vertical centre line of the face.

Comments on Example 3-4

1. This example demonstrates why engineers usually use gauge pressures instead of absolute pressures when evaluating forces. Like the coffer dam, most structures are built in the atmosphere and the gauge pressure force on one side is zero. The *net* force is therefore equal to the gauge pressure force on the other side. Similar comments apply to the centre of pressure; it is the centre of *gauge* pressure that is 3.33 m below the water surface on the outside face of the dam.
2. This example could be dealt with more easily using Eqs (3-22) and (3-24). It is easy to verify that the gauge pressure force is equal to the product of the area of the wetted surface bd_0 and the gauge pressure $\tfrac{1}{2}\rho g d_0$ at its centroid. Also the depth $\bar{d}$ of the centre of pressure is the ratio of the second and first moments of area of the wetted surface about the water line axis, namely $\tfrac{1}{3}bd_0^{3}$ and $\tfrac{1}{2}bd_0^{2}$.
3. In a complete design of a coffer dam, it is also necessary to take account of forces imposed by the water as a consequence of its motion. These include drag forces, wave forces and scouring action that threatens to undermine the dam.

Example 3-5: Sluice gate Determine the horizontal and vertical components of the hydrostatic force on the radial gate shown in Fig. 3-14a. Also determine the moment of these forces about the axis of the gate. The specific weight of water is 9.81 kN/m^3 and the channel containing the water is 3 m wide (normal to the page).

SOLUTION The *horizontal* component of the hydrostatic force on the curved surface of the gate is the same as the force that would be experienced by a vertical surface of the same depth. The area of the equivalent vertical surface of width $b = 3$ m is $bR = 4.5 \text{ m}^2$, the depth of its centroid is $d_c + \tfrac{1}{2}R = 3.25$ m, and the gauge pressure wd at this depth is 31.9 kPa. Equation (3-22) shows that the net horizontal thrust is $F_H = 4.5 \times 31.9 \text{ kN} = 143.5 \text{ kN}$.

The second moment of area I_{Gx} of the rectangular projected surface

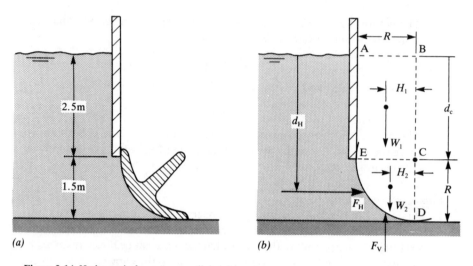

Figure 3-14 Hydrostatic forces on a radial sluice gate.

about a horizontal axis through its centroid is $bR^3/12$, i.e. $0.84\,\text{m}^4$. The second moment of area about an axis in the water surface is therefore

$$I_x = (0.84 + 4.5 \times 3.25^2)\,\text{m}^4 = 48.4\,\text{m}^4$$

The depth of the line of action of the horizontal thrust follows from Eq. (3-24). It is $d_H = 48.4/14.6\,\text{m} = 3.31\,\text{m}$.

The *vertical* component of the force on the gate could be found by integrating an expression such as (3-16). However, it can be found much more easily by noting that it must be equal to the weight of water that could occupy the region ABCDE in Fig. 3-14 if the gate was removed.

The volume of the region ABCE is $bRd_c = 11.25\,\text{m}^3$. The weight of water that could be contained within it is $W_1 = 110.4\,\text{kN}$ and its centre of mass would be a distance $H_1 = \frac{1}{2}R = 0.75\,\text{m}$ from vertical line BC. The volume of the region CDE is $\frac{1}{4}\pi R^2 b = 5.30\,\text{m}^3$ and the weight of water that could be contained within it is $W_2 = 52.0\,\text{kN}$. The centre of mass would be a distance $H_2 = 4R/3\pi = 0.637\,\text{m}$ from the vertical line CD. The upthrust on the gate is the sum of W_1 and W_2 namely $162.4\,\text{kN}$.

The counterclockwise moment exerted by the hydrostatic thrust about the pivot C is

$$M = F_H(d_H - d_c) - W_1 H_1 - W_2 H_2 = 155\,\text{Nm} \qquad (3\text{-}53)$$

Comments on Example 3-5
1. The self weight of the gate will also act in the counterclockwise direction unless it is counterbalanced. Therefore a clockwise moment must be

provided to maintain equilibrium. For example, the gate might be supported from above or it might have a large counterbalancing weight.

2. In this particular example each elemental pressure force on the surface of the gate is radial and therefore has no moment about C. It follows that the resultant force will have no moment about C, i.e. it too is radial. This makes it possible to obtain a solution in a simpler manner that does not involve a need to determine the distance H_2.

Example 3-6: Sluice gate control Figure 3-15 depicts the gauge pressure distribution on the face of a vertical rectangular sluice gate in a free surface flow. By assuming that the distribution satisfies

$$p - p_{AT} = \rho g d \left\{ 1 - \left(\frac{d}{H} \right)^N \right\} \tag{3-54}$$

estimate the magnitude and location of the resultant horizontal force on the gate.

SOLUTION The net pressure force on an elemental strip of width b and height δd is $\delta F_x = (p - p_{AT})b \, \delta d$. Therefore the total horizontal force is

$$F = \int_0^H \rho g b \left(d - \frac{d^{N+1}}{H^N} \right) dd = \tfrac{1}{2} \rho g b H^2 \left(\frac{N}{N+2} \right) \tag{3-55}$$

The moment of the elemental force about the y-axis (normal to the page in

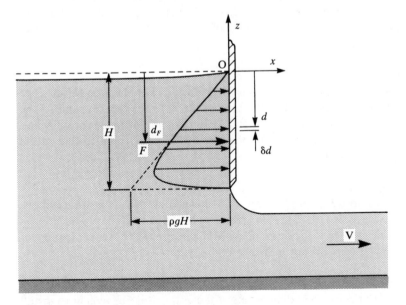

Figure 3-15 Pressure force on an open sluice gate.

the free surface) is $d \, \delta F = (p - p_{AT})bd \, \delta d$. Therefore the moment of the total force is

$$d_F F = \int_0^H \rho g b \left(d^2 - \frac{d^{N+2}}{H^N} \right) dd = \tfrac{1}{3}\rho g b H^3 \left(\frac{N}{N+3} \right) \qquad (3\text{-}56)$$

and so the depth of the centre of pressure is

$$d_F = \frac{2H}{3} \left(\frac{N+2}{N+3} \right) \qquad (3\text{-}57)$$

Comments on Example 3-6

1. In the case of a closed gate, the liquid is stationary and hydrostatic conditions prevail. By inspection, the various equations are asymptotic to the usual results when $N \rightarrow \infty$.
2. In the more general case, $N \neq \infty$ and the magnitude of the resultant force and the depth of its line of action are both reduced. Further information is presented in Example 7-4 (p. 189) and in Problem 12 at the end of Chapter Eight.

Example 3-7: Buoyancy of a storage vessel A cylindrical oil tank floats in the sea with its axis vertical. It is open at the top and the bottom. When it contains no oil, the buoyancy of the vessel and its external floatation tanks is counterbalanced by a tensile force of 50 MN in the fixing spar AB shown in Fig. 3-16. Estimate (a) the height of the oil surface above sea level and (b) the force in the fixing spar when the 120 m long, 30 m internal diameter tank contains 50×10^6 kg crude oil of relative density 0.89. The density of the sea water is $1020 \, \text{kg/m}^3$.

SOLUTION

(a) Since the density of the oil is $890 \, \text{kg/m}^3$, its volume is $56\,180 \, \text{m}^3$. The cross-sectional area of the tank is $900\pi/4 \, \text{m}^2 = 706.9 \, \text{m}^2$ and so the height of the oil column is $56\,180/706.9 \, \text{m} = 79.47 \, \text{m}$. At the oil–water interface, the hydrostatic pressure in the oil is equal to that in the water, that is

$$\rho_0 g H_0 = \rho_w g H_w \qquad (3\text{-}58)$$

Therefore $H_w = 69.34 \, \text{m}$ and so the elevation of the oil surface above sea-level is $(79.47 - 69.34)$ metres $\approx 10.1 \, \text{m}$.

(b) The force in the spar is still 50 MN because the contents of an open-ended tank do not influence its buoyancy.

Comments on Example 3-7

1. To reduce the external dimensions of the vessel, it is convenient to install the buoyancy tanks internally. However, their buoyancy in oil differs from

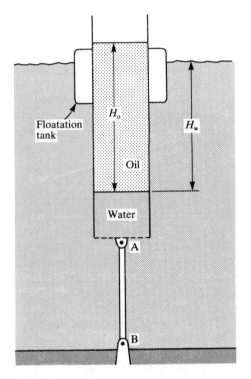

Figure 3-16 Floating storage vessel (open at top and bottom).

that in water and so their dimensions and position must be chosen very carefully if it is desired to maintain a constant tension in the fixing spar.
2. The response of the tank to surface waves is a major design consideration. In practice this will be deduced from model tests because current theoretical knowledge is inadequate for the simulation of such complex flows.

Example 3-8: Stability of a pontoon The 14 m wide pontoon shown in Fig. 3-17 is 42 m long and its mass, including all fittings, is 900 Mg. Its moment of inertia about a longitudinal axis through its centre of mass, 3.2 m above the base, is 25 000 Mg m². The pontoon carries a 45 m long prestressed concrete bridge girder which has a mass of 700 Mg. The moment of inertia of the beam about a longitudinal axis through its centre of mass, 9.35 m above the pontoon base, is 8000 Mg m². The girder is to be carried to the side of a bridge over a fresh-water estuary where it will be lifted into position. Investigate the stability of this arrangement.

SOLUTION The total mass is 1600 Mg and so the submerged volume in fresh water (density $= 1$ Mg/m³) is $V = 1600$ m³. Since the pontoon is 42 m

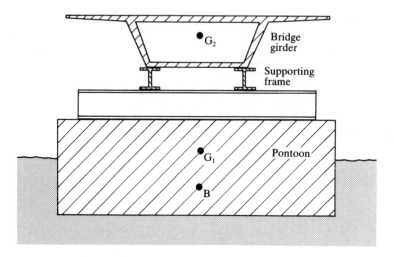

Figure 3-17 Stability of a pontoon.

long and 14 m wide, the submerged depth is $1600/(42 \times 14)$metres $=$ 2.72 m. The height of the centre of buoyancy B above the base is one half of the buoyant depth, namely 1.36 m.

The water-line section is 42 m long and 14 m wide. Its second moment of area about a central longitudinal axis is $I_{WL} = 42 \times 14^3/12 \, \text{m}^4$ $= 9600 \, \text{m}^4$. Therefore the height of the metacentre above the centre of buoyancy is $I_{WL}/V = 6.00$ m, and so the height of the metacentre above the pontoon base is $6.00 + 1.36$ m $= 7.36$ m.

The centre of mass of the whole structure is a distance z_G above the pontoon base where

$$z_G = \frac{(900 \times 3.2) + (700 \times 9.35)}{1600} \, \text{m} = 5.89 \, \text{m}$$

Therefore the metacentre is $(7.36 - 5.89)$ metres $= 1.47$ m above the centre of mass and so the pontoon and girder are stable.

Period of roll The centre of mass of the pontoon and the girder are respectively 2.78 m below and 3.37 m above the overall centre of mass. Using the parallel axes theorem (3-25), the combined moment of inertia about the overall centre of mass is therefore $I_G = (25\,000 + 900 \times 2.78^2) + (800 + 700 \times 3.37^2) \, \text{Mg}\,\text{m}^2$. That is $I_G = 47\,900 \, \text{Mg}\,\text{m}^2$. Since the axis of roll roughly coincides with an axis through the centre of mass, Eq. (3-27) shows that the period of roll is

$$T \simeq 2\pi \sqrt{\frac{I_G}{WH}} \simeq 9.0 \, \text{s} \tag{3-59}$$

Comments on Example 3-8

1. A considerable amount of effort is hidden in the specification of this example. It is important to determine the total mass and its distribution as accurately as possible.
2. The actual period of roll should be quite close to the predicted value, but the oscillation will be fairly strongly damped by non-hydrostatic forces developed in the fluid as a result of its motion.
3. Sometimes, water is allowed into a pontoon to act as ballast while a bridge beam is floated into position. If the water is then pumped out to allow the deck to rise, the least stable configuration occurs when the deck reaches its highest point.

PROBLEMS

1 The internal and external diameters of the pipe shown in Fig. 3-18 are 80 mm and 84 mm respectively. The pipe is closed at both ends and contains fluid at an absolute pressure of 1 MPa. The pressure in the surroundings is 0.1 MPa.
(a) Estimate the axial and circumferential stresses in the pipe wall.
(b) Estimate the axial extension of the 14 m long pipe if the Young's modulus and Poisson's ratio of the pipe material are 200 GPa and 0.28 respectively.

$$[(a)\ 9\,\text{MPa},\ 18\,\text{MPa}\ (b)\ 0.277\,\text{mm}]$$

2 A 50 mm bore pipeline connects two reservoirs in which the water surface levels are respectively 25 m and 5 m above the elevation datum. Estimate the net pressure force on a closed valve preventing flow along the pipeline.

$$[385\,\text{N}]$$

3 The inclined manometer shown in Fig. 3-19 is used to determine the pressure in a water pipeline. Estimate the gauge pressure in the pipeline when $H_1 = 450\,\text{mm}$ and $H_2 = 40\,\text{mm}$. The manometric fluid is mercury and its relative density is 13.6.

$$[922\,\text{Pa}]$$

4 Two closed tanks containing water and air are mounted as shown in Fig. 3-20. Initially, valves A and B are open to atmosphere and valves C and D are closed.
(a) Describe the consequences of closing valves A and B and then opening valve C.

[Water falls; hydrostatic water pressure distribution develops]

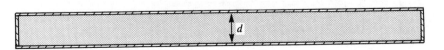

Figure 3-18 Water-filled pipe closed at both ends.

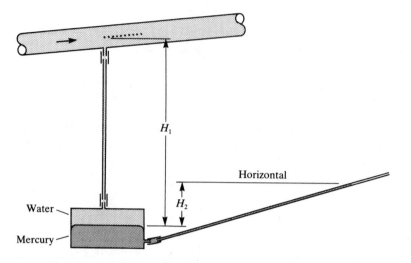

Figure 3-19 Use of an inclined manometer.

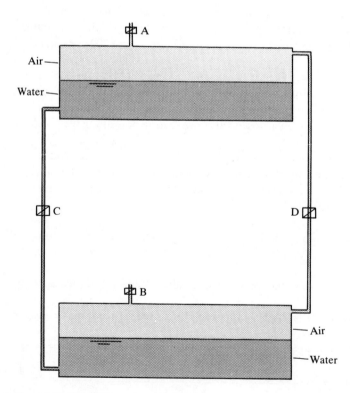

Figure 3-20 Equalization of pressures in tanks.

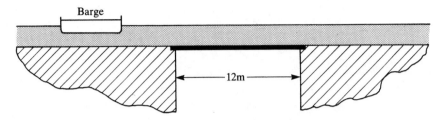

Figure 3-21 Canal supported on a bridge.

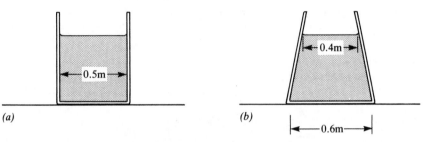

Figure 3-22 Hydrostatic forces on the bases of containers.

(b) Describe the consequences of subsequently closing valve C and opening valve D.

[Air rises; hydrostatic air pressure distribution develops]

5 The water in an 8 m wide canal is 2.5 m deep where it passes over a 12 m span bridge (Fig. 3-21).
  (a) Estimate the hydrostatic force on the bridge as a 30 Mg barge approaches.
  (b) Estimate the hydrostatic force on the bridge when the barge is directly overhead.

[(a) 2.35 MN, (b) 2.35 MN]

6 The water depths in the tanks shown in Fig. 3-22 are both equal to 0.5 m. Tank (a) is cylindrical and its internal diameter is 0.5 m. Tank (b) is conical and its internal diameter at the base is 0.6 m. For each tank, determine (i) the weight of water and (ii) the hydrostatic pressure force on the base.

[(a) 963 N, 963 N, (b) 976 N, 1387 N]

7 The water depth alongside the L-shaped dam in Fig. 3-23 is $H = 4$ m. Estimate the minimum possible width L of the horizontal limb if the dam does not overturn. Neglect seepage and the self weight of the dam.

[2.31 m]

8 The radius of the dam shown in Fig. 3-24 is $R = 4$ m and the water depth is also 4 m. Estimate the minimum possible width L of the horizontal support if the dam does not overturn. Neglect seepage and the self weight of the

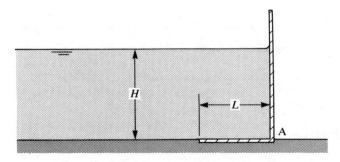

Figure 3-23 Stability of a L-shaped dam.

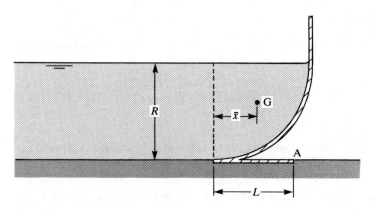

Figure 3-24 Stability of a curved dam.

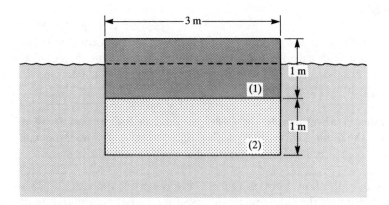

Figure 3-25 Stability of a top-heavy, floating object.

dam. The distance $\bar{x}$ to the centroid of the quadrant is $4R/3\pi$. (*Hint:* See comment 2 on Example 3-5.)

[2.55 m]

9 An object floats at the free surface between two liquids, 80 per cent by volume being above the interface. If the densities of the liquids are $850 \, \text{kg/m}^3$ and $1250 \, \text{kg/m}^3$, what is the average density of the object?

[930 kg/m³]

10 A spherical gas bubble of 5 mm diameter is released from a diving suit at a water depth of 100 m. Estimate the buoyancy force acting on the bubble. Also estimate the buoyancy force when the bubble has risen to a depth of 1 m below the surface at which the atmospheric pressure is 100 kPa. Assume the density of the seawater to be $1026 \, \text{kg/m}^3$ and neglect changes in the gas temperature.

[5.27 mN, 53.0 mN]

11 The rectangular object shown in Fig. 3-25 is $8 \, \text{m} \times 3 \, \text{m}$ in plan. The densities of the two materials are $\rho_1 = 1000 \, \text{kg/m}^3$ and $\rho_2 = 600 \, \text{kg/m}^3$. Prove that the object can float stably in water ($\rho = 1000 \, \text{kg/m}^3$) in this position.

12 Estimate the frequency of oscillation of the object in Fig. 3-25 when it rocks back and forth after being turned upside down.

[0.0967 Hz]

FURTHER READING

Denton, E.J. (1974) *Buoyancy in Marine Animals*, Oxford University Press.

Hammitt, F.G. (1980) *Cavitation and Multiphase Flow Phenomena*, McGraw-Hill.

Holman, J.P. (1984) *Experimental Methods for Engineers*, 4th edn, McGraw-Hill.

Huey, L.J. (1978) A yaw-insensitive static pressure probe, *J. Fluids Engng, Trans. ASME*, **100**, 229–231.

Petrucci, L.G.D. (1979) *Measurement of Pressure*, L & A Press.

Trevena, D.H. (1987) *Cavitation and Tension in Liquids*, Adam Hilger.

FOUR

FLUID SHEAR

4-1 LAMINAR AND TURBULENT FLOWS

The origin of shear stresses in fluids was described in section 2-7 where they were shown to be related to differences between velocities on adjacent streamlines. Unlike fluid pressure, there are no shear stresses in a stationary fluid—because there are no velocity differences to induce them. Shear stresses are a result of fluid *flow* and, other things being equal, they increase with the speed of flow.

Throughout this chapter, attention is focused on *Newtonian* fluids such as water and air. As described in section 2.7, these are fluids in which shear stresses are found to be directly proportional to rates of shear strain. In other types of fluid, more complex (non-linear) relationships exist, and there can even be a limiting (yield) stress below which the substance behaves as a solid.

In comparison with normal stresses (i.e. pressure) shear stresses are usually very small. Indeed they are tiny in the case of sufficiently slow flows. Nevertheless they can have a significant impact because they are the only source of resistance to flow. In a steady flow along a pipe for instance, the total force resulting from small shear stresses on the large internal surface of the pipe wall is often sufficient to make us use pumps to maintain satisfactory rates of flow. Likewise, some of the aerodynamic drag forces experienced by road vehicles have their origins in fluid shear.

The fluid pressure at a point in a stationary fluid was shown in section 3-2 to be the same in all directions, and variations in this behaviour were said to be small even in moving fluids. The same is not true for fluid shear. In stationary

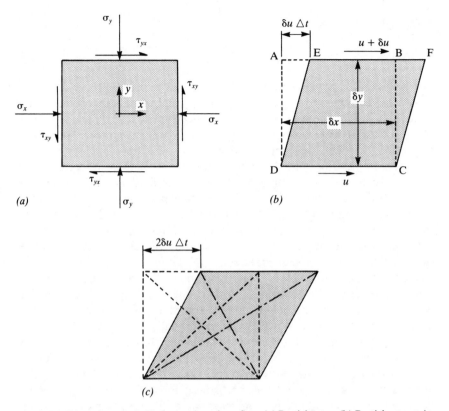

Figure 4-1 Distortion of a fluid element in a shear flow. (a) Particle at t_0. (b) Particle at $t_0 + \Delta t$. (c) Particle at $t_0 + 2\Delta t$.

fluids and in hypothetical inviscid flows, the shear stresses are zero. In real flows, however, shear stresses vary significantly with direction. A simple example will suffice to demonstrate this.

Consider the element of fluid shown at successive times in Fig. 4-1 as it flows parallel to the x-axis, constrained by the sides of a channel perhaps. The velocity of the upper surface exceeds that of the lower surface by an amount δu and so the element distorts as it moves downstream. The degree of distortion increases continuously as the element moves downstream, but the shear stress remains constant (contrast this with a solid where increasing distortion implies increasing shear stress). The shear strain at the instant $t_0 + \Delta t$ is $(x_E - x_A)/\delta y$ and so the rate of increase of shear strain in the interval Δt is

$$\dot{\gamma} = \frac{(x_E - x_A)/\delta y}{\Delta t} = \frac{\delta u}{\delta y} \tag{4-1}$$

For a Newtonian fluid, the shear stress is proportional to the rate of increase of shear strain. That is,

$$\tau_{yx} = \mu \frac{du}{dy} \tag{4-2}$$

in which μ is the dynamic viscosity and u is the velocity in the x-direction.

In this example, there need be no component of velocity normal to the page. If there isn't, there will be no rate of shear strain in any plane normal to the x-axis and so there will be no components of shear stress in these planes. By inspection, therefore, the components of shear stress are different in different directions at the same point in the flow (and they are in most other flows too).

It is worth noting in passing that the particular fluid motion considered in this example is *rotational*. That is, individual fluid particles rotate as they move downstream. This may be verified by observing the rotation of both diagonals of the element relative to their original orientation (Fig. 4-1c).

Equation (4-2) is valid only for Newtonian fluids and, as a complete expression for the effective shear stress, is valid only for laminar flows. In the case of turbulent flows, it is not reasonable to imagine particles flowing in an orderly manner parallel to an axis. Even in a highly constrained environment such as a flow along a straight pipe, the various fluid particles jostle for position as they move downstream, continually swapping places in a seemingly random fashion. One consequence of this is that slow-moving particles from the outer regions of flow jump into mid-stream where they have to be accelerated to the local velocity. Likewise, faster-moving particles arriving in the outer regions have to be slowed down. The overall effect of this behaviour is equivalent to a huge increase in the effective shear stress, and the additional contributions are commonly known as *Reynolds stresses*.

We shall not attempt to analyse Reynolds stresses in this book because of their complexity. However, this is a fascinating area of research and much has been learnt in recent years especially about local 'bursts' of particularly strong turbulence within flows. In some of the simplest models of homogeneous turbulence (i.e. no 'bursts') the existence of a kinematic eddy viscosity ε is hypothesized, enabling a turbulent flow version of (4-2) to be written as

$$\tau_{yx} = (\mu + \rho\varepsilon) \frac{d\bar{u}}{dy} \tag{4-3}$$

in which $\bar{u}$ denotes the average velocity at a point after allowing for random turbulent fluctuations. The major complication arising in all turbulence models is that the Reynolds stresses—and hence parameters such as the kinematic eddy viscosity—are functions of the flow behaviour, not properties of the fluid. Unlike the dynamic viscosity, the eddy viscosity and its counterparts are not constants, even in the flow of a Newtonian fluid.

4-2 BOUNDARY LAYERS

Because of the difficulties involved in analysing turbulent or even laminar flows, we often prefer to neglect shear stresses altogether and analyse instead the equivalent flow of an inviscid fluid. This naturally yields useful results when the true influence of fluid shear is small and so we need to have a clear idea of the types of flow for which this is so. Generally speaking, the effects of fluid shear are greatest close to solid surfaces past which the fluid is attempting to slide. The no-slip condition (section 2-7) requires that there is no tangential velocity component at the surface itself and so large velocity gradients can arise—close to the surface of an aircraft wing or to the inside wall of a pipe for instance.

In many external flows (section 1-2) but few internal ones, it is reasonable to analyse the flow in two regions, one very close to solid surfaces (where fluid shear matters) and the other remote from these surfaces (where it doesn't). To illustrate this approach, let us consider one of nature's most fascinating and elegant flows, waves on the surface of water.

When oscillatory waves propagate in deep water—an ocean, say—the individual fluid particles follow almost circular paths as shown in Fig. 4-2a. The orbital motion of adjacent particles is out of phase, and this is what produces the wave effect.

The magnitude of the orbital motion decreases exponentially with depth and very little movement is induced at depths exceeding about one wavelength below the surface. This is therefore an example of a flow where there is no significant tangential movement adjacent to a solid surface and, as a consequence, there is no region of significant fluid shear. It is therefore not surprising that experiment shows a very close correlation with inviscid flow theory. Incidentally, the speed of the wave—$(gL/2\pi)^{1/2}$ where L is the wavelength—is typically much larger than the speeds of the individual particles. More important, the velocity gradients associated with the particle movements are small and so shear stresses are small. In particular, they are far smaller than the inertia forces and gravitational forces that dominate the flow behaviour.

When similar waves propagate in less deep water—nearer to a coastline perhaps—the region of significant particle motion may extend as far as the bed. In this case, inviscid flow theory predicts that individual particles will follow elliptical paths and that particles at the bed will simply oscillate back and forth along the bed. Once again, experiment shows that the theory is accurate in the upper regions of flow. However, it breaks down in the lower regions because the no-slip condition prevents any motion at the bed itself. Just above the bed, the tangential movements of the particles imply velocity gradients that predictably involve significant shear stresses. For obvious reasons, the lower region is known as a *boundary layer*. There is no precise upper limit to the boundary layer (see Fig. 4-2b), but we usually take it to be

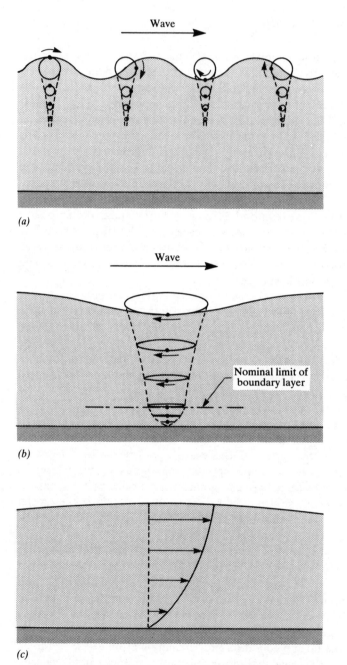

Figure 4-2 Free surface waves over a horizontal bed. (a) Short waves. (b) Intermediate waves. (c) Long waves.

the position at which the velocities predicted by inviscid flow theory differ by 1 per cent from reality. Within the boundary layer, the fluid shear might cause either laminar or turbulent conditions. The same holds true outside the boundary layer, but there the shear stresses are negligible anyway.

When the waves propagate into a region of shallow water—very close to a beach, say—the boundary layer extends through the whole flow and there is no region approximating to inviscid conditions. All the flow is influenced by shear stresses, either laminar or turbulent or both.

Figure 4-3 shows four different examples of fluid flows involving significant shear stresses. In the first, boundary layers gradually increase in thickness along the sides of an aerofoil and subsequently form a narrow wake behind it. In the second, similar boundary layers form on the sides of a poorly streamlined body and abruptly 'separate' from the solid surface to form a large wake of relatively slowly moving fluid. The conditions in the wake are similar to those downstream of a jet of fluid emerging into an otherwise stationary fluid (Fig. 4-3c).

The fourth example in the figure shows boundary layers developing on the sides of a pipe or channel with a streamlined entrance region. The gradual thickening of the layers causes them to merge and the conditions further downstream are wholly boundary layer type flow. That is, they are either laminar or turbulent or both. Experiment shows that the length of pipe or channel into which the inviscid core extends is rarely greater than 50 diameters and is usually far less because most entrance regions are poorly streamlined.

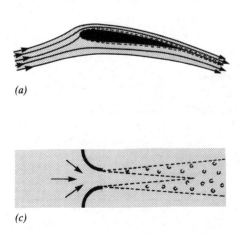

(a)

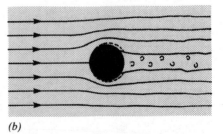

(b)

(c)

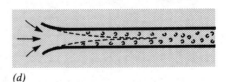

(d)

Figure 4-3 Typical boundary layer flows.

4-2-1 Development of a Boundary Layer in a Steady Flow

Figure 4-4 depicts developing boundary layers on the opposite surfaces of a flat plate aligned parallel to a uniform stream. This configuration is of relatively little interest in its own right, but it illustrates some general features that are also found in flows such as those in Fig. 4-3.

The most obvious feature is that the thickness of the boundary layer increases continuously downstream. This complicates the analysis of external flows because the rate of increase in thickness is not known, being dependent upon the outer flow field. However, it tends to simplify the analysis of internal flows because pipes and channels (e.g. rivers) are often much longer than the region before the boundary layers merge (Fig. 4-3d). In this case it is reasonable to neglect the special features of the entrance regions and to pretend that boundary layer type flow exists everywhere.

For external flows the thickness h of a *laminar* boundary layer at a distance x from the leading edge of a flat plate was shown by Blasius to satisfy

$$\frac{h}{x} = 4.93 \left(\frac{Vx}{\mu} \right)^{-1/2} \tag{4-4}$$

where V denotes the free stream velocity and $v \equiv \mu/\rho$ is the fluid kinematic viscosity. The ratio Vx/v is a Reynolds number (section 1-2). The thickness of a *turbulent* boundary layer is more difficult to predict, but for Reynolds numbers less than about 10^7 the ratio h/x is approximately proportional to $Re^{-1/5}$. Sufficiently far downstream, the boundary layer will separate from the plate and cause a wake-region. The thickness of this and the location at which it begins are strongly dependent upon the condition in the external flow—as Fig. 4-3 illustrates.

A second important feature of Fig. 4-4 is that the upstream part of the

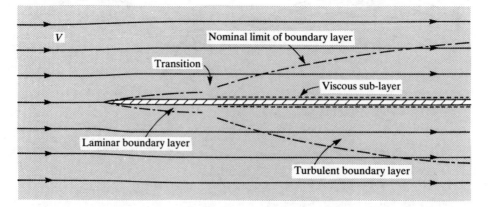

Figure 4-4 Boundary layer over a smooth, flat plate at zero incidence.

boundary layer is laminar and the downstream portion is turbulent. This is always the case unless large disturbances exist for some reason, in which case the whole of the boundary layer will be turbulent. In the absence of large disturbances, the length L_1 of the laminar boundary layer will usually satisfy

$$0.5 \times 10^6 < \frac{VL_1}{v} < 2 \times 10^6 \qquad (4\text{-}5)$$

in which VL_1/v is a Reynolds number as above.

Allen (1981) has given evidence that in general the location of the transition from laminar to turbulent conditions in a boundary layer might be more closely dependent upon the local skin friction coefficient $C_F \equiv \tau_w / \frac{1}{2} \rho V^2$ than on the Reynolds number. If so, analytical and numerical studies of flows such as those in Fig. 4-3a and b will be improved. However, Eq. (4-4) is likely to be preferred in engineering design practice because it is difficult to estimate the values of shear stress needed to apply Allen's criterion.

The existence of the laminar region is often unimportant in large-scale engineering design because its length is typically small. In water flowing at 5 m/s, for example, L_1 is shown by Eq. (4-4) to lie somewhere between 0.1 m and 0.4 m. These values are small in comparison with the lengths of full-scale ships or sewers. However, they are not small in comparison with the size of fish or otters. Annoyingly, they are often not small in comparison with model-scale ships, either. Thus engineers attempting to deduce the correct dynamics of full-scale vessels from model tests must take special precautions to induce an early transition to turbulence in their models.

A third important feature illustrated in Fig. 4-4 is that a turbulent boundary layer can have a viscous sub-layer very close to the solid surface. If so, the conditions in this layer will be similar to those in a wholly laminar boundary layer. Also there will be a thin layer just outside it in which the conditions are similar to those in the transitional region shown in the figure. In both cases, the portion of the transition region that bounds the laminar region is almost laminar itself. Further away it is a mixture of laminar and turbulent conditions with more and more turbulent pockets as we approach the fully turbulent region.

In a turbulent boundary layer over a rough surface, the viscous sub-layer is effectively non-existent. Wakes form behind individual bits of surface roughness and the transition region extends to the boundary surface. In the case of an extremely rough surface, even the transition layer is effectively absent, and the fully developed turbulence extends between the individual roughnesses.

The final feature to be noted from Fig. 4-4 is that the streamlines are almost parallel to the solid surface even when they become subsumed in the boundary layer. This feature is also displayed in the various examples in Fig. 4-3. It is of great importance from an analytical point of view because it enables us to regard the conditions in most boundary layers as locally uniaxial.

Whenever the boundary layer remains close to the boundary—i.e. upstream of separation points such as those in Fig. 4-3b and c—the direction of the mean flow is known a priori.

4-3 LAMINAR FLOWS

In general, laminar flows may be strongly three-dimensional and unsteady—random waves in a highly viscous fluid, for instance. However, for present purposes, we shall consider only steady, uniaxial flow. Figure 4-5 depicts three possible velocity distributions in a steady laminar flow between two parallel horizontal plates, one of which is moving with a steady velocity V in the positive x-direction. To investigate the reasons for the different velocity distributions, we consider the forces acting on the fluid element depicted in Fig. 4-6. For a unit width of flow (normal to the page) the pressure forces on the left- and right-hand faces are $p \, \delta y$ and $-\{p + (\partial p/\partial x) \, \delta x\} \, \delta y$ respectively and the shear forces on the lower and upper surfaces are $-\tau \, \delta x$ and $\{\tau + (\partial \tau/\partial y) \, \delta y\} \, \delta x$ respectively. Since the flow is steady, these forces are in equilibrium and so

$$-\frac{\partial p}{\partial x} \, \delta x \, \delta y + \frac{\partial \tau}{\partial y} \, \delta y \, \delta x = 0 \tag{4-6}$$

For a Newtonian fluid, the shear stress in a uniaxial laminar flow satisfies

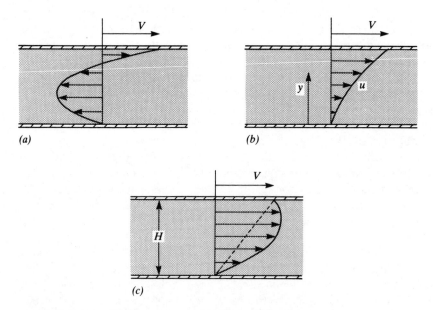

(a)   *(b)*

(c)

Figure 4-5 Velocity distributions in a laminar flow between parallel plates.

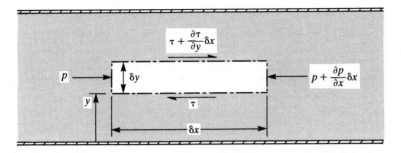

Figure 4-6 Equilibrium of a laminar flow between parallel plates. (Element shown at a particular instant. It subsequently distorts.)

$\tau = \mu(\partial u/\partial y)$ (section 4-1) and so Eq. (4-6) shows that the pressure gradient $\partial p/\partial x$ is given by

$$\frac{\partial p}{\partial x} = \mu \frac{\partial^2 u}{\partial y^2} \qquad (4\text{-}7)$$

After integrating twice with respect to y, we obtain

$$\frac{1}{2}\frac{\partial p}{\partial x} y^2 = \mu u + K_1 y + K_2 \qquad (4\text{-}8)$$

in which K_1 and K_2 are constants. These can be found by noting that $u = 0$ at $y = 0$ and that $u = V$ at $y = H$, and so

$$\frac{\partial p}{\partial x} = \frac{2\mu(u - Vy/H)}{y(y - H)} \qquad (4\text{-}9)$$

The velocity distribution is obtained by rearranging this expression to give

$$u = \frac{Vy}{H} + \frac{y(y - H)}{2\mu}\frac{\partial p}{\partial x} \qquad (4\text{-}10)$$

Each of the velocity distributions depicted in Fig. 4-5 satisfies Eq. (4-10). They are obtained with different pressure gradients.

(a) $\partial p/\partial x = 0$ When there is no variation in pressure in the direction of flow, the velocity distribution is linear, namely $u/V = y/H$. In this case, every particle behaves in the manner depicted in Fig. 4-1 and all of them distort at the same rate.

(b) $\partial p/\partial x < 0$ When the pressure decreases in the direction of flow, all particles travel more quickly than they would with a linear velocity distribution. As shown in Fig. 4-5c, the additional contribution is greatest at the mid-point between the plates. In the special case where both plates are stationary, $V = 0$ and Eq. (4-10) shows that the velocity distribution is parabolic. This case is analogous to a laminar flow in a circular-section pipe (see section 2-7 and Example 4-1).

(c) $\partial p/\partial x > 0$ When the pressure increases in the direction of flow, all particles travel more slowly than with $\partial p/\partial x = 0$ as shown in Fig. 4-5b. With a sufficiently large pressure gradient, most of the flow will be in the negative x-direction even though that close to the moving plate will still be in the positive x-direction. This condition occurs in many lubricating flows.

Lateral variations in pressure: When integrating Eq. (4-7), it was implicitly assumed that the pressure does not vary within any particular cross-section (x=constant). That is, the pressure varies with x but not with y. The justification for this assumption is deferred until Chapter Ten, but in essence it is that a lateral pressure gradient would imply the existence of lateral accelerations and yet the latter are absent in a parallel flow.

4-3-1 Laminar Boundary Layers

If the negative pressure gradient in Fig. 4-5c was slightly smaller, the velocity gradient $\partial u/\partial y$ at the moving plate would become zero and there would be no shear stress at the plate (Fig. 4-7a). In this case no force would be required to maintain its motion. All the effort involved in maintaining the fluid motion would be provided by the pressure gradient.

Another way of looking at this condition would be to argue that the flow conditions could remain unchanged if the upper plate was removed. This would lead to the free surface flow condition shown in Fig. 4-7b. In practice, of course, this would not be a realistic possibility in a horizontal channel because there would be no way of providing the required pressure gradient. However, it is effectively possible in a sloping channel because gravity can provide the necessary driving force. This condition is considered in Problem 3 at the end of this chapter.

A third way of viewing the flow is to imagine that the upper plate is removed and replaced by an inviscid flow-field as shown in Fig. 4-7c. Once again, the conditions could not be sustained downstream because the pressure gradient would have to exist in the inviscid flow-field as well as in the laminar flowfield. There would therefore be accelerations in the inviscid flow. Nevertheless, something similar to this could undoubtedly exist for a very short distance (or time). The depicted condition may therefore be regarded as an approximate example of a laminar boundary layer bordering an inviscid flow. In particular, it illustrates how the viscous and inviscid flow-fields can merge at an interface where the velocity gradient is zero.

4-3-2 Skin Friction

In each case shown in Fig. 4-7, there is a velocity gradient at the lower plate and therefore a shear stress $\tau_w = \mu(\partial u/\partial y)_w$. This is sometimes used to define a

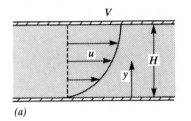

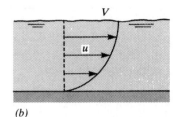

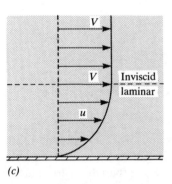

Figure 4-7 Alternative interpretations of a velocity distribution with $\partial u/\partial y = 0$ at $y = H$. (a) $\tau = 0$ at upper plate. (b) Free-surface flow. (c) Boundary layer flow.

skin friction coefficient C_F by

$$C_F \equiv \frac{\tau_w}{\frac{1}{2}\rho V^2} \qquad (4\text{-}11)$$

For any particular free stream velocity the skin friction coefficient is proportional to the wall shear stress. However, C_F is of relatively little use in laminar flows because it is rarely constant (or even approximately so). It is of far greater use in turbulent flows, where it may often be regarded as approximately constant.

4-4 TURBULENT FLOWS

In the majority of flows of practical interest, especially those at large scale, the conditions in a boundary layer or wake are more likely to be turbulent than laminar. Individual fluid particles follow rather erratic paths akin to those of autumn leaves fluttering groundwards in a breeze. The overall direction of flow is well defined, but the instantaneous direction of motion of any particular particle is highly unpredictable. Moreover, individual particles distort and rotate equally erratically.

The seemingly random movements of large-scale particles in a turbulent

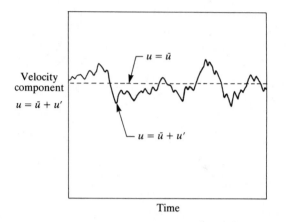

Figure 4-8 Velocity history at a point in a 'steady' turbulent flow.

flow are loosely analogous to the random motions of molecules at a microscopic level. The length and time scales associated with the turbulent randomness are hugely greater than those associated with molecular randomness, but the two phenomena have much in common. In particular, the methods used to define a velocity at a point in a continuum (section 2-2) can be applied to turbulent flows simply by choosing sufficiently large intervals of space and time. However, some turbulent motions are so large that they should not strictly be neglected in comparison with typical dimensions of pipes, channels, vehicles and buildings, etc.

Figure 4-8 illustrates the signal that might be recorded by a sufficiently sensitive velocity probe at a point in a turbulent flow-field. The signal fluctuates about a mean value $\bar{u}$ and the fluctuations u' appear random.

Similar results will be obtained for probes measuring the velocity components in the y- and z-directions and so

$$u = \bar{u} + u' \qquad v = \bar{v} + v' \qquad w = \bar{w} + w' \qquad (4\text{-}12)$$

In a typical boundary layer, the flow is predominantly uniaxial and two of the *mean* velocity components—$\bar{v}$ and $\bar{w}$, say—are almost zero. However, the turbulent fluctuations in these directions do not disappear. Therefore

$$u = \bar{u} + u' \qquad v \simeq v' \qquad w \simeq w' \qquad (4\text{-}13)$$

4-4-1 Shear Stress in a Turbulent Boundary Layer

In a uniaxial *laminar* flow of a Newtonian fluid, the shear stress is $\tau_{yx} = \mu \, \partial u / \partial y$. The stress exists because molecular diffusion involves movements normal to the direction of mean flow. When the velocities on adjacent streamlines are unequal, molecules moving from a slow-moving streamline to a faster moving one will on average require to be accelerated a little. Likewise, molecules moving in the opposite direction will on average require to be slowed down.

In a uniaxial turbulent flow, the same effect occurs, but it is dwarfed by the equivalent phenomenon due to turbulent eddies behaving in an analogous manner. In this case, large mass transfers occur normal to the *mean* streamlines and the shear stress is composed of both laminar and turbulent contributions, i.e.

$$\tau_{yx} = \bar{\tau}_{yx} + \tau'_{yx} \qquad (4\text{-}14)$$

in which τ'_{yx} is often called a *Reynolds stress*. It is sometimes alternatively called an *eddy stress*, *apparent stress* or *virtual stress* and it can be shown to satisfy $\tau'_{yx} = -\rho \overline{u'v'}$. Thus

$$\tau_{yx} = \mu \frac{\partial \bar{u}}{\partial y} - \rho \overline{u'v'} \qquad (4\text{-}15)$$

Figure 4-9 depicts typical velocity and shear stress distributions in a turbulent boundary layer over a smooth surface. The velocity gradient $\partial \bar{u}/\partial y$ is very large close to the wall and reduces with increasing distance from the wall. The total shear stress varies similarly but the relative magnitudes of the laminar and turbulent contributions vary markedly. Throughout most of the boundary

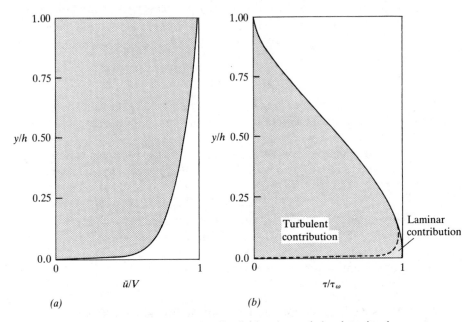

Figure 4-9 Typical velocity and shear stress distributions in a turbulent boundary layer over a smooth surface. (h=nominal thickness of the boundary layer). (a) Velocity distribution. (b) Shear stress distribution.

layer, τ'_{yx} is the dominant contribution, but it is much smaller than $\bar{\tau}_{yx}$ in the viscous sub-layer very close to the wall (see also Figure 4-4).

Since the magnitudes of the turbulent fluctuations vary unpredictably, Eq. (4-15) cannot be used directly to analyse turbulent flow behaviour. However, numerous empirical models of turbulence have been developed and it is possible to simulate many types of flow reasonably well. Some of the simplest models make use of a *kinematic eddy viscosity* ε defined to satisfy

$$\tau'_{yx} = \rho\varepsilon\frac{\partial\bar{u}}{\partial y} \tag{4-16}$$

Since $\mu = \rho v$ where v is the fluid kinematic viscosity, the total shear stress at any distance y from the wall may be written as

$$\tau_{yx} = \rho(v + \varepsilon)\frac{\partial\bar{u}}{\partial y} \tag{4-17}$$

If ε, like v, was a fluid property that could be regarded as a constant, Eq. (4-17) would be no more difficult to use than its laminar flow counterpart (4-2). Unfortunately, however, ε is not a constant and it is not even a fluid property; it is dependent upon u' and v'.

Usually, turbulence models attempt to relate u' and v' to the mean velocity gradients, and in the case of boundary layer flows only $\partial\bar{u}/\partial y$ is important. An early model developed by Prandtl using physical reasoning hypothesises that the kinematic eddy viscosity ε is proportional to $|\partial\bar{u}/\partial y|$ and to the square of the distance y from the boundary wall, namely

$$\varepsilon \propto y^2 \left|\frac{\partial\bar{u}}{\partial y}\right| \tag{4-18}$$

The influence of the distance from the wall is important because the freedom of fluid particles to move in a random-like turbulent manner is greatly restricted close to the wall.

Often, turbulence models also involve a parameter u_τ known variously as a *friction velocity*, a *wall-friction velocity* or a *wall-shear velocity* and defined as

$$u_\tau \equiv (\tau_w/\rho)^{1/2} \tag{4-19}$$

The parameter is introduced because the ratio τ_w/ρ frequently occurs in equations describing boundary layers. Its various names reflect the fact that it has the dimensions of velocity, but that its magnitude is determined by the wall shear stress. In some books, the symbol is written as u^*.

4-4-2 Velocity Distribution in a Turbulent Boundary Layer

Figure 4-9*b* shows that the laminar contribution to the overall shear stress is detectable for a significant distance from the wall. Nevertheless, turbulent shear stresses dominate throughout most of the boundary layer. Within a zone of significant thickness in the neighbourhood of the wall($y/h<0.2$, say, in the figure) the shear stress is nearly constant and we may continue to write $\tau \approx \tau_w$ as a first approximation. In this case, the kinematic eddy viscosity relationship (4-18) may be substituted into Eq. (4-16) to give

$$\tau_w = \rho K^2 y^2 \left(\frac{d\bar{u}}{dy}\right)^2 \tag{4-20}$$

in which K is a constant. Since $\tau_w/\rho = u_\tau^2$, this may be written as

$$\frac{d\bar{u}}{dy} = \frac{u_\tau}{K} \cdot \frac{1}{y} \tag{4-21}$$

and integrated with respect to y to give the velocity distribution within this zone of the boundary layer, namely

$$\boxed{\bar{u} = \frac{u_\tau}{K}\ln(y) + C_1} \tag{4-22}$$

in which C_1 is a constant of integration.

Although this relationship has been obtained in a less than rigorous manner (the expression (4-18) has simply been assumed for instance) it is found to give good agreement with experiment. Remarkably, the agreement persists throughout almost the whole of the boundary layer (about 85 per cent) even though the expression has been derived only for the region in which the shear stress is approximately constant. For the purposes of this book, it will be assumed to be valid throughout the boundary layer (except in the viscous sub-layer very close to a smooth wall—see Figs 4-4 and 4-9b).

The constant K in Eq. (4-22) is found experimentally to be about 0.4. The constant C_1 is usually eliminated by substituting $u = V$ at the outer limit of the boundary layer where $y = h$. In this case, we obtain the so-called *velocity-defect law*

$$\frac{V-\bar{u}}{u_\tau} = \frac{1}{K}\ln\left(\frac{h}{y}\right) \approx 2.5\ln\left(\frac{h}{y}\right) \tag{4-23}$$

This simple expression is sufficiently accurate for many practical purposes. However, Hinze (1975) shows that better agreement with experiment is obtained throughout most of the boundary layer with the slightly different expression

$$\frac{V-\bar{u}}{u_\tau} = 2.44\ln\left(\frac{h}{y}\right) + 2.5 \tag{4-24}$$

4-4-3 Smooth Walls

All solid surfaces appear rough when viewed with sufficiently large magnification. Nevertheless, they do not 'feel' rough to the mainstream flow unless the size of the roughness is significant in comparison with the thickness of the viscous sub-layer.

Detailed experiments show that disturbances exist sporadically in this layer when pockets of intense turbulence move towards the wall, but the overall behaviour is sensibly laminar. Within this thin zone (shown as less than 1 per cent of the boundary layer thickness in Fig. 4-9b) we may write $\tau = \mu \, d\bar{u}/dy$ and, since $\tau \simeq \tau_w$,

$$\frac{d\bar{u}}{dy} = \frac{\tau}{\mu} = \frac{\rho \tau_w}{\mu \rho} = \frac{u_\tau{}^2}{v} \tag{4-25}$$

On integrating with respect to y, we obtain a linear distribution of velocity in the viscous sub-layer, namely

$$\bar{u} = \frac{u_\tau{}^2}{v} y \tag{4-26}$$

in which the constant of integration is omitted because $\bar{u} = 0$ at the wall ($y = 0$).

Experimental measurements show that this expression is highly accurate for distances from the wall satisfying $u_\tau y/v < 5$ (that is when $u < 5 u_\tau$) and that it is a good approximation within $u_\tau y/v < 8$. There is then a transitional region in which deviations from the laminar distribution become increasingly large until the logarithmic turbulent distribution (4-22) takes over when $u_\tau y/v$ exceeds about 30.

Although neither equation is valid within the transitional region, it is instructive to extrapolate them both into this region to meet at the point 1 shown in Fig. 4-10. At this position, the laminar Eq. (4-26) gives $y_1 = v\bar{u}_1/u_\tau{}^2$ and so the turbulent Eq. (4-22) may be written as

$$\bar{u}_1 = \frac{u_\tau}{K} \ln \left(\frac{v\bar{u}_1}{u_\tau{}^2} \right) + C_1 \tag{4-27}$$

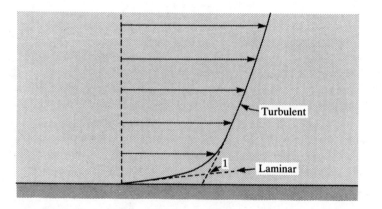

Figure 4-10 Velocity distribution in a turbulent boundary layer close to a smooth surface.

The advantage of this procedure is that it enables us to develop a slightly more general expression than (4-22) for the special case of turbulent flows over smooth walls. Using (4-27) to eliminate C_1 from (4-22), we obtain after a little manipulation

$$\frac{\bar{u}}{u_\tau} = \frac{1}{K} \ln\left(\frac{yu_\tau}{v}\right) + C_2 \tag{4-28}$$

At first sight, this is no more general than (4-22) because it involves nothing more than a change in the constant of integration, namely $C_2 = C_1 - 1/K \ln(u_\tau/v)$. However, it is a major advance in practice because C_2 turns out to be almost universally constant, independent of the particular flow. Various values have been obtained in different sets of experiments, but the following expression lies within the range suggested by most workers:

$$\frac{\bar{u}}{u_\tau} \simeq 2.5 \ln\left(\frac{yu_\tau}{v}\right) + 5.5 \tag{4-29}$$

A simple method of deriving Eq. (4-28) based solely on the requirements of physical similarity is presented in Example 5-7.

Skin friction The wall-shear velocity u_τ is defined by Eq. (4-19) as a function of the wall shear stress. It may therefore be used to deduce the local skin friction coefficient C_F using

$$C_F \equiv \frac{\tau_w}{\frac{1}{2}\rho V^2} = 2\left(\frac{u_\tau}{V}\right)^2 \tag{4-30}$$

For flow in pipes and channels—where the boundary layer extends throughout the region of flow—it is usual to define a skin friction coefficient, f, in terms of the overall mean velocity $\bar{V}$ rather than the free stream velocity V at the outer limit of the boundary layer. Thus

$$f \equiv \frac{\tau_w}{\frac{1}{2}\rho \bar{V}^2} = 2\left(\frac{u_\tau}{\bar{V}}\right)^2 \tag{4-31}$$

By integrating Eq. (4-29) over the whole of the boundary layer (see Example 4-8, p. 110), it can be shown that

$$\sqrt{\frac{2}{f}} \simeq 2.5 \ln\left(Re\sqrt{\frac{f}{2}}\right) + 3 \tag{4-32}$$

where $Re = \bar{V}h/v$ is a Reynolds number and h denotes the thickness of the boundary layer. An equivalent expression can be derived similarly for the flow in a circular-section pipe, but this is not strictly valid because (4-29) applies to two-dimensional, not axi-symmetric, flows. It is therefore necessary to adjust the resulting constants of integration slightly to obtain good correlation with experiment. After so doing, we obtain

$$\sqrt{\frac{1}{4f}} \simeq 2 \log (Re\sqrt{4f}) - 0.8 \qquad (4\text{-}33)$$

in which $Re = \bar{V}D/\nu$, where D denotes the pipe diameter. The change to base 10 logarithms was traditionally made for computational convenience. The expression is written as a function of '$4f$' rather than 'f' to simplify comparisons with other literature.

4-4-4 Rough Walls

In a turbulent boundary layer over a sufficiently rough wall, the viscous sub-layer on the solid surfaces is not thick enough to prevent the individual roughnesses from penetrating into the turbulent region of flow. In this case, it is reasonable to imagine that the zone of turbulence extends all the way to the wall. For this condition, the constant C_1 can be eliminated from Eq. (4-22) by writing $u = u_k$ at $y = k_s$ to give

$$\frac{\bar{u}}{u_\tau} = \frac{1}{K} \ln \left(\frac{y}{k_s} \right) + \frac{u_k}{u_\tau} \qquad (4\text{-}34)$$

in which k_s is the characteristic roughness size. Experiments show that the substitutions $K = 0.4$ and $u_k = 8.5 u_\tau$ lead to a universally valid approximation

$$\frac{\bar{u}}{u_\tau} \simeq 2.5 \ln \left(\frac{y}{k_s} \right) + 8.5 \qquad (4\text{-}35)$$

Strictly, it is not valid to use the single parameter k_s to denote the size of numerous three-dimensional roughnesses. Nor is it possible to find a simple way of defining the position of the origin $y = 0$. In practice, however, these complications do not cause significant difficulty. For example, common sense shows that the precise location of $y = 0$ cannot be important provided that the roughness size is small in comparison with the overall thickness of the boundary layer.

Skin friction Like its smooth-wall counterpart (4-29), Eq. (4-35) can be used to deduce the skin friction coefficient for a turbulent boundary layer. After carrying out the necessary integration, we obtain

$$\sqrt{\frac{2}{f}} \simeq 2.5 \ln \left(\frac{h}{k_s} \right) + 6 \qquad (4\text{-}36)$$

in which h denotes the thickness of the boundary layer and $(1 - k_s/h)$ has been approximated to unity. The corresponding relationship for steady flow in a circular-section pipe of diameter D is found by experiment to be

$$\sqrt{\frac{1}{4f}} \simeq 2 \log \left(\frac{D}{k_s} \right) + 1.14 \qquad (4\text{-}37)$$

when the roughnesses are sufficiently uniform.

4-4-5 Walls of Intermediate Roughness

In most boundary layer flows and internal flows, the roughnesses are too large to cause smooth-wall conditions, but too small to cause rough-wall conditions. Therefore, both viscosity and surface roughness contribute significantly to the degree of turbulence and hence to the skin friction.

For commercially available pipes of circular cross-section, the most widely acknowledged expression for skin friction is the Colebrook–White equation, namely

$$\sqrt{\frac{1}{4f}} = -2\log\left\{\frac{k_s}{3.72D} + \frac{2.51}{Re\sqrt{4f}}\right\} \qquad (4\text{-}38)$$

This is asymptotic to the corresponding smooth-wall equation (4-33) when the Reynolds number and the roughness size are sufficiently small. It is asymptotic to the rough-wall equation (4-37) when they are sufficiently large. As a consequence, it may be regarded as a universal expression for turbulent skin friction in a circular-section pipe for any Reynolds number and any roughness size (provided that the latter does not exceed, say, 10 per cent of the pipe diameter).

A disadvantage of the Colebrook–White equation is that it cannot be solved explicitly for the skin friction coefficient f (because this appears both inside and outside the logarithm). Accordingly, several people have proposed the use of alternative expressions that do not have this drawback. A particularly accurate one proposed by Jain (1976) is

$$\sqrt{\frac{1}{4f}} = -2\log\left\{\frac{k_s}{3.72D} + \frac{5.72}{Re^{0.9}}\right\} \qquad (4\text{-}39)$$

This agrees very closely with the Colebrook–White expression for most values of k_s and Re of practical interest.

A warning note is merited before leaving this topic. The Colebrook–White expression is only one of a wide range of possible expressions that are asymptotic to the smooth-wall and rough-wall relationships. It has gained widespread acceptance because it approximates the behaviour of flows in many commercial pipes quite well. However, it should not be used indiscriminately for pipes made from non-traditional materials until their resistance characteristics have been assessed experimentally. The expression is highly unsuitable for representing the famous data obtained by Nikuradse in artificially roughened pipes.

4-5 APPLICATIONS OF FLUID SHEAR

Example 4-1: Poiseuille flow Show that the axial pressure gradient dp/dx in a steady laminar flow along a straight, horizontal pipe of radius R is

$-4\mu V_0/R^2$ where μ is the dynamic viscosity of the fluid and V_0 is the velocity on the pipe axis. Hence find the shear stress at the wall of a 0.2 m diameter pipe conveying a crude oil of density 925 kg/m^3 and dynamic viscosity 0.1 Pa s when the velocity on the axis is 1.5 m/s.

SOLUTION Consider the elemental cylindrical core of fluid of radius r and length δx shown in the centre of the pipe in Fig. 4-11. The pressure and shear stresses acting on the element parallel to the axis are shown in their mathematically positive directions. The net pressure force in the positive x-direction is $-\pi r^2 (dp/dx)\,\delta x$ and the shear force on the lateral surfaces of circumference $2\pi r$ and length δx is $2\pi r\,\delta x\,\tau$.

As the element moves along the pipe, it will distort, but none of the fluid particles will accelerate. Therefore the net force acting on the element must be zero, and so

$$2\pi r\,\delta x\tau - \pi r^2 \frac{dp}{dx}\,\delta x = 0 \tag{4-40}$$

i.e.
$$\tau = \frac{r}{2}\frac{dp}{dx} \tag{4-41}$$

Using the Newtonian relationship between the shear stress and the velocity gradient, $\tau = \mu\,du/dr$, we obtain

$$\frac{du}{dr} = \frac{r}{2\mu}\frac{dp}{dx} \tag{4-42}$$

The pressure gradient dp/dx is constant in a steady pipe flow. Therefore this expression may be integrated to give

$$u - V_0 = \frac{r^2}{4\mu}\frac{dp}{dx} \tag{4-43}$$

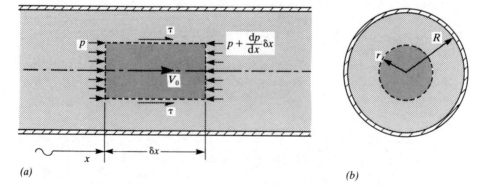

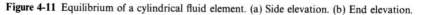

(a)

(b)

Figure 4-11 Equilibrium of a cylindrical fluid element. (a) Side elevation. (b) End elevation.

in which V_0 denotes the velocity on the axis ($r=0$). At the pipe wall, $r=R$ and the velocity u is zero, and so

$$\frac{dp}{dx} = -\frac{4\mu V_0}{R^2} \qquad (4\text{-}44)$$

The shear stress at the pipe wall is shown by Eq. (4-41) to be $\tau_w = \frac{1}{2}R\,(dp/dx)$. Therefore, using 4-44,

$$\tau_w = -\frac{2\mu V_0}{R} \qquad (4\text{-}45)$$

For the numerical values given in the example, we obtain $\tau_w = -3\,\text{Pa}$.

Comments on Example 4-1
1. The minus signs in Eq. (4-44) and (4-45) indicate that the pressure decreases in the direction of flow and that the shear stress opposes the motion. Both of these results are in agreement with common sense.
2. Equation (4-41) shows that the maximum magnitude of the shear stress occurs at the wall (where r is a maximum). Even here, it is only 3 Pa and so it is tiny in comparison with typical normal stresses. For example atmospheric pressure is typically about 10^5 Pa.
3. Equation (4-43) shows that the velocity distribution in the pipe is paraboloidal. It is shown in Example 7-1 (p. 185) that this implies that the mean velocity of flow $\bar{V}$ is only half the maximum velocity V_0.
4. The Reynolds number is defined by Eq. (1-1) as $Re = \rho d V/\mu$. For flows in pipes, it is conventional to choose the pipe diameter and the mean velocity as the representative parameters. In this example we obtain $Re = 1388$. Since this is smaller than 2000, experience shows that the flow in the pipe will be laminar as assumed, not turbulent.

Example 4-2: Shaft end bearing Figure 4-12a depicts the end bearing of a 0.4 m diameter shaft rotating at $\Omega = 600\,\text{rev/min}$ over a 1 mm thick layer of oil with a viscosity of 0.4 Pa s. Estimate the power required to overcome frictional resistance in the bearing.

SOLUTION The tangential velocity of the bearing at a radius r is $V = r\Omega$ and so the velocity gradient in the oil at this radius is $V/h = r\Omega/h$. The corresponding shear stress is $\mu r\Omega/h$ and the contribution of an elemental annulus (Fig. 4-12b) to the torque about the central axis is

$$\delta T = \frac{\mu r\Omega}{h} \times 2\pi r\,\delta r \times r = \frac{2\pi\mu\Omega}{h}r^3\,\delta r \qquad (4\text{-}46)$$

The power required to overcome this torque is $\delta\dot{W} = \Omega\,\delta T$ and so the total power requirement is

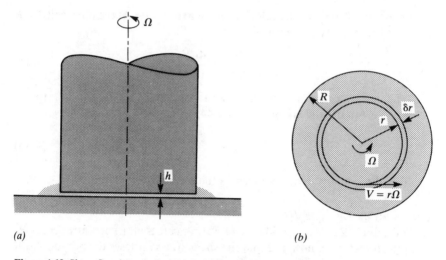

(a)   (b)

Figure 4-12 Shear flow in a shaft end bearing. (a) Side elevation. (b) End elevation.

$$\dot{W} = \frac{2\pi\mu\Omega^2}{h} \int_0^R r^3\,dr = \frac{\pi\mu\Omega^2 R^4}{2h} \tag{4-47}$$

Using the numerical values stated above and noting that 600 rev/min is 20πrad/sec, we obtain $\dot{W} = 3.97$ kW.

Comment on Example 4-2 In practice the fluid particles will not simply follow the circular paths implied in this analysis. The rotation of the shaft will induce a radial pressure gradient which will induce an inward component of flow over the base plate. This will be accompanied by an outward component of flow over the end of the shaft.

Example 4-3: Laminar skin friction force on a flat plate Estimate the skin friction force per unit width on the upper surface of a 2 m long flat plate aligned parallel to a uniform stream of air moving at $V = 10$ m/s. Assume that the boundary layer is laminar and that its thickness h at a distance x from the leading edge of the plate is given by

$$h = \sqrt{\frac{30v}{V}x} \tag{4-48}$$

Also assume that the shear stress varies linearly across the width of the boundary layer at any section. The density of the air is $\rho = 1.23$ kg/m^3 and its dynamic viscosity is $\mu = 17.9$ Pa s. The plate is 8 m wide normal to the page (Fig. 4-13).

SOLUTION The linear stress distribution may be written as

$$\tau = \tau_{\text{w}}\left(1 - \frac{y}{h}\right) \tag{4-49}$$

which is a maximum at the wall ($y = 0$) and zero at the outer edge of the boundary layer ($y = h$). In a laminar flow of air, $\tau = \mu\,du/dy$ and so

$$\frac{du}{dy} = \frac{\tau_{\text{w}}}{\mu}\left(1 - \frac{y}{h}\right) \tag{4-50}$$

This may be integrated to give the velocity distribution

$$u = \frac{\tau_{\text{w}}}{\mu}\left(y - \frac{y^2}{2h}\right) \tag{4-51}$$

where the constant of integration is omitted because $u = 0$ at $y = 0$.

Equation (4-51) shows that the velocity distribution is paraboloidal. For present purposes, however, its main use is to provide an expression for the wall shear stress as a function of the boundary layer thickness. Since $u = V$ at $y = h$, we obtain

$$\tau_{\text{w}} = \frac{2\mu V}{h} \tag{4-52}$$

The shear force on an elemental length δx of the plate of width b is $\delta F_x = \tau_{\text{w}} b \,\delta x$ and the total shear force on the plate is

$$F_x = \int_0^L \tau_{\text{w}} b \,dx = 2b\mu V \int_0^L \frac{dx}{h} \tag{4-53}$$

On substituting (4-48) for the thickness h, we obtain

$$F_x = 0.365 b\rho^{1/2}\mu^{1/2}V^{3/2}\int_0^L \frac{dx}{x^{1/2}} = 0.730 b(\rho\mu L V^3)^{1/2} \tag{4-54}$$

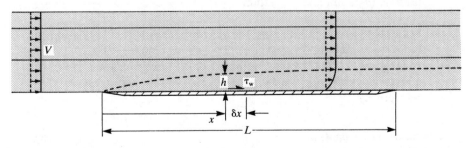

Figure 4-13 Growth of a boundary layer on a flat plate at zero incidence (boundary layer thickness exaggerated).

For the particular numerical values given above, the total skin friction force on the surface of the plate is $F_x = 1.23\,N$.

Comments on Example 4-3

1. It is usual to introduce an average skin friction coefficient $\bar{C}_F$ defined to satisfy $F_x = \frac{1}{2}\bar{C}_F\rho bLV^2$. Using (4-54) we obtain

$$\bar{C}_F = 1.46\,(Re)^{-1/2} \tag{4-55}$$

where $Re = LV/v$ is a Reynolds number. This is actually an overestimate, however. Using a more accurate analysis due to Balsius, we obtain

$$\bar{C}_F = 1.328\,(Re)^{-1/2} \tag{4-56}$$

which for the present example gives $\bar{C}_F = 0.001\,13$.

2. The average skin friction coefficient in a *turbulent* boundary layer over a flat plate can be estimated in a manner similar to that used above. However, the necessary integrations tend to be rather complex if a logarithmic velocity distribution is assumed. Also the coefficients in the resulting expressions are found by experiment to be in need of adjustment. Schlichting (1979) gives an alternative expression

$$\bar{C}_F = \frac{0.455}{(\log Re)^{2.58}} - \frac{C}{Re} \tag{4-57}$$

which is the Prandtl–Schlichting formula for turbulent flows over smooth plates at zero incidence. The value of the constant C depends upon the ratio L_1/L where L_1 denotes the length of the laminar boundary layer upstream of the transition region (Fig. 4-4). For sufficiently short laminar regions on long plates, C may be approximated to zero.

Example 4-4: Deterioration of pipes Estimate the skin friction coefficient in a 1 m diameter smooth-walled pipe conveying water ($v = 1.2\,mm^2/s$) with a mean velocity of 1.8 m/s. Also estimate the skin friction coefficient in later years when its characteristic roughness size increases to (a) 0.1 mm, (b) 1 mm and (c) 10 mm.

SOLUTION For a smooth pipe, the appropriate skin friction relationship is Eq. (4-33). Using $Re = D\bar{V}/v = 1.5 \times 10^6$, we obtain $f = 0.002\,72$. For a rough pipe, Eq. (4-37) should be used. For the three roughness sizes given above, it yields (a) $f = 0.002\,99$, (b) $f = 0.004\,90$ and (c) $f = 0.009\,46$ respectively.

Comments on Example 4-4

1. Few real pipes resist this process. Even plastic pipes tend to increase in roughness over the years as a result of deposition. The process is more

severe in metal pipes because of corrosion and it is quite important in sewers because of biological activity. In the case of long sea outfalls, saltwater intrusion can bring with it marine life such as barnacles that can be virtually impossible to dislodge by realistic means.

2. In the solution it has been assumed that the rough-wall law is appropriate even for roughnesses as small as 0.1 mm. To investigate the validity of this assumption, we should estimate the thickness of the viscous sub-layer. This is stated in section 4-4-3 to extend to a distance of approximately $y = 5v/u_\tau$ from a wall. Using $u_\tau^2 = \tau_w/\rho = \frac{1}{2}f\bar{V}^2$ and substituting $f = 0.00272$, we obtain a thickness of about 0.09 mm. Although the thickness reduces when the skin friction increases, only the two larger roughness sizes significantly exceed the thickness of the viscous sub-layer. For the first case, Jain's expression (Eq. 4-39) indicates that the smooth and rough effects are both important. A better estimate of the skin friction coefficient is $f \approx 0.00327$.

Example 4-5: Free surface flow in a channel Water flows steadily down a wide channel at a constant depth of $d = 0.6$ m and at a mean velocity of $\bar{V} = 2$ m/s. If the channel slopes downhill at a rate of 1 in 500, estimate (a) the characteristic roughness size of the channel bed and (b) the mean velocity on a subsequent occasion when the depth of flow is 1.2 m. Assume that the rough wall relationship (4-35) is applicable. The specific weight of water is 9810 N/m^3.

SOLUTION Consider the block of fluid of length L and width b normal to the page in Fig. 4-14a. The volume of the fluid is bdL and its weight is $\rho gbdL$. The component of this force in the direction of flow is $\rho gbdL \sin \theta$.

The shear stress τ_w at the bed acts on an area bL and so the force resisting the motion is $\tau_w bL$. Since equilibrium conditions prevail, the two forces are equal and so

$$\tau_w bL = \rho gbdL \sin \theta \qquad (4\text{-}58)$$

Now $\tau_w/\rho = \frac{1}{2}f\bar{V}^2$ and so we may write

$$\tfrac{1}{2}f\bar{V}^2 = gd \sin \theta \qquad (4\text{-}59)$$

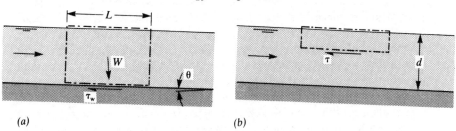

(a) (b)

Figure 4-14 Equilibrium conditions in uniform channel flows.

(a) For the particular numerical values pertaining at a depth of 0.6 m, this yields $f = 0.005\,89$, and it follows from Eq. (4-36) that the characteristic roughness size is $k_s = 4.15$ mm.

(b) When the depth is 1.2 m, Eq. (4-36) gives $f = 0.004\,92$ since k_s is unchanged. In this case, Eq. (4-59) yields $\bar{V} = 3.09$ m/s.

Comments on Example 4-5

1. It is typical of free surface flows that a higher mean velocity is obtained when the depth is increased. Thus a doubling of the depth results in more than a doubling of the rate of flow (assuming uniform flow conditions in both cases). In this particular instance, doubling the depth leads to more than a threefold increase in the rate of flow.

2. Consider the forces on a block of fluid such as that shown in Fig. 4-14b. A similar expression to (4-58) is obtained, thus showing that the shear stress varies linearly with the depth (see Problem 2 at the end of this chapter). Nevertheless, the use of Eq. (4-36) is found to be acceptable in practice (i.e. through experiments).

Example 4-6: Laminar and turbulent regimes At a typical section in a steady flow of water ($\mu = 1.1$ mPa s, $\rho = 1000$ kg/m³) along a wide channel, the depth is $d = 2$ m and the mean velocity is $\bar{V} = 3$ m/s. Estimate the shear stress at the bed assuming (a) inviscid, (b) laminar, (c) smooth wall turbulent and (d) rough wall ($k_s = 50$ mm) turbulent conditions.

SOLUTION

(a) *Inviscid*: There are no shear stresses in an inviscid flow. Hence $\tau_w = 0$.

(b) *Laminar*: Using Eq. (4-52), and assuming that $\bar{V} \approx \frac{2}{3}V$ for a parabolic velocity distribution, we obtain $\tau_w = 3\mu \bar{V}/d = 4.95$ Pa.

(c) *Smooth-wall turbulent*: Using Eq. (4-32) and noting that $Re = d\bar{V}/\mu = 5.45 \times 10^6$, we obtain $f = 0.001\,83$. Hence $\tau_w = \frac{1}{2}f\rho \bar{V}^2 = 8.25$ Pa.

(d) *Rough-wall turbulent*: Using Eq. (4-36), we obtain $f = 0.006\,83$ and so $\tau_w = 38.8$ Pa.

Comments on Example 4-6

1. Inviscid conditions are nearly always assumed when considering phenomena taking place in a short length of channel. Other forces are usually vastly greater than shear forces in regions of rapidly varied flow.

2. When skin friction may not be neglected—in a long channel reach for instance—we need to know whether to assume laminar or turbulent conditions. For a channel flow, experiments show that laminar flows prevail when the Reynolds number $Re = \rho d\bar{V}/\mu$ is less than about 3000. Turbulent conditions are almost certain to exist in normal circumstances at significantly higher Reynolds numbers. The relatively low shear stress predicted for laminar flow in this particular example is therefore irrelevant (since $Re \approx 5.45 \times 10^6$).

3. For turbulent flows the value of k_s must first be estimated as accurately as possible. Then the smooth-wall and rough-wall friction coefficients should be estimated from Eqs (4-32) and (4-36). The larger of the two values obtained will be the more realistic (for a correctly chosen k_s).

4. In practice the estimation of k_s can be very difficult in channels, especially natural ones (rivers, etc.). Therefore a completely different method of estimating channel resistance is commonly used in practice, based on the empirical Manning equation.

Example 4-7: Natural gas supply line The mean velocity at any cross-section in the final 2 km of a 1 m diameter pipeline delivering natural gas ($\gamma = 1.31$, $R = 520 \, \text{J/kg K}$) is found to be

$$\bar{V} = (A - Bx)^{-1/2} \tag{4-60}$$

where $A = 5 \times 10^{-4} \text{s}^2/\text{m}^2$ and $B = 2 \times 10^{-7} \text{s}^2/\text{m}^3$. The skin friction coefficient is $f = 0.0075$ and the rate of flow of the gas is $\dot{m} = 50 \, \text{kg/s}$. Estimate the total skin friction resistance in this part of the pipeline.

SOLUTION The shear stress at any position is $\tau_w = \frac{1}{2} f \rho \bar{V}^2$ and the shear force δF in an elemental length δx of the pipeline of circumference πD is $\tau_w \pi D \, \delta x$, i.e.

$$\delta F = \frac{1}{2} \pi D f \rho \bar{V}^2 \, \delta x \tag{4-61}$$

The mass flux along the pipeline ($\dot{m} = \rho a \bar{V}$) is constant and so the product $\rho \bar{V}$ is also constant. Writing this as $\dot{m}/a$, we obtain

$$\delta F = \frac{1}{2} \pi D f \frac{\dot{m}}{a} \bar{V} \, \delta x = \frac{\frac{1}{2} \pi D f \dot{m}/a}{(A - Bx)^{1/2}} \, \delta x \tag{4-62}$$

On integrating, the total skin friction resistance (drag) along the length $L = 2 \, \text{km}$ of the pipeline is found to be

$$F = \frac{1}{2} \pi D f \frac{\dot{m}}{a} \left[-\frac{2}{B} (A - Bx)^{1/2} \right]_0^L = 92.7 \, \text{kN} \tag{4-63}$$

Comments on Example 4-7

1. Equation (4-60) indicates that the mean velocity increases as the gas flows downstream (e.g. $\bar{V} = 44.7 \, \text{m/s}$, $57.7 \, \text{m/s}$ and $100 \, \text{m/s}$ at $x = 0$, 1 km and 2 km respectively). This is a direct consequence of the skin friction force which induces a decreasing pressure—and therefore a decreasing density—downstream.

2. By inspection, Eq. (4-60) yields $\bar{V} = \infty$ at $x = 2.5 \, \text{km}$ and no solution is obtainable for larger values of x. It would therefore have been a nonsense to pose the identical question with the origin for x a few hundred metres further upstream. The same rate of flow could then be maintained only with a higher upstream density and hence with a higher upstream pressure.

Example 4-8: Skin friction in a smooth-wall turbulent flow By integrating the velocity distribution (4-29) for turbulent flow over a smooth wall, show that the skin friction coefficient f satisfies Eq. (4-32).

SOLUTION The distance travelled in an elemental time δt by a particle moving with a velocity $\bar{u}$ is $\bar{u} \, \delta t$. Therefore the volume of fluid crossing an elemental strip δy of an imaginary stationary surface of width b (normal to the page in Fig. 4-15) is $b \, \delta y \, \bar{u} \, \delta t$, and the total volume of fluid crossing the imaginary surface in the time δt is $b \, \delta t \int_0^h \bar{u} \, dy$.

Suppose that the same total volume crossed the imaginary surface, but with all particles travelling at the same velocity $\bar{V}$. In this case, each would travel a distance $\bar{V} \, \delta t$ and that total volume could be expressed as $bh\bar{V} \, \delta t$.

By equating these two expressions for the volume, we deduce that the mean velocity satisfies

$$h\bar{V} = \int_0^h \bar{u} \, dy \tag{4-64}$$

For the particular case of a turbulent boundary layer over a smooth plate, the velocity distribution in the turbulent zone satisfies (4-29) and so

$$h\bar{V} \simeq \int_\delta^h u_\tau \left\{ 2.5 \ln \left(\frac{y u_\tau}{v} \right) + 5.5 \right\} dy \tag{4-65}$$

in which the lower limit δ is used instead of zero because the velocity distribution is invalid at tiny values of y. On integrating this expression

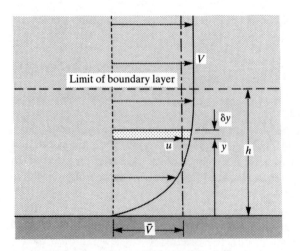

Figure 4-15 Mean velocity $\bar{V}$ in a boundary layer.

and noting that $\int \ln(x) = x \ln(x) - x$, we obtain

$$h\bar{V} \simeq u_\tau \left[2.5y \ln \left(\frac{yu_\tau}{v} \right) + 3y \right]_\delta^h \qquad (4\text{-}66)$$

Therefore, neglecting terms in δ, the mean velocity is given by

$$\frac{\bar{V}}{u_\tau} \simeq 2.5 \ln \left(\frac{hu_\tau}{v} \right) + 3 \qquad (4\text{-}67)$$

In developing this expression into the form used in Eq. (4-32) we note that

$$u_\tau = \sqrt{\frac{\tau_w}{\rho}} = \sqrt{\frac{\frac{1}{2} f \rho \bar{V}^2}{\rho}} = \bar{V} \sqrt{\frac{f}{2}} \qquad (4\text{-}68)$$

Equation (4-67) therefore reduces to the required form, namely

$$\sqrt{\frac{2}{f}} \simeq 2.5 \ln \left(\frac{h\bar{V}}{v} \sqrt{\frac{f}{2}} \right) + 3 \qquad (4\text{-}69)$$

Comments on Example 4-8

1. Because the viscous sub-layer is so thin in comparison with the overall thickness of the boundary layer, only tiny errors are introduced by neglecting its contribution to the total rate of flow. Similarly the value of δ implied in the development of Eq. (4-67) is sufficiently small for its contribution to be negligible.
2. The use of Eq. (4-69) in practice is slightly cumbersome because the skin friction coefficient f appears both inside and outside the logarithm. To illustrate this point consider water ($v = 1\,\text{mm}^2/\text{s}$) flowing with a mean velocity $\bar{V} = 3\,\text{m/s}$ in a boundary layer of thickness $h = 2\,\text{m}$. Readers will find it easy to *verify* that $f \approx 0.001\,81$, but they would find it more difficult to deduce this value *ab initio*. An iterative method of solution must be used.

Example 4-9: Colebrook–White equation The general expression (4-22) for the velocity distribution in a turbulent boundary layer can be expressed in the form

$$\frac{\bar{u}}{u_\tau} = \frac{1}{K} \ln \left(\frac{y}{L} \right) \qquad (4\text{-}70)$$

in which L is an unspecified constant with the dimensions of a length. An inspection of Eqs (4-28) and (4-34) reveals that $L \propto v/u_\tau$ in a smooth-wall flow and $L \propto k_s$ in a rough-wall flow. The linear combination

$$L = A_1 \frac{v}{u_\tau} + A_2 k_s \qquad (4\text{-}71)$$

is therefore a plausible approximation for the value of the constant in a boundary layer alongside a wall of intermediate roughness, A_1 and A_2 being dimensionless constants.

Show that this approximation leads to a skin friction formula with the same form as the Colebrook–White equation (4-38).

SOLUTION The solution closely follows that of Example 4-8. The distance travelled by a particle in an elemental time δt is $\bar{u}\,\delta t$, and so the volume of fluid crossing an elemental annular section of radius $r = R - y$ and width δy is $2\pi(R-y)\,\delta y\,\bar{u}\,\delta t$ (see Fig. 4-16). The total volume crossing an imaginary surface in the cross-section is $2\pi\,\delta t \int_0^R \bar{u}(R-y)\,dy$. By comparing this with the volume $\pi R^2 \bar{V}\,\delta t$ that would cross the section if all particles travelled at the same speed $\bar{V}$, we deduce that the mean velocity is

$$\bar{V} = \frac{2}{R^2}\int_0^R \bar{u}(R-y)\,dy \tag{4-72}$$

For the particular velocity distribution described by Eq. (4-70) this leads to

$$\bar{V} = \frac{2u_\tau}{R^2 K}\int_\delta^R \left\{ R\ln\left(\frac{y}{L}\right) - y\ln\left(\frac{y}{L}\right)\right\}dy \tag{4-73}$$

in which the lower limit of integration is changed to δ to exclude negative velocities close to the wall where Eq. (4-70) is invalid. After integration and a little manipulation, we obtain

$$\bar{V} = \frac{u_\tau}{K}\ln\left(\frac{D}{3L}\right) \tag{4-74}$$

in which $D = 2R$ is the pipe diameter. Using Eq. (4-31) we obtain

$$\sqrt{\frac{2}{f}} = \frac{\bar{V}}{u_\tau} = \frac{1}{K}\ln\left(\frac{D}{3L}\right) = -\frac{1}{K}\ln\left(\frac{3L}{D}\right) \tag{4-75}$$

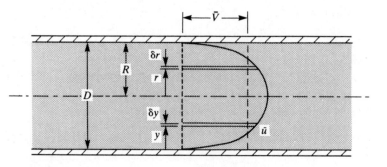

Figure 4-16 Velocity distribution in a turbulent pipe flow.

and so, for the intermediate-roughness expression (4-71),

$$\sqrt{\frac{2}{f}} = -\frac{1}{K}\ln\left\{3A_1\frac{v}{u_\tau} + 3A_2\frac{k_s}{D}\right\} \tag{4-76}$$

Using Eq. (4-31) again and noting that $D\bar{V}/v$ is the Reynolds number Re, we obtain

$$\sqrt{\frac{2}{f}} = -\frac{1}{K}\ln\left\{\frac{3\sqrt{2}A_1}{Re\sqrt{f}} + 3A_2\frac{k_s}{D}\right\} \tag{4-77}$$

which is a valid functional form of the Colebrook–White equation.

Comment on Example 4-9 The usual form of the Colebrook–White equation (4-38) can be obtained by changing to base 10 logarithms and choosing the coefficients A_1 and A_2 appropriately. The value of this analysis (suggested to the author by G.D. Matthew at the University of Aberdeen) is that it links the resulting expression to a simple combination of the parameters governing the smooth-wall and rough-wall velocity distributions.

4-6 THREE-DIMENSIONAL FLOWS

Throughout most of this chapter, attention has been focused on flows predominantly parallel to a single axis. In general, however, flows can be three-dimensional and it is useful to develop relationships between the shear stress and the velocity gradients.

Consider the elemental cuboid shown in Fig. 4-17. The normal and shear stresses exerted *on* the element *by* the surrounding fluid are shown at the centres of the visible faces. Equal and opposite stresses act on the other three faces. In the limit, as the dimensions of the element tend to zero, the nine independent stresses define the stress tensor at a point, namely

$$\boldsymbol{\sigma} = \begin{pmatrix} \sigma_x & \tau_{xy} & \tau_{xz} \\ \tau_{yx} & \sigma_y & \tau_{yz} \\ \tau_{zx} & \tau_{zy} & \sigma_z \end{pmatrix} \tag{4-78}$$

Normal stresses In a stationary fluid, the three normal stresses σ_x, σ_y and σ_z are all equal to the thermodynamic pressure p. In a moving incompressible fluid, the three are unequal, but their average is the thermodynamic pressure. They are:

$$\sigma_x = p - 2\mu\frac{\partial u}{\partial x} \qquad \sigma_y = p - 2\mu\frac{\partial v}{\partial y} \qquad \sigma_z = p - 2\mu\frac{\partial w}{\partial z} \tag{4-79}$$

Slightly more complex expressions apply in compressible flows, and the mean of σ_x, σ_y and σ_z is not quite equal to the pressure.

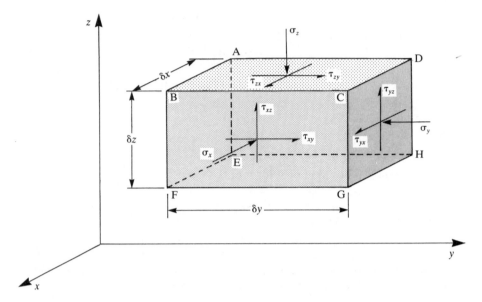

Figure 4-17 Surface stresses on an elemental cuboid. (Stresses illustrated only on positive faces.)

In practice the difference between the pressure and any one of the normal stresses is usually very small. For engineering purposes it is common practice to neglect variations of normal stress with direction whenever the Reynolds number (see section 1-1) is sufficiently large.

It should be noted that the sign convention used for normal stresses in fluid mechanics, namely *compression is positive*, is the opposite of the convention most commonly used in solid mechanics.

Shear stresses Figure 4-17 illustrates the sign convention used for shear stresses exerted *on* the element *by* the surrounding fluid. On *positive* faces of the element (i.e. where the outward normal is in the positive coordinate direction), positive shear stresses also act in the *positive* coordinate direction. On *negative* faces, they act in the *negative* coordinate direction. This convention is the same as that most commonly used in solid mechanics.

Although there are six components of shear stress in the stress tensor, only three of them are independent because $\tau_{xy} = \tau_{yx}$, $\tau_{yz} = \tau_{zy}$ and $\tau_{zx} = \tau_{xz}$. This can be demonstrated quite easily by considering the rotational equilibrium of the element. For example, the only unbalanced forces tending to cause the element to rotate about the edge EA are $\tau_{xy}\,\delta y\,\delta z$ on the face BCGF and $\tau_{yx}\,\delta x\,\delta z$ on the face CDHG. All other normal and shear forces tending to cause rotation are counterbalanced by equal forces on the opposite faces. The moments due to the two unbalanced forces are $\tau_{xy}\,\delta x\,\delta y\,\delta z$ and $-\tau_{yx}\,\delta x\,\delta y\,\delta z$ respectively.

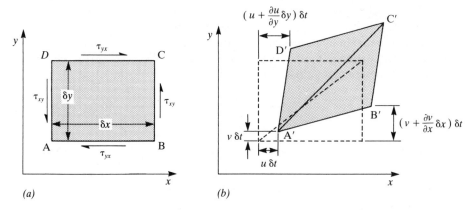

Figure 4-18 General displacement of a fluid element. (a) Applied stresses. (b) Displaced fluid element.

Since the element does not accelerate at an infinite rate about EA, the net moment must be zero and so $\tau_{xy} = \tau_{yx}$.

The *laminar* shear stress in a Newtonian fluid flowing parallel to a single axis satisfies Eq. (2-8) in which the stress is related to the rate of shear strain, i.e. the velocity gradient. In order to develop equivalent relationships applicable to three-dimensional flows, we consider the equilibrium of the element shown in Fig. 4-17. Attention is focused on shear stresses tending to distort the element in the (x-y) plane, and these are shown again in Fig. 4-18a. During a small time δt, the element moves to a new position whose projection is shown in Fig. 4-18b. The movement is composed of (1) translational displacement, namely a change in the position of the centroid, (2) rotation and (3) distortion. The translational and rotational displacements are directly related to the linear and angular velocities of the element and the distortion is caused by shear forces.

Let the velocity of a fluid particle at the point A be $V_A = (u, v, w)$. Then the velocities of particles at the points B and D are

$$V_B = \left(u + \frac{\partial u}{\partial x}\delta x, \ v + \frac{\partial v}{\partial x}\delta x, \ w + \frac{\partial w}{\partial x}\delta x \right) \tag{4-80}$$

and

$$V_D = \left(u + \frac{\partial u}{\partial y}\delta y, \ v + \frac{\partial v}{\partial y}\delta y, \ w + \frac{\partial w}{\partial y}\delta y \right) \tag{4-81}$$

respectively. During the interval δt, particles initially at the points A and B move in the y-direction by $v \ \delta t$ and $[v + (\partial v/\partial x) \ \delta x] \ \delta t$ respectively. Since the movement at B exceeds that at A by $(\partial v/\partial x) \ \delta x \ \delta t$, the side AB of the element has rotated anticlockwise through an angle $(\partial v/\partial x) \ \delta t$. Similarly, particles at A and D move in the x-direction by $u \ \delta t$ and $[u + (\partial u/\partial y) \ \delta y] \ \delta t$ respectively and so the side AD has rotated clockwise through an angle $(\partial u/\partial y) \ \delta t$.

The net shear strain (angular distortion) in the $(x\text{-}y)$ plane is therefore $(\partial u/\partial y + \partial v/\partial x)\,\delta t$ and the *rate* of shear strain is $(\partial u/\partial y + \partial v/\partial x)$.

For Newtonian fluids, experiments show that the laminar shear stress is directly proportional to the rate of shear strain. Together with similar expressions describing the distortion about the x and y axes, the relationship is

$$\tau_{xy} = \tau_{yx} = \mu\left(\frac{\partial u}{\partial y} + \frac{\partial v}{\partial x}\right)$$

$$\tau_{yz} = \tau_{zy} = \mu\left(\frac{\partial v}{\partial z} + \frac{\partial w}{\partial y}\right) \qquad (4\text{-}82)$$

$$\tau_{zx} = \tau_{xz} = \mu\left(\frac{\partial w}{\partial x} + \frac{\partial u}{\partial z}\right)$$

in which μ is the dynamic viscosity of the fluid.

In a *turbulent* flow, the fluctuating components of velocity give rise to Reynolds stresses such as those described in section 4-4-1 for uniaxial flows. These act *in addition* to the laminar contributions and they influence the normal stress components as well as the shear stress components.

The Reynolds stresses at a point are

$$\sigma' = \begin{pmatrix} -\rho\overline{u'^2} & -\rho\overline{u'v'} & -\rho\overline{u'w'} \\ -\rho\overline{v'u'} & -\rho\overline{v'^2} & -\rho\overline{v'w'} \\ -\rho\overline{w'u'} & -\rho\overline{w'v'} & -\rho\overline{w'^2} \end{pmatrix} \qquad (4\text{-}83)$$

in which, for example, $\overline{u'v'}$ denotes the mean value of the product $u'v'$. In practice, these expressions are rarely used explicitly. We usually attempt to relate them to the local mean velocity gradients through some form of turbulence model that can be handled conveniently in numerical computations.

4-6-1 Rotation

Figure 4-18 shows a net *rotation* of the element about the z-axis. Since the figure is obtained by viewing the element of Fig. 4-17 in the *negative z-direction*, we define anticlockwise rotations as positive (so that the right-hand screw rule will apply in the positive z-direction). The sides AB and AD have rotated through angles of

$$\frac{\partial v}{\partial x}\,\delta t \qquad \text{and} \qquad -\frac{\partial u}{\partial y}\,\delta t \qquad (4\text{-}84)$$

respectively in the interval δt. The mean angular rotation of the element is half the algebraic sum of these rotations and its angular velocity is

$$\Omega = \tfrac{1}{2}\left(\frac{\partial v}{\partial x} - \frac{\partial u}{\partial y}\right) \qquad (4\text{-}85)$$

It is shown in Chapter Ten that the equations of fluid motion adopt a particularly simple form when individual fluid particles do not rotate, that is when $\Omega = 0$. Such flows are aptly termed *irrotational*.

The expression $(\partial v/\partial x - \partial u/\partial y)$ is the component of *vorticity* about the z-axis. It is simply twice the angular velocity. The three components of vorticity are

$$\xi \equiv \frac{\partial w}{\partial y} - \frac{\partial v}{\partial z}$$

$$\eta \equiv \frac{\partial u}{\partial z} - \frac{\partial w}{\partial x} \qquad (4\text{-}86)$$

$$\zeta \equiv \frac{\partial v}{\partial x} - \frac{\partial u}{\partial y}$$

PROBLEMS

1 Determine the values of the constant pressure gradient dp/dx that will lead to $du/dy = 0$ (and hence to $\tau = 0$) at (a) the upper plate and (b) the lower plate in Fig. 4-7. Use $\mu = 0.1$ Pa s, $H = 2$ mm and $V = 20$ m/s.

$$[-1 \text{ MPa/m}, +1 \text{ MPa/m}]$$

2 Show that the shear stress varies linearly with depth in the steady flow of a liquid along an open channel. (*Hint*: Consider the equilibrium of the block of liquid highlighted in Fig. 4-14b.)

3 Show that the velocity distribution in a laminar flow of a Newtonian liquid in an open channel of slope α is

$$u = \frac{g \sin \alpha}{2v} (d^2 + 2dy - y^2) \qquad (4\text{-}87)$$

where d is the depth of flow and v is the kinematic viscosity. (*Hint*: Follow the analysis in section 4-3 and assume that $du/dy = 0$ at the free surface.)

4 Water ($v = 1.2$ mm^2/s) flows off an airport runway at a depth of 1.5 mm and a surface velocity of 50 mm/s. Estimate the gradient of the runway.

$$[1 \text{ in } 368]$$

5 Figure 4-19 depicts a 100 mm diameter shaft rotating in a 175 mm long journal bearing. The 0.5 mm gap is full of oil with a viscosity of $\mu = 0.6$ Pa s and the shaft rotates at 500 rev/min. Estimate the power required to maintain the rotation.

$$[452 \text{ W}]$$

6 In sufficiently small capillaries, the red corpuscles have to distort to squeeze along the vessel and the conditions are broadly analogous to those shown in Fig. 4-20. By considering the equilibrium of a corpuscle, show that the shear stress τ_w on its sides is $\frac{1}{4}D\partial p/\partial x$ where $\partial p/\partial x$ is the pressure

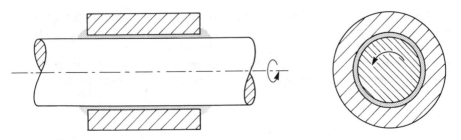

Figure 4-19 Journal bearing.

gradient in the annular region around the corpuscle. Hence determine the pressure difference needed to maintain the flow of a single corpuscle of length 4 μm and diameter 5 μm at a speed of 0.1 mm/s in a 5.05 μm diameter capillary if the viscosity of the blood plasma is 0.005 Pa s.

[31.4 Pa]

7 The velocity distribution in the turbulent portion of a boundary layer over a smooth surface satisfies Eq. (4-29) when $u_\tau y/v > 30$. Very close to the wall ($u_\tau y/v < 5$), the velocity distribution satisfies the viscous sub-layer equation (4-26). By extrapolating both of these relationships beyond their regions of strict validity, show that they both predict the same velocity at a distance $y = 11.64v/u_\tau$ from the wall.

8 Evaluate this distance ($11.6\, v/u_\tau$) for each of the following flows:

(a) water ($\rho = 1000\,\text{kg/m}^3$, $v = 1.2\,\text{mm}^2/\text{s}$) when $\tau_w = 3\,\text{Pa}$;

[0.254 mm]

(b) water when $\bar{V} = 2\,\text{m/s}$ and $f = 0.008$;

[0.110 mm]

(c) air ($\rho = 1.15\,\text{kg/m}^3$, $v = 15\,\text{mm}^2/\text{s}$) when $\tau_w = 3\,\text{Pa}$;

[0.108 mm]

(d) air when $\bar{V} = 40\,\text{m/s}$ and $f = 0.006$.

[0.079 mm]

9 Blasius suggested the following explicit expression for the skin friction coefficient in turbulent, smooth-walled pipe flows at sufficiently low Reynolds numbers:

$$f = 0.079\,(Re)^{-1/4} \qquad (4\text{-}88)$$

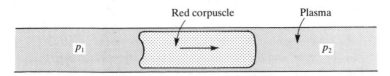

Figure 4-20 Blood flow in a capillary. (Highly idealized.)

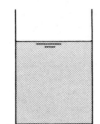

Figure 4-21 Optimum channel shapes. The areas of flow are equal. Which gives the greater resistance?

Show that the value of the friction coefficient f obtained from this expression differs by approximately 2.3 per cent, 1.2 per cent and 14.2 per cent from the values predicted by Eq. (4-33) at Reynolds numbers of $\bar{V}D/v = 10^4$, 10^5 and 10^6 respectively.

10 Water flows at a constant depth d along a rectangular channel of width B and slope 1 in 400. The skin friction coefficient is $f = 0.01$. Estimate the mean velocity of flow assuming (a) $B = 8\,\text{m}$, $d = 1\,\text{m}$, (b) $B = 4\,\text{m}$, $d = 2\,\text{m}$ and (c) $B = 2\,\text{m}$, $d = 4\,\text{m}$. Can you find an $8\,\text{m}^2$ rectangular shape better than (b)?

[1.98 m/s, 2.21 m/s, 1.98 m/s, no]

11 In a sufficiently smooth pipe, the influence of roughnesses is negligible in comparison with the influence of the viscous sub-layer. The reverse holds true for sufficiently rough pipes. Jain's equation, Eq. (4-39), is often used when neither of these extreme conditions obtains. Estimate the friction coefficient for a gas flow at 20 m/s in a 50 mm diameter pipe of roughness $k_s = 0.1\,\text{mm}$. Take $\rho = 1\,\text{kg/m}^3$ and $\mu = 20 \times 10^{-6}\,\text{Pa s}$ and assume (a) smooth-wall conditions—Eq. (4-88), (b) Jain's expression and (c) rough-wall conditions—Eq. (4-37).

[0.0053, 0.0063, 0.0058]

12 Water ($v = 1.2\,\text{mm}^2/\text{s}$) flows steadily along an 8 m long domestic heating pipe of 10 mm diameter. The pipe roughness is estimated to be 0.06 mm and the mean velocity of flow is 2 m/s. Using the most suitable expression for the resistance to flow, estimate the pressure head difference between the upstream and downstream ends of the pipe.

[6.04 m]

FURTHER READING

Allen, J. (1981) Significance of the boundary shear stress/momentum-flux ratio, *Proc. Instn Civ. Engrs*, Part 2, **71**, 269–271.

Bradshaw, P., Cebeci, T. and Whitelaw, J.H. (1981) *Engineering Calculation Methods for Turbulent Flow*, Academic Press.

Chang, P.K. (1976) *Control of flow separation*, Hemisphere.

Hinze, J.O. (1975) *Turbulence*, 2nd edn, McGraw-Hill.

Jain, A.K. (1976) Accurate explicit equation for friction factor, *J. Hydr. Engng*, ASCE, **102**, 674–677.

Prandtl, L. (1927) The generation of vortices in fluids of small viscosity, *J. R. Aero. Soc.*, **31**, 721–741.

Schlichting, H. (1979) *Boundary Layer Theory*, 7th edn, McGraw-Hill.

FIVE

DIMENSIONAL ANALYSIS

5-1 THE NEED FOR EMPIRICAL DATA

Most of this book is slanted towards the quantification of fluid flows based on mathematical statements of the natural laws. However, we rarely manage to complete analyses of fluid flows by means of theoretical arguments alone. Even seemingly simple topics such as steady flows in rivers are found to be intractable, partly because of the complex geometry, but mostly because of our failure to find an exact mathematical representation of turbulence. Nevertheless, we are able to design pipe networks, ships and gas turbines, etc. with a high degree of success. To achieve this, we make use of *experimental* evidence to bridge gaps in theoretical knowledge. In effect, we admit our ignorance and observe how nature solves the problem.

At first sight it might appear that designs based on experimental measurement are not really 'designs' at all. Indeed, if we could not estimate, say, the performance characteristics of an aircraft except by actually building and flying it, this criticism would be entirely valid. However, the interpretation of experimental evidence is carried out in such a way that the principal characteristics of the aircraft behaviour can be predicted from measurements made with a small model. Similarly, by recording appropriate data for a pump operating at a given speed, we can infer its probable behaviour at a wide range of other speeds without actually buying and fitting alternative motors.

In the interpretation of experimental measurements, use is made of the fact that natural processes do not exhibit any preferred scales for the measurement of physical parameters. The fundamental characteristics of natural phenomena can always be described in terms of *ratios*. For example, the extent to which fluid compressibility influences a steady flow depends upon the ratio of the particle velocity and the sonic velocity (section 2-6-2). The purpose of this chapter is to describe how the relevant ratios are identified and utilized. We shall see that it is possible to glean a great deal of valuable information from the proper interpretation of carefully devised experiments.

5-2 QUANTITIES AND DIMENSIONS

The nature of physical *quantities* such as pressure and viscosity can be expressed in terms of fundamental *dimensions*, the most important in fluid mechanics being mass $[M]$, length $[L]$, time $[T]$ and temperature $[\Theta]$. For example the dimensions of acceleration, density and specific heat are $[L/T^2]$, $[M/L^3]$ and $[L^2/\Theta T^2]$ respectively.

Because force is more readily understandable than mass, some engineers prefer to use the dimensions $[F, L, T, \Theta]$ instead of $[M, L, T, \Theta]$. It does not really matter which choice is made and examples of both usages are given herein. However, only three of the four dimensions of mechanics are *independent* because Newton's second law of motion shows that

$$[F] = [MLT^{-2}] \qquad (5\text{-}1)$$

In the SI system, the units used to quantify the dimensions force, mass, length, time and temperature are N, kg, m, s and K respectively. Since $1\,\text{N} = 1\,\text{kg}\,\text{m/s}^2$, only four of these are independent.

5-2-1 Dimensionless Groups

Whenever we set about describing a physical phenomenon, we tend to make use of quantities such as pressure, velocity and diameter, etc. When we are lucky enough to be able to deduce mathematical relationships between the various quantities—using the natural laws, for instance—we usually end up with an equation that is most conveniently interpreted in dimensional terms. A simple example will illustrate this point.

One of the most famous equations in fluid mechanics is the so-called Bernoulli equation which relates the pressure p, velocity V, density ρ, elevation z and gravitational acceleration g. Provided that the datum for the elevation is chosen suitably, the equation can be written in the familiar form:

$$\frac{p}{\rho g} + \frac{V^2}{2g} + z = 0 \qquad (5\text{-}2)$$

In this form, the equation relates three expressions $(p/\rho g)$, $(V^2/2g)$ and z, each of which has the dimensions of length $[L]$. Although Eq. (5-2) is one of the commonest ways of writing the Bernoulli equation, it is by no means the simplest. Indeed there are several possible ways of simplifying it. For example, dividing each term of the equation by V^2/g leads to

$$\frac{p}{\rho V^2} + \tfrac{1}{2} + \frac{gz}{V^2} = 0 \qquad (5\text{-}3)$$

which is an equation relating only two parameters $(p/\rho V^2)$ and (gz/V^2), each of which is dimensionless. By inspection, therefore, this particular relationship between five dimensional quantities $(p, V, \rho, z$ and $g)$ can be expressed as a functional relationship between just two dimensionless parameters, namely

$$\left(\frac{p}{\rho V^2}\right) = f\left(\frac{gz}{V^2}\right) \qquad (5\text{-}4)$$

We shall see that simplifications such as this are by no means unusual. Indeed in section 5.3 the identical expression is derived without any reference to the Bernoulli equation itself.

5-2-2 Buckingham Π Theorems

Two important theorems in dimensional analysis are illustrated (but not proven) by the preceding derivation. Of general application in any physical analysis (not only fluid mechanics), they are as follows:

1. All natural phenomena can be described in terms of dimensionless ratios.
2. The number of physical quantities exceeds the number of *independent* dimensionless ratios by the number of fundamental dimensions.

These are known as the Buckingham Π theorems and the dimensionless ratios are usually called *dimensionless groups* or Π *groups*.

The first theorem follows from the need for dimensional homogeneity of physical equations. We can always produce dimensionless ratios from a dimensional equation by dividing throughout by one of the terms. Although this is by no means the only way of achieving this objective, it happens to be the way used in the preceding section.

The second theorem is also verified above. There are 5 physical quantities and 3 fundamental dimensions $[M, L, T]$. In accordance with the second theorem, there are 5 minus 3, namely 2, independent dimensionless groups.

5-3 IDENTIFICATION OF DIMENSIONLESS GROUPS

In the above example, the dimensionless groups are obtained from an equation that is known to describe a particular type of flow. Usually, however, we cannot proceed in this manner. The great advantage of dimensional analysis is that it enables us to develop the functional forms of equations describing natural phenomena even when we have no prior knowledge of their forms.

Suppose that we wish to describe any natural phenomenon that involves only a stress p, a density ρ, a velocity V, a length l and an acceleration g (due to gravity, say). Two such flows are depicted in Fig. 5-1. In general, the relationship between these quantities may be written as

$$f_1(p, \rho, V, l, g) = 0 \qquad (5\text{-}5)$$

in which f_1 denotes an unknown function. The dimensions of these quantities are

$$
\begin{array}{ccccc}
p & \rho & V & l & g \\
[M/LT^2] & [M/L^3] & [L/T] & [L] & [L/T^2]
\end{array}
\qquad (5\text{-}6)
$$

In total, there are 5 quantities and 3 dimensions and so there are 2 independent dimensionless groups, Π_1 and Π_2. Therefore, Eq. (5-5) can also be written as

$$f_2(\Pi_1, \Pi_2) = 0 \qquad (5\text{-}7)$$

At this stage, experienced engineers can immediately write down the result (5-13) simply because they have been through this process many times before. Until such experience is gained, however, it is necessary to continue with a formal development of the equations. The following procedure involves the use of certain conventions that are not strictly necessary, but are nevertheless strongly recommended.

Mathematically, *any* two independent dimensionless groups formed from the five quantities may be used to describe the phenomenon. However, for engineering convenience, it is highly desirable to organize the development in such a way that we all end up with the *same* two groups. There are several methods of achieving this objective, but in fluid mechanics the most important contribution derives from the way in which the independence of the groups is ensured.

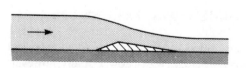
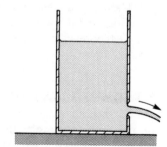

Figure 5-1 Flows involving only p, ρ, V, l and g.

The quantities ρ, l and V appear in most analyses of fluid flows. They involve all three of the dimensions $[M, L, T]$ and so can be combined in a dimensionless manner with *any* other quantity that depends only on these three dimensions. In the present example one dimensionless group will be obtained by combining ρ, l, and V with p and the second will be formed by combining ρ, l and V with g. The two groups will necessarily be independent of each other because one will contain p but not g while the other will contain g but not p. The groups are

$$\Pi_1 = \Pi_1(p, \rho, l, V) \qquad (5\text{-}8)$$

and $$\Pi_2 = \Pi_2(g, \rho, l, V) \qquad (5\text{-}9)$$

To determine the algebraic form of each group we express it as, typically,

$$\Pi_1 = p\rho^a l^b V^c \qquad (5\text{-}10)$$

in which a, b and c are numerical powers. Since Π_1 is dimensionless, the dimensions of the various quantities given in (5-6) must satisfy

$$[0] = \left[\frac{M}{LT^2}\right]\left[\frac{M}{L^3}\right]^a [L]^b \left[\frac{L}{T}\right]^c \qquad (5\text{-}11)$$

On equating the powers of $[M]$, $[L]$ and $[T]$ we obtain

$$[M]: \qquad 0 = 1 + a \qquad (5.12a)$$

$$[L]: \qquad 0 = -1 - 3a + b + c \qquad (5.12b)$$

$$[T]: \qquad 0 = -2 - c \qquad (5.12c]$$

which are three linear simultaneous equations for which the solution is $a = -1$, $b = 0$ and $c = -2$. By substituting these values into (5-10) we find that the first group is $\Pi_1 = p/\rho V^2$. In a similar manner it can be shown that $\Pi_2 = gl/V^2$ and so the functional relationship (5-7) is

$$f_2\left(\frac{p}{\rho V^2}, \frac{gl}{V^2}\right) = 0 \qquad (5\text{-}13)$$

or $$\boxed{\frac{p}{\rho V^2} = f_3\left(\frac{gl}{V^2}\right)} \qquad (5\text{-}14)$$

which is identical to Eq. (5-4) if we interpret the general length l as an elevation z.

Let us recapitulate briefly on what has been achieved. We started with a dimensional statement (5-1) that simply identifies five parameters believed to be involved in some physical phenomenon. We have ended up with a dimensionless statement (5-14) indicating that, irrespective of the particular form of the phenomenon, it can be regarded as a two-parameter event. That is, *all* phenomena involving *any* stress, density, velocity, length and acceleration can be represented by the two-parameter Eq. (5-14). While we grapple with five parameters, Nature contents herself with just two. Dimensional analysis enables us to make the same simplification.

It is worth noting that dimensional analysis does not enable us to go the whole way and thereby eliminate the need for conventional analysis using the natural laws. No amount of manipulation would enable us to get from Eq. (5-14) to anything like Eq. (5-2) or (5-3). The particular form of the function f_3 depends upon the phenomenon under investigation; we must find it by other means such as conventional analysis or experimental measurements.

5-3-1 Alternative Dimensionless Groups

Equation (5-14) is by no means the only possible dimensionless relationship involving the five quantities p, ρ, V, l, and g. For example, if ρ, l and g are chosen as the repeated quantities instead of ρ, l and V, the relationship obtained is

$$\frac{p}{\rho g l} = f_4\left(\frac{V}{\sqrt{gl}}\right) \tag{5-15}$$

None of the various possible relationships is 'better' or 'more accurate' than any of the others. However, some are more convenient than others in particular applications.

It is not essential (merely helpful) to make use of repeated quantities when deriving the dimensionless groups. All that is necessary is to make sure that the groups in the functional relationship collectively involve all the physical quantities while being mutually independent. Nevertheless, experience shows that the relationships obtained with ρ, l and V as repeated quantities are often the most useful in fluid mechanics.

The method presented above for obtaining a dimensionless group corresponding to a set of quantities is rarely used explicitly because it is usually possible to develop the group more simply. For example, by inspection of Eq. (5-11), it is obvious that $a = -1$ because the final group must not involve $[M]$. Similarly $c = -2$ because the group must not involve $[T]$. Therefore the group must contain the ratio $p/\rho V^2$, and an inspection of the dimensions of this ratio shows that the group cannot also include the quantity l. Barr (1984) gives a concise matrix method of determining Π-groups.

5-4 GENERAL PHENOMENA INVOLVING MECHANICAL QUANTITIES

An appropriate functional equation is now developed for any phenomenon involving the quantities stress p, density ρ, velocity V, acceleration g, viscosity

μ, surface tension σ, time t and lengths l, l_1 and l_2. Although many of these quantities are denoted by symbols that have a special meaning throughout the rest of this book, they can be regarded more generally in this article. For example, the stress p could represent a shear stress in a solid body and the acceleration g could denote the acceleration of that body.

The dimensional relationship between these quantities is

$$f_1(p, \rho, V, g, \mu, \sigma, t, l, l_1, l_2) = 0 \tag{5-16}$$

and their dimensions in the $[FLT\Theta]$ system are

$$
\begin{array}{cccccccccc}
p & \rho & V & g & \mu & \sigma & t & l & l_1 & l_2 \\
\left[\dfrac{F}{L^2}\right] & \left[\dfrac{FT^2}{L^4}\right] & \left[\dfrac{L}{T}\right] & \left[\dfrac{L}{T^2}\right] & \left[\dfrac{FT}{L^2}\right] & \left[\dfrac{F}{L}\right] & [T] & [L] & [L] & [L]
\end{array} \tag{5-17}
$$

Since the ten quantities involve only three dimensions, there are seven independent dimensionless groups (i.e. $10-3$). By choosing ρ, V and l as the repeated quantities, we obtain

$$\Pi_1(p, \rho, l, V) = p/\rho V^2 \tag{5-18}$$

$$\Pi_2(g, \rho, l, V) = gl/V^2 \tag{5-19}$$

$$\Pi_3(\mu, \rho, l, V) = \mu/\rho l V \tag{5-20}$$

$$\Pi_4(\sigma, \rho, l, V) = \sigma/\rho l V^2 \tag{5-21}$$

$$\Pi_5(t, \rho, l, V) = tV/l \tag{5-22}$$

$$\Pi_6(l_1, \rho, l, V) = l_1/l \tag{5-23}$$

$$\Pi_7(l_2, \rho, l, V) = l_2/l \tag{5-24}$$

Each of these seven groups is obtained in the manner described above. They are necessarily independent of one another because each contains a quantity which appears in no other group. Therefore a general expression describing all phenomena involving these ten quantities is

$$f_2\left(\frac{p}{\rho V^2}, \frac{gl}{V^2}, \frac{\mu}{\rho l V}, \frac{\sigma}{\rho l V^2}, \frac{tV}{l}, \frac{l_1}{l}, \frac{l_2}{l}\right) = 0 \tag{5-25}$$

By inspection, the simplification achieved in this case is less sweeping than in the previous case, but it should be realized that few phenomena involve so many parameters. It is usually possible to neglect the influence of sufficient parameters to reduce the functional expression to a manageable size (e.g. to three dimensionless groups).

5-5 GENERAL PHENOMENA INVOLVING THERMODYNAMIC QUANTITIES

In processes involving heat transfers, the relevant quantities might include a rate of heat transfer $\dot{Q}$, a specific heat capacity c_p, a thermal conductivity λ and a temperature T (which will normally be a temperature *difference* causing a heat flux). These quantities are additional to the mechanical quantities considered above, and neither list is exhaustive. To avoid an unnecessarily long derivation, let us omit some mechanical quantities and consider a phenomenon which can be described by

$$f_1(p, \mu, V, l, \dot{Q}, c_p, \lambda, T) = 0 \tag{5-26}$$

Two such flows are depicted in Fig. 5-2. In the $[MLT\Theta]$ system, the dimensions of these quantities are

$$
\begin{array}{cccccccc}
\rho & \mu & V & l & \dot{Q} & c_p & \lambda & T \\
\left[\dfrac{M}{L^3}\right] & \left[\dfrac{M}{LT}\right] & \left[\dfrac{L}{T}\right] & [L] & \left[\dfrac{ML^2}{T^3}\right] & \left[\dfrac{L^2}{T^2\Theta}\right] & \left[\dfrac{ML}{T^3\Theta}\right] & [\Theta]
\end{array} \tag{5-27}
$$

Since the eight quantities involve four dimensions, there are four (i.e. $8-4$) dimensionless groups. The set of repeated quantities must contain four members (one per dimension). By choosing T in addition to the usual trio ρ, l and V we obtain

$$\Pi_1(\mu, \rho, l, V, T) = \mu/\rho l V \tag{5-28}$$

$$\Pi_2(\dot{Q}, \rho, l, V, T) = \dot{Q}/\rho l^2 V^3 \tag{5-29}$$

$$\Pi_3(c_p, \rho, l, V, T) = c_p T/V^2 \tag{5-30}$$

$$\Pi_4(\lambda, \rho, l, V, T) = \lambda T/\rho l V^3 \tag{5-31}$$

and so a possible functional expression for such phenomena is

$$f_2\left(\frac{\mu}{\rho l V}, \frac{\dot{Q}}{\rho l^2 V^3}, \frac{c_p T}{V^2}, \frac{\lambda T}{\rho l V^3}\right) = 0 \tag{5-32}$$

Manipulation of dimensionless groups In practice the expression is rarely used in this form. It is often modified by (1) replacing the second term by the ratio of itself and the fourth term, and (2) replacing the third term by itself multiplied by the first term and divided by the fourth. These manipulations yield

$$f_3\left(\frac{\mu}{\rho l V}, \frac{\dot{Q}}{\lambda l T}, \frac{c_p \mu}{\lambda}, \frac{\lambda T}{\rho l V^3}\right) = 0 \tag{5-33}$$

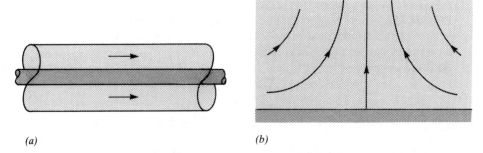

(a)   (b)

Figure 5-2 Flows involving heat transfers. (a) Forced convection past a hot core. (b) Natural convection over a hot surface.

The *reason* for these changes is that the resulting groups are more familiar to experienced engineers (see below). The mathematical *justification* is that no loss of generality results; the new groups are independent of one another (none can be derived by a combination of others).

In the presentation of experimental results, we strive to find sets of groups in which some members have little influence on the phenomenon under investigation. For example suppose that the group $c_p\mu/\lambda$ has little influence in a particular flow (see Example 5-4, p. 141). In this case, Eq. (5-33) may be approximated by

$$f_3\left(\frac{\mu}{\rho l V}, \frac{\dot{Q}}{\lambda l T}, \frac{c_p\mu}{\lambda}, \frac{\lambda T}{\rho l V^3}\right) = 0 \qquad (5\text{-}34)$$

Notice that although the manipulation of groups sometimes helps us to simplify functional relationships, it does not *in itself* result in a reduction in the number of independent groups (unless erroneous substitutions are inadvertently made).

Heat as a dimension In some flows there is very little interaction between thermal and mechanical effects even though heat transfers exist. For example, moderate heat flows to or from water flowing along a pipe or channel may have negligible influence on the rate of flow or on the pressure gradient. Experienced analysts can take advantage of this knowledge by ignoring known relationships between thermodynamic and mechanical quantities. They use *five* independent dimensions. $[M, L, T, \Theta, H]$ and hence obtain one fewer dimensionless group than would otherwise be the case. Here $[H]$ has the dimensions of energy and it is used for 'thermal' quantities such as heat flux, thermal conductivity, specific heat capacities and internal energy. It is not used for 'mechanical' quantities such as work, density, shear stress and velocity.

Great caution must be exercised when making use of this type of simplification; it is all too easy to step outside the range of approximate validity. Until appropriate experience has been gained it is wise to avoid the use of additional dimensions and to allow experiment to indicate which of the resulting groups are unimportant.

5-6 COMMON DIMENSIONLESS GROUPS

Most of the groups developed above occur so commonly that they are easily recognized by experienced engineers. In many cases, the groups have been named in honour of the person who first proposed their use or who was the first to make extensive use of them.

Reynolds number *(Re)* The Reynolds number is defined in section 1-1. It is

$$Re \equiv \frac{\rho l V}{\mu} = \frac{lV}{\nu} \tag{5-35}$$

in which l and V denote a length and a velocity that characterize the flow under investigation. In flows along circular section pipes, for example, l is nearly always chosen as the pipe diameter and V as the mean velocity.

In *any* steady flow (not just along a pipe) the Reynolds number is a measure of the relative importance of inertia and viscous forces. The components of acceleration of a fluid particle along and normal to a streamline are $V \partial V/\partial s$ and V^2/R (see Chapters Nine and Ten). Both of these scale with V^2/l where V and l are the characteristic quantities. The mass of a particle scales with ρl^3 and so the product mass × acceleration—i.e. the 'inertia force'—scales with $\rho l^2 V^2$. The viscous stress is typified by $\tau = \mu \, \partial u/\partial r$ which scales with $\mu V/l$ and so the viscous force (stress × area) scales with $\mu l V$. Therefore the *ratio* of the inertia and viscous forces scales with $\rho l^2 V^2/\mu l V$, that is with $\rho l V/\mu$, the Reynolds number.

In a turbulent flow, the effective shear stress is dominated by contributions such as $\overline{\rho u' v'}$ due to fluctuating components of velocity (sections 4.4 and 4.6). These contributions—the so-called Reynolds stresses—scale with ρV^2 and so the ratio of Reynolds stresses and viscous stresses scales with $\rho V^2/(\mu V/l)$, that is with the Reynolds number $\rho l V/\mu$. This is an important result because it demonstrates that Reynolds stresses dominate laminar stresses at high Reynolds numbers and that the converse must be true at low Reynolds numbers. That is, high Reynolds number flows will normally be turbulent and low Reynolds number flows will normally be laminar.

Froude number (Fr) In some flows, notably stratified and/or free surface flows, gravitational forces are important. These scale with $g\rho l^3$ where g is the gravitational force per unit mass. Therefore the *ratio* of inertia and gravitational forces scales with V^2/gl, which is the square of the Froude number,

$$Fr \equiv \frac{V}{\sqrt{gl}} \qquad (5\text{-}36)$$

Some confusion exists in the literature regarding this definition. Occasionally the Froude number is defined as V^2/gl and this differs from (5-36) at all values except unity. The second definition has merit because it follows the convention of defining dimensionless groups as the ratio of a specified force and an inertia force whenever this is reasonably possible. Nevertheless, (5-36) is the more commonly accepted definition.

Weber number (*We*) The units of the surface tension coefficient σ are force per unit length. The force therefore scales with σl and a logical definition for the Weber number is

$$We \equiv \frac{\rho l V^2}{\sigma} \qquad (5\text{-}37)$$

Even more confusion exists regarding this definition. Not only is the square root sometimes preferred, but the reciprocals of both possibilities are also used. Great care must be taken in the interpretation of numerical values quoted in the literature.

Mach number (*Ma*) The Mach number is defined in section 2-6-2 as

$$Ma \equiv \frac{V}{c} \qquad (5\text{-}38)$$

in which V is a local fluid particle velocity and c is the local sonic velocity.

Strouhal number (*St*) The Strouhal number, defined typically by

$$St \equiv \frac{Nl}{V} \qquad (5\text{-}39)$$

in which N is a frequency (dimension $[1/T]$), is useful in flows in which periodic behaviour is exhibited. In more general unsteady flows the combination Vt/l is more useful and is also sometimes referred to as a Strouhal number.

Prandtl number (*Pr*) The Prandtl number is a fluid property, defined in section 2-9 as a function of other properties, namely

$$Pr \equiv \frac{\mu c_p}{\lambda} \qquad (5\text{-}40)$$

Nusselt number (Nu) In heat transfers through fluids, both conduction and convection processes exist. Although convection implies *mass* (not heat) transfers, its effect is to increase the heat interactions at the fluid boundaries. The Nusselt number is the ratio of the actual heat flux $\dot{Q}$ to the flux that would occur across a surface of area l^2 due to *conduction* with a temperature gradient of $\Delta T/l$, i.e.

$$Nu \equiv \frac{\dot{Q}}{\lambda l \, \Delta T} \tag{5-41}$$

Sometimes the *Stanton number* is used in preference to the Nusselt number. It is equal to $Nu/RePr$ and it is commonly denoted by the same symbol as the Strouhal number, namely St.

Stress and Force Coefficients In Eqs (5-14) and (5-25), a *stress* (denoted by p) appears as a dimensionless ratio with the product ρV^2. Although this is a logical ratio, we usually choose to introduce a coefficient of $\frac{1}{2}$ to give a closer correlation with, say, the Bernoulli and energy equations. Thus, for example, skin friction coefficients are conventionally defined as

$$C_F \equiv \frac{\tau_w}{\frac{1}{2}\rho V^2} \qquad \text{or} \qquad f \equiv \frac{\tau_w}{\frac{1}{2}\rho \overline{V}^2} \tag{5-42}$$

Force coefficients (e.g. drag and lift) are defined similarly, using

$$C_D \equiv \frac{F}{\frac{1}{2}\rho l^2 V^2} \tag{5-43}$$

Work and power Work, heat and energy all have the dimensions of force × length. They may be expressed as dimensionless ratios by, typically, $W/\rho l^3 V^2$. Power is a *rate* of work or heat transfer. It has the dimensions of force × velocity and a suitable dimensionless form is $\dot{W}/\rho l^2 V^3$.

5-7 PHYSICAL SIMILARITY

One of the most common applications of dimensional analysis is the interpretation of experimental data. Typically, we might wish to present a vast quantity of data in a concise form or we might want to use measurements from a model to predict the behaviour of a prototype. In either case, we must begin by deciding which physical quantities can independently influence the phenomenon.

A useful way of achieving this objective is to imagine that we are preparing an instruction manual enabling a technician to create the phenomenon to be measured. We would first describe the geometrical configuration, then the materials to be used (especially the fluid) and finally sufficient parameters to ensure that the desired flow conditions would be established. Once the technician has carried out the instructions we can come along with our instruments and take measurements as required.

As a simple example we might instruct our technician to build a smooth-walled, rectangular channel 10 m long and 0.3 m wide and to support it with its base sloping at 5 per cent to the horizontal. We could then ask her to devise some method of delivering water to the upper end of the channel at a constant rate of 0.1 m³/s and to make suitable arrangements for its disposal at the other end without obstructing the flow along the channel. On completion of this task (Fig. 5-3) we could set about measuring the depth, the velocity of flow and the gradient of the free surface, etc.

The parameters used to tell the technician what is required are the *independent* parameters that appear in lists such as Eq. (5-16). The additional parameters that we are able to measure are *dependent* on this basic set.

To ensure complete *physical similarity* between a model and a prototype or between models constructed and operated by different technicians, we must ensure that the instructions are complete but not over-specified. It would be nonsense to give the technician all of the above instructions and additionally to require her to achieve a particular depth of flow (unless she is allowed to insert weirs and sluices, etc.). In general, it is convenient to consider geometric, kinematic, dynamic and thermodynamic requirements in turn.

5-7-1 Geometric Similarity—Involves [L] Only

Geometrical configurations can be described by means of suitable lengths alone. Some objects, such as spheres, can be described very simply. Others, like trees, are more complex. Strictly, every possible independent length should be specified, but it is intuitively obvious that extreme detail may often be disregarded. In deciding whether a particular shape should be modelled, it may be necessary to consider the whole phenomenon. For instance, quite small details can have a pronounced influence on the drag force on a vehicle. Conversely, quite large features may have little effect if they are in a 'dead' region of flow.

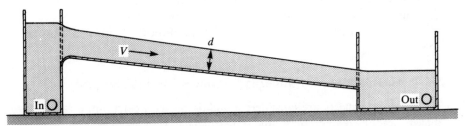

Figure 5-3 Flow along a laboratory channel.

Usually, geometrical parameters refer to solid surfaces. Occasionally however it is desirable to specify the flow conditions by a geometrical parameter (as in Example 5-3 on p. 139). We could, for instance, have asked our technician to reproduce a particular depth of flow at the mid-point of the channel instead of specifying a particular rate of flow.

5-7-2 Kinematic Similarity—Involves $[L, T]$ Only

Once the geometrical backcloth has been established, the *kinematic* conditions (i.e. the state of motion) may be specified. For rigid bodies it is usually sufficient to specify the velocity and acceleration, but further temporal derivatives such as the rate of change of acceleration are occasionally needed. When a body rotates in relation to the reference frame, angular as well as linear contributions must be specified.

Steady, continuous *fluid* motion can usually be specified sufficiently by a rate of flow. When greater detail is required, the magnitude and direction of the velocity may be specified at a suitable point on each independent streamline.

An unsteady, continuous flow may be specified by velocity and acceleration histories on independent streamlines only when the influence of pressure waves is negligible. In other cases, the specification of kinematic conditions can become as complex as the geometrical description of a tree. Happily, however, it is usually possible to regard flow parameters as dependent variables in these cases.

5-7-3 Dynamic Similarity—Involves $[M, L, T]$ Only

Dynamic similarity deals with forces. The force required to accelerate a body depends upon its inertial mass, and we usually allow for this by introducing the mass density ρ. In the case of fluids, the density is nearly always stipulated because of its influence on turbulence and/or because individual particles accelerate through constrictions and around obstructions even in steady flows. For similar reasons, the viscosity of a flowing fluid is usually considered (although its influence might be neglected subsequently in a simplified analysis). Field effects are normally stipulated by means of parameters such as g, the gravitational force per unit mass. In flows with a free liquid surface, account may also have to be taken of surface tension forces.

When necessary, temporal rates of change of these parameters must be specified.

5-7-4 Thermodynamic Similarity—Involves $[M, L, T, \Theta]$ Only

When thermal effects are considered, additional parameters may be introduced, notably temperatures and rates of heat transfer. The ability of a substance to conduct heat can be important. So can its specific heat capacities because these determine the changes in temperature corresponding to given amounts of heat or work.

Not all of these parameters will necessarily be independent. For instance, when the temperature gradient in a solid is specified then the rate of heat conduction can be calculated for any particular value of thermal conductivity (or vice versa). A similar comment applies to film heat transfer coefficients at the interfaces between solids and fluids.

5-7-5 Pure Substances

The state of any particular pure substance can be specified completely by any two independent properties. Therefore no more than two such properties should appear in the set of independent quantities. For example, if the density and temperature of air are both specified, we cannot also choose the pressure, viscosity and specific heat capacities, etc.; these follow automatically.

Notice that the above comment applies only to any *particular* pure substance. When different fluids are to be used in the 'model' and 'prototype', this simplification does not apply.

A different simplification can often be used for *general* pure substances. In the case of perfect gases, for example, only four members of the set $p, \rho, T, R, \gamma, c_p, c_v, u, h, s$ can be specified independently. The remainder can be deduced from relationships given in Chapter Two.

5-8 APPLICATIONS OF DIMENSIONAL ANALYSIS

Example 5-1: Skin friction in pipe flows Figure 5-4 depicts a steady flow along a rough-walled pipe of constant diameter. Derive a dimensionless expression that can be used for the presentation of experimental data relating to the skin friction between the fluid and the pipe. In the figure, the shear stress τ_w is shown in the direction in which it acts on the wall; it acts in the opposite direction on the fluid.

SOLUTION The parameters that we shall assume are needed to specify the state of flow are

$$
\begin{array}{lll}
\text{Geometric:} & D, l, l_1, l_2, \ldots & \\
\text{Kinematic:} & \bar{V} & \text{(5-44)} \\
\text{Dynamic:} & \rho, \mu &
\end{array}
$$

The lengths l, l_1, and l_2 typify the roughness of the pipe wall. At this stage of the development we cannot know in how much detail they must be specified (see comment 1). The axial length of the pipe is omitted from the list of geometric quantities because it is not plausible to suppose that this influences the shear stress remote from the ends. The velocity field is

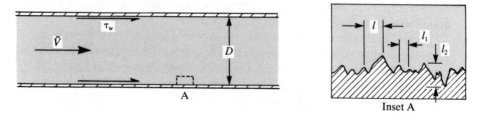

Figure 5-4 Skin friction in a pipe flow. (The shear stress τ_w is shown in the direction in which it acts on the pipe wall.)

characterized by the mean velocity of flow, $\bar{V}$, and dynamic effects are assumed to be due to inertia and viscosity alone. Gravitational forces may induce a hydrostatic-type pressure distribution, but this will have negligible influence on the local fluid motions.

Since the flow is presumed to be specified by the quantities (5-44), the wall shear stress τ_w must depend upon these parameters alone, i.e.

$$\tau_w = f_1(\rho, \mu, \bar{V}, D, l, l_1, l_2, \ldots) \tag{5-45}$$

which may be written equally generally as

$$f_2(\tau_w, \rho, \mu, \bar{V}, D, l, l_1, l_2, \ldots) = 0 \tag{5-46}$$

After manipulating this expression in the manner described in section 5-4, we obtain

$$f_3\left(\frac{\tau_w}{\rho \bar{V}^2}, \frac{\mu}{\rho D \bar{V}}, \frac{l}{D}, \frac{l_1}{D}, \frac{l_2}{D}, \ldots\right) = 0 \tag{5-47}$$

and so a suitable dimensionless expression for the skin friction is

$$f = f_4\left(Re, \frac{l}{D}, \frac{l_1}{D}, \frac{l_2}{D}, \ldots\right) \tag{5-48}$$

in which the skin friction coefficient f and the Reynolds number Re are defined by (5-42) and (5-35) respectively.

Comments on Example 5-1

1. In practice, experiments show that at low Reynolds numbers the skin friction coefficient is independent of the ratios l/D, l_1/D, etc. provided that they are sufficiently small. That is, (5-48) simplifies to $f = f(Re)$. At high values of Re, however, the coefficient f becomes independent of the Reynolds number. In this regime, (5-48) is usually written as $f = f(k_s/D)$ in which k_s is a characteristic size that depends upon all of l, l_1, l_2, etc.
2. Equation (5-48) has been derived for a steady uniform flow in a pipe of constant diameter. However, experiments show that it may also be applied

with good accuracy to flows in which particles experience gradual accelerations due to unsteadiness, compressibility or changes in area.

Example 5-2: Model of an aerogenerator An aerogenerator (Fig. 5-5) is designed to produce 3 MW while rotating at 60 rev/min in a 50 km/h wind. To test the design, a ⅓-scale model is built in a location where wind speeds of 20 km/h are expected. At what rotational speed should the model be operated and what power is it expected to generate?

SOLUTION The independent quantities needed to represent the flow are

$$\begin{array}{lll} \text{Geometric:} & D, l, l_1, \ldots & \\ \text{Kinematic:} & V, N & (5\text{-}49) \\ \text{Dynamic:} & \rho, \mu & \end{array}$$

The geometrical quantities include the blade diameter D and as many other lengths l, l_1, etc. as are needed to specify the object. The velocity of the wind V and the angular velocity of rotation N are both included in the kinematic set because they can vary independently. The power generated

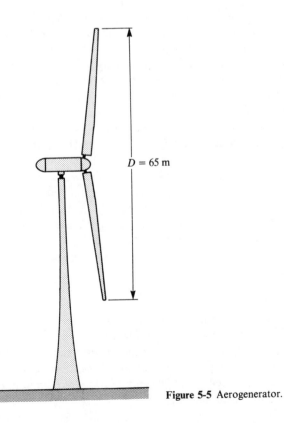

Figure 5-5 Aerogenerator.

by the machine is a function of the flow conditions, and so

$$\dot{W} = f_1(\rho, \mu, V, N, D, l, l_1, \ldots) \tag{5-50}$$

With the usual set of repeated quantities—ρ, D and V—we obtain

$$\frac{\dot{W}}{\rho D^2 V^3} = f_2 \left(Re, St, \frac{l}{D}, \frac{l_1}{D}, \ldots \right) \tag{5-51}$$

in which $Re = \rho D V / \mu$ is a Reynolds number and $St = ND/V$ is a Strouhal number.

The next step in the development is crucial for all modelling applications. We note that (5-51) can be applied equally well to either the prototype or the model and, in particular, that the function f_2 must be the same for both. The importance of this observation is that it can lead to the possibility of equating the dimensionless ratio $\dot{W}/\rho D^2 V^3$ for the prototype and the model. By writing (5-51) for the prototype P and for the model M, and forming their ratio, we obtain

$$\frac{(\dot{W}/\rho D^2 V^3)_P}{(\dot{W}/\rho D^2 V^3)_M} = \frac{f_2(Re, St, l/D, l_1/D, \ldots)_P}{f_2(Re, St, l/D, l_1/D, \ldots)_M} \tag{5-52}$$

If the model is truly geometrically similar to the prototype, then the ratios l/D, l_1/D, etc. will be the same for both machines. Therefore if we can also arrange that $Re_M = Re_P$ and that $St_M = St_P$ then the right-hand side of (5-52) will become equal to unity. This will necessarily follow even though we know nothing at all about the form of the function f_2.

Unfortunately, our hopes are dashed immediately because the statement of the problem requires that $Re_P = 12.5 Re_M$. Nevertheless, we notice that the Strouhal numbers can be made equal by simply choosing the rotational speed of the model to satisfy $(ND/V)_M = (ND/V)_P$, that is by choosing $N_M = 120$ rev/min. To proceed further, we call upon engineering judgement and argue that since the Reynolds numbers are very large (approximately 7×10^7 and 6×10^6 respectively), the fact that they are different is unimportant (see comment 2). In this case, it must be nearly correct to disregard this parameter and hence to equate the right-hand side of (5-52) to unity. Therefore we argue that when $N_M = 120$ rev/min, $(\dot{W}/\rho D^2 V^3)_M \approx (\dot{W}/\rho D^2 V^3)_P$ and so the expected power output from the model is about 10.2 kW.

Comments on Example 5-2

1. The above reasoning is typical of dimensional analysis applied to models. It is rare to find that all of the independent dimensionless groups can be made equal for the model and the prototype, but it is nevertheless usually possible to assume an approximate identity for the dependent dimensionless group. Of course, the use of engineering judgement is possible only after appropriate experience has been gained.

2. In this particular case, the relevant experience is that flow patterns tend to become independent of the Reynolds number when it is sufficiently large— as indicated in the comments on Example 5-1. However, experience also warns against over-confidence because the efficiency of the aerogenerator will be influenced by factors such as the detailed flow patterns over the surfaces of the blade. It is unlikely that these will be reproduced as faithfully as the overall flow patterns.

Example 5-3: Free surface flow over a spillway A $\frac{1}{25}$-scale model is used to investigate the pressure distribution on the surface of a spillway at various rates of flow. How should the measurements be interpreted? The spillway is shown in Fig. 5-6.

SOLUTION The following parameters are assumed to be sufficient to specify the flow:

$$\begin{array}{ll} \text{Geometric:} & H, l, l_1, \dots \\ \text{Dynamic:} & \rho, \mu, g, \sigma \end{array} \qquad (5\text{-}53)$$

No kinematic parameters are specified because they are all dependent upon the chosen set of independent parameters. Once we have chosen the height of the undisturbed free surface above the spillway crest, nature will choose, say, the velocity at the crest. The geometrical quantities l, l_1, etc. specify the shape of the spillway itself.

Let q denote the volumetric rate of flow per unit width (normal to the page) over the spillway. Since it is a dependent parameter,

$$q = f_1(\rho, \mu, g, \sigma, H, l, l_1, \dots) \qquad (5\text{-}54)$$

The usual repeated quantities cannot be used in this case because no velocity is specified explicitly. Using ρ, H and g instead, we obtain

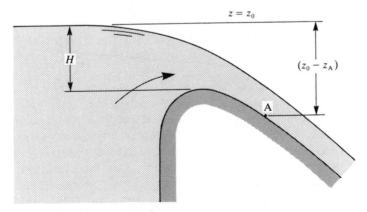

Figure 5-6 Free surface flow over a spillway.

$$\frac{q}{g^{1/2}H^{3/2}}=f_2\left(\frac{\mu}{\rho g^{1/2}H^{3/2}},\frac{\sigma}{\rho g H^2},\frac{l}{H},\frac{l_1}{H},\cdots\right) \qquad (5\text{-}55)$$

Usually neither of the first two groups in the function f_2 is the same in the model as it is in the prototype because the same liquid (water) and the same gravitational field are used in both cases. However, experience shows that their influence on the flow is not important unless the model is too small. Therefore with a sufficiently large and geometrically correct model, the right-hand side of (5-55) has nearly the same value for the two cases. It follows that $(q^2/gH^3)_M \approx (q^2/gH^3)_P$ or, for a $\frac{1}{25}$-scale model, $q_P \approx 125 q_M$.

The absolute pressure at any point in the flow field depends upon the ambient pressure at the free surface. However, the pressure *difference* Δp between any two points must be a function of the independent quantities (5-53) alone. Following the same procedure as before, we obtain

$$\frac{\Delta p}{\rho g H}=f_3\left(\frac{\mu}{\rho g^{1/2}H^{3/2}},\frac{\sigma}{\rho g H^2},\frac{l}{H},\frac{l_1}{H},\cdots\right) \qquad (5\text{-}56)$$

and so $(\Delta p/\rho g H)_M \approx (\Delta p/\rho g H)_P$. Any convenient reference pressure within the liquid may be used to form a pressure difference with the measured values on the spillway surface. In practice, the most convenient is undoubtedly the pressure at the free surface and so Δp is usually interpreted as $(p-p_{AT})$, namely the gauge pressure. For the $\frac{1}{25}$-scale model, $(p-p_{AT})_P \approx 25(p-p_{AT})_M$.

Comments on Example 5-3

1. In practice, the velocity of the flow over a spillway can be so high that the gauge pressures are negative over a large part of its surface. There is therefore a significant risk of the development of cavitation and of consequential damage (section 2-4-1). The easiest (and wisest) method of investigating this effect is to operate the model in normal atmospheric conditions and to infer the full-scale gauge pressure distribution in the manner described above. It is then a simple matter to inspect the predicted pressures to see whether any are less than (or near to) the pressure at which cavitation is likely to occur.

2. A less credible procedure can be devised in which cavitation is induced in the model. After first determining the pressures p_{CM} and p_{CP} at which cavitation will begin in typical samples of the model and prototype liquids, we could choose to operate the model with an ambient pressure p_{AM} that satisfies $(p_{CP}-p_{AT})=25(p_{CM}-p_{AM})$. The onset of cavitation (but not its subsequent behaviour) should then be the same in the model as in the prototype. In practice, however, this procedure is fraught with difficulties. Despite its visual advantages, it should not be relied upon.

3. Readers are strongly advised to consult Kenn and Garrod (1981) before

designing any spillway model. If highly sheared flows can be developed, the preceding development is inappropriate because the influence of viscosity must not be neglected. Very low local pressures inside individual vortices can cause cavitation that will not be predicted by the above model. In such cases, accurate modelling is possible only if the balance between the viscous forces (proportional to $\mu l V$) and the surface tension forces (proportional to σl) is modelled correctly. The ratio of these is $\mu V/\sigma$ and so the velocity scale must be unity if the same fluid is to be used in the model and the prototype.

Example 5-4: Heat pipe Figure 5-7 depicts a copper heat pipe of diameter D containing pure water in its saturation condition where the liquid and vapour (steam) phases coexist in thermodynamic equilibrium. When steady-state conditions prevail, one end of the pipe is heated at a rate $\dot{Q}$ and the other end is cooled at a similar rate. In operation, capillary action causes liquid to travel through the wick to the hot end of the pipe where it is evaporated because of the heat supply. It passes to the other end of the tube and is condensed by cooling. Develop a suitable dimensionless expression for the interpretation of experimental measurements of the heat flux and the temperature difference between the heat source and sink.

SOLUTION The physical conditions can be fully specified by the parameters

$$
\begin{array}{lll}
\text{Geometric:} & D, l, l_1, \ldots & \\
\text{Dynamic:} & \rho, g & \text{(5-57)} \\
\text{Thermodynamic:} & T_1, \Delta T &
\end{array}
$$

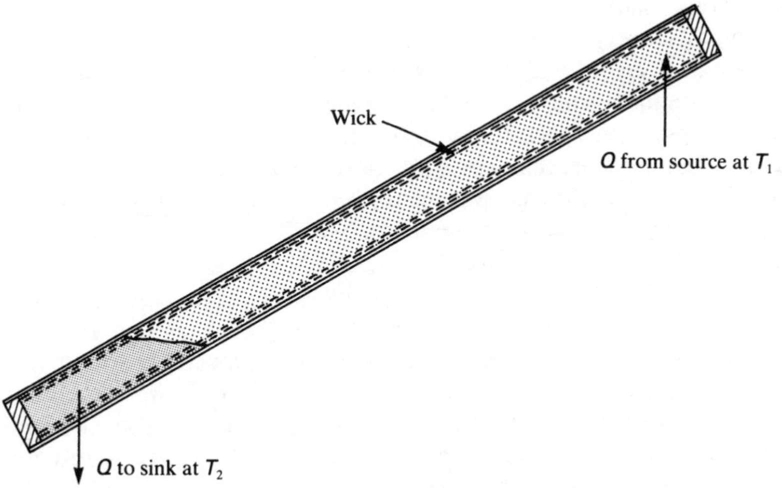

Figure 5-7 Heat pipe containing a pure fluid substance in liquid and vapour forms.

together with stipulations about the materials of which the tube and wick are made and a statement that the fluid is pure water in its saturated condition. The geometrical parameters l, l_1, etc. include information about the size of the pipe and its elevation at every point as well as a complete description of the wick. However, they do not include the position of the free surface; it is more convenient to allow this to be a dependent parameter. The density of the steam ρ is chosen to characterize the mass of fluid and its thermodynamic state. Since the fluid is in its saturated condition, all other properties must be regarded as dependent quantities. Therefore, for example, the fluid temperature, viscosity and specific heat capacities are not included in the list of independent thermodynamic quantities. However, the absolute temperature of the source T_1 is included as well as the temperature difference between the source and the sink $\Delta T \equiv T_1 - T_2$ because these parameters can be varied independently.

The heat flux is dependent upon the above parameters only. Therefore

$$\dot{Q} = f_1 \left(T_1, \Delta T, \rho, g, D, l_1, \ldots \right) \tag{5-58}$$

Using ρ, D, g and ΔT as the repeated quantities, we obtain

$$\frac{\dot{Q}}{\rho g^{3/2} D^{7/2}} = f_2 \left(\frac{T_1}{\Delta T}, \frac{l}{D}, \frac{l_1}{D}, \ldots \right) \tag{5-59}$$

and so the experimental results should give the same result for all geometrically similar pipes if they are plotted as a graph of $\dot{Q}/\rho g^{3/2} D^{7/2}$ against $\Delta T/T_1$.

Comments on Example 5-4

1. Heat pipes are very efficient devices for cooling high-temperature sources. The heat flux can be hundreds of times greater than the corresponding flux owing to conduction along a solid pipe of the same diameter. However, any particular pipe can operate only over a restricted range of temperatures because the fluid must remain saturated. If the working fluid is water, for example, its temperature must lie between the triple point temperature of $0.01\,°C$ and the critical point temperature $374.15\,°C$. Different fluids are used for higher or lower temperature ranges.

2. To minimize the length of the liquid column between the free surface and the heat sink, the volume of liquid in the pipe should be kept small. In practice, this limits the upper range of temperatures at which the device can operate successfully because all of the fluid will vaporize before the temperature rises to the critical point value. The operation of the pipe would be less effective in any case at the higher temperatures because the enthalpy of evaporation (sometimes called the latent heat of vaporization) reduces as the temperature increases.

Example 5-5: Speed of propagation of free surface waves Surface waves in the sea and in lakes, etc. are found to travel at a wide range of velocities. Determine the principal parameters likely to influence wavespeeds and hence deduce appropriate relationships for different types of waves.

SOLUTION If we use the notation depicted in Fig. 5-8 to describe the geometry then the principal parameters that may influence the wavespeed are

$$
\begin{array}{ll}
\text{Geometric:} & d, L, H \\
\text{Dynamic:} & \rho, g, \mu, \sigma
\end{array}
\tag{5-60}
$$

Additional geometrical quantities could be used to describe the shape of the surface profile, but it is reasonable to suppose that the overall wavespeed will depend primarily upon the overall size and not upon the detailed shape. No kinematic parameters are included because the velocities and accelerations cannot be varied *independently* of the above quantities.

Short waves When the wavelength L is smaller than about twice the depth, experiments show that the particle motions close to the bed are very small. In this case, the depth may logically be discarded from the quantities in (5-60). Assuming that surface tension effects may also be neglected (i.e. restricting the analysis to wavelengths and liquid depths of typical engineering interest for the present), the wavespeed V_w may be expressed as

$$
V_w = f_1(\rho, g, \mu, L, H)
\tag{5-61}
$$

Using ρ, g and L as repeated quantities, we obtain

$$
\frac{V_w}{g^{1/2}L^{1/2}} = f_2\left(\frac{\mu}{\rho g^{1/2}L^{3/2}}, \frac{H}{L}\right)
\tag{5-62}
$$

When the liquid is water or some other low-viscosity substance, the first term

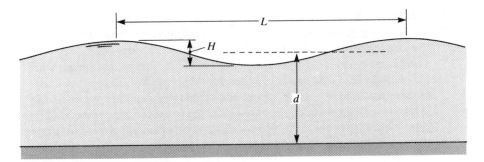

Figure 5-8 Free surface waves over a horizontal bed.

on the right-hand side of (5-62) has little influence. When, additionally, the ratio H/L is small, we expect the right-hand side to be approximately constant because the frequency of all small oscillations is independent of their amplitude. Therefore for small-amplitude short waves in a low viscosity liquid

$$V_w^2/gL \simeq \text{constant} \tag{5-63}$$

Long waves When the wavelength is more than about 50 times the depth, the particle motions close to the bed are similar to those at the surface (except for boundary layer effects). Experiments show that surface disturbances are strongly influenced by the proximity of the bed and that the wavelength is unimportant. It follows that the wavelength L should be discarded from the quantities in (5-60), but the above equations are reproduced unchanged except for this replacement. In particular, for small-amplitude, long waves in a low viscosity liquid,

$$V_w^2/gd \simeq \text{constant} \tag{5-64}$$

Intermediate waves When L/d exceeds 2 but is less than 50 then neither the depth nor the wavelength should be discarded from the quantities in (5-60). In this case, the dimensionless group d/L appears on the right-hand side of Eq. (5-62) and it is not possible to obtain a simple relationship such as (5-63) or (5-64).

Capillary waves When the wavelength is small, it is necessary to take account of surface tension forces which resist the increases in the free surface area that accompany deviations from the horizontal. With tiny wavelengths—a few millimetres in water—the surface tension forces are much larger than the gravitational forces. For a low-viscosity fluid and small amplitude waves the functional expression is

$$V_w \simeq f_3(\rho, \sigma, L) \tag{5-65}$$

and so
$$V_w^2 \rho L/\sigma \simeq \text{constant} \tag{5-66}$$

Comments on Example 5-5

1. As usual, dimensional analysis can give no information about the values of the constants in (5-63), (5-64) and (5-66). These must be found by experiment or by analytical methods. The appropriate relationships are $V_w^2 = gL/2\pi$, $V_w^2 = gd$ and $V_w^2 = 2\pi\sigma/\rho L$ respectively.
2. Notice that, unlike gravity waves, the velocity of capillary waves varies inversely with the wavelength; the shorter the wave, the greater its velocity. This feature helps the surfaces of lochs and other inland waters to attain their glass-like smoothness in very calm conditions.

Example 5-6: Shock waves in a railway tunnel Two ventilation shafts are to be provided in a main-line railway tunnel to ensure that passengers will not experience aural discomfort as a result of shock waves generated when a train enters the tunnel. The optimum positions of the shafts along the tunnel and their required cross-sectional areas are to be determined from an inexpensive physical model which is to be built in a 35 m long laboratory. Choose appropriate length and time scales for the model and indicate how the results should be interpreted. The full-size tunnel is 1800 m long, its cross-sectional area is $35\,m^2$ and the maximum train speed is 250 km/h.

SOLUTION The air velocities generated by a train in a tunnel are much smaller than the train speed. Therefore the Mach number is very small and compressibility effects have little influence on the steady components of the airflows. Nevertheless, account must be taken of compressibility because shock waves are present. Using the speed of sound c for this purpose, the independent parameters specifying the overall phenomenon may be chosen as

$$\begin{array}{lll} \text{Geometric:} & D, l, l_1, l_2, \ldots & \\ \text{Kinematic:} & V, c, t & (5\text{-}67) \\ \text{Dynamic:} & \rho, \mu & \end{array}$$

in which V is the train speed and l, l_1 and l_2 are relevant lengths other than the tunnel diameter D. The time t is included as an independent parameter because the conditions in the tunnel are highly time dependent.

The magnitudes of the pressure fluctuations in the tunnel are dependent upon these parameters. Using ρ, D and V as repeated variables, a suitable functional relationship is

$$\frac{\Delta p}{\frac{1}{2}\rho V^2} = f\left(\frac{\rho D V}{\mu}, \frac{V}{c}, \frac{Vt}{D}, \frac{l}{D}, \frac{l_1}{D}, \frac{l_2}{D}, \ldots\right) \qquad (5\text{-}68)$$

The first two terms on the right-hand side of this expression are the Reynolds number and the Mach number, and in practice we cannot reproduce both of them inexpensively. Since pressure wave action is expected to be more important than frictional resistance, we shall discount the Reynolds number and design the model to suit the remaining parameters.

The speed of sound in air will be the same in the model as in the prototype (at the same temperature) and so the Mach number V/c can be reproduced only by having the same train speed in the model as at full scale. The model length is dictated by the size of the available laboratory, and a scale of 1 : 100 should give sufficient space for the necessary acceleration and deceleration tracks for the vehicle.

The required time scale can be deduced from the dimensionless group Vt/D. For this to have the same value in the model and the prototype, the time scale must be the same as the length scale, namely 1 : 100.

Subject to inaccuracies resulting from the failure to model the Reynolds number correctly, the pressure histories in the model should be the same as those occurring a hundred times more slowly at full scale.

Comments on Example 5-6

1. With a length scale of $1:100$ the Reynolds number $Re = \rho D V / \mu$ will be 100 times smaller in the model than in the prototype. The influence of skin friction will therefore be different in the two cases and this is important because the train speed is so high. In practice, more accurate results can be obtained by using a geometrically distorted model in which the radial length scale exceeds the axial length scale. This adjustment has the advantage that the detailed nature of complex flows close to, say, the base of a ventilation shaft or the nose of the train is reproduced more accurately. The disadvantage is that the ratio L/D that governs the overall pressure drop due to friction is not scaled correctly.
2. Neither the distorted model nor the undistorted model will reproduce correctly the relative magnitudes of the pressure fluctuations due to shock waves and skin friction. It is therefore of considerable advantage to have available a computer program that is capable of simulating the flows at both scales. After using the model to verify the accuracy of the computed predictions at the small scale, confidence can be placed in the corresponding predictions for the full-scale flows.
3. The need to reproduce full-scale velocities in the model is highly demanding, especially for a train speed of 250 km/h. Nevertheless attempts have been made by experimenters to build such models.

Example 5-7: Law of the wall Deduce the logarithmic expression (4-28) describing the velocity distribution in the constant-stress region of a turbulent boundary layer over a smooth wall using the principle of physical similarity.

SOLUTION Figure 5-9 illustrates a typical velocity distribution close to the wall. In the region of approximately constant shear stress, $\tau \approx \tau_w$ and the local velocity component $\bar{u}$ at a distance y from the wall may be assumed to be a function of the following parameters:

$$\left. \begin{array}{ll} \text{Geometric:} & y \\ \text{Dynamic:} & \rho, \mu, \tau_w \end{array} \right\} \tag{5-69}$$

The shear stress is approximately equal to the Reynolds stress $-\rho \overline{v'u'}$ —and so there is no need to stipulate the kinematic parameters u' and v' independently. A suitable functional relationship for $\bar{u}$ is

$$\bar{u} = f_1(y, \rho, \mu, u_\tau) \tag{5-70}$$

in which the replacement of the wall-shear stress by the wall-shear velocity

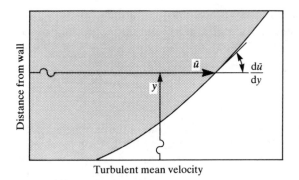

Figure 5-9 Velocity distribution within the constant-stress region close to a smooth wall.

$u_\tau = \sqrt{\tau_w/\rho}$ involves no loss of generality. By choosing y, ρ and $\bar{u}$ as repeated variables and forming dimensionless groups in the usual manner, we obtain

$$\frac{\bar{u}}{u_\tau} = f_2\left(\frac{yu_\tau}{v}\right) = f_2(\eta) \tag{5-71}$$

in which $v = \mu/\rho$ is the kinematic viscosity and the dimensionless parameter $\eta \equiv yu_\tau/v$ is introduced for convenience. Equation (5-71) is a correct functional form of Eq. (4-28). However, we can proceed further by differentiating it to obtain the velocity gradient, namely

$$\frac{d\bar{u}}{dy} = \frac{u_\tau}{v}\frac{df_2(\eta)}{d\eta} \tag{5-72}$$

Now the velocity gradient in the constant-stress region is determined primarily by the local turbulence and is nearly independent of the fluid viscosity (since $\tau \approx -\rho\overline{v'u'}$). Therefore the right-hand side of Eq. (5-72) must also be independent of the viscosity. It follows that $df_2/d\eta$ must be directly proportional to v and so it is inversely proportional to η. That is $df_2/d\eta = C_1/\eta$ in which C_1 is a constant. On integration, this yields $f_2 = C_1 \ln(\eta) + C_2$ or, in the original notation,

$$\frac{\bar{u}}{u_\tau} = C_1 \ln\left(\frac{yu_\tau}{v}\right) + C_2 \tag{5-73}$$

Comments on Example 5-7
1. It is remarkable that the approximate form of the function f_2 can be deduced from physical reasoning. Usually, only functional relationships such as (5-71) can be obtained.
2. By following similar reasoning the corresponding relationship for rough

walls can be obtained. In this case the characteristic wall-roughness size k_s takes over the role played by the kinetic viscosity v in the above development.

PROBLEMS

1 Show that all possible dimensionless combinations of the parameters ρ, D, V and μ are powers of the Reynolds number $Re = \rho D V/\mu$. Each parameter has its usual meaning.

2 Show that all possible dimensionless combinations of the parameters F, D, V and μ are powers of $(F/\mu D V)$. Each parameter has its usual meaning and F denotes a force.

3 Show that all possible dimensionless combinations of the parameters $\dot{Q}$, λ, c_p, T and g are powers of $(g\dot{Q}/\lambda c_p T^2)$. The parameters are respectively a rate of heat transfer, a thermal conductivity, a specific heat capacity, a temperature and an acceleration.

4 The drag force F on a submerged sphere in a steady flow of liquid is believed to depend only upon its diameter D and velocity V in addition to the density ρ and viscosity μ of the liquid. Assuming this to be true, show that the force can be expressed as

$$F = \tfrac{1}{2} C_D \rho D^2 V^2 \qquad (5\text{-}74)$$

where C_D is a drag coefficient that is dependent solely upon the Reynolds number.

5 A liquid flows steadily down a channel with a heated bed as shown in Fig. 5-10. The temperature rise of the liquid ΔT_1 is believed to depend upon the slope α and roughness k_s of the channel, the gravitational acceleration g, the upstream depth d, the excess of the bed temperature over the incoming liquid temperature ΔT_B and the density ρ, viscosity μ and specific heat capacity c_v of the liquid. Assuming this to be so and choosing ρ, g, d and ΔT_B as repeated variables, show that

$$\Delta T_1 = \Delta T_B f\left(\frac{\mu}{\rho g^{1/2} d^{3/2}}, \frac{c_v \Delta T_B}{gd}, \frac{k_s}{d}, \alpha\right) \qquad (5\text{-}75)$$

6 An air bubble rising through a liquid at a steady speed may have any of the

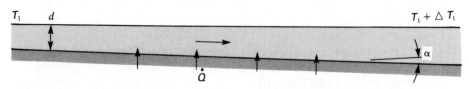

Figure 5-10 Flow over a heated bed.

Figure 5-11 Air bubbles rising through a liquid.

shapes shown in Fig. 5-11. Explain how the shape of the bubble can be influenced independently by (*a*) liquid viscosity and density, (*b*) air viscosity and density, (*c*) liquid–air surface tension, (*d*) gravitational acceleration, (*e*) volume of air.

Also explain why the pressure and temperature of the air and liquid and the velocity of the bubble should not be included as additional independent variables. (*Suggestion*: Observe bubbles in shampoo bottles in a supermarket.)

7 Figure 5-12 depicts two possible surface profiles for the flow over a broad-crested weir in a channel. Explain how, for any particular weir, the shape of the surface profile is dependent upon (*a*) the upstream depth and velocity, (*b*) the liquid density and viscosity and (*c*) the gravitational acceleration. In what circumstances should the surface tension of the liquid be introduced as an additional parameter?

8 Natural gas flows steadily along a horizontal pipeline. Explain how the rate of heat transfer through the walls of a 1 km length of the pipe can be influenced independently by (*a*) the upstream pressure, temperature and velocity, (*b*) the diameter, thickness, roughness and thermal conductivity of the pipe and (*c*) the temperature of the surroundings. Also explain why the viscosity and thermal conductivity of the gas and the flow conditions at the downstream end of the section should not be introduced as additional independent parameters.

9 A rifle bullet travels through initially still air at approximately twice the speed of sound. Explain how the deceleration of a particular bullet at an instant when it is travelling horizontally can be influenced independently by (*a*) the density and temperature of the air and (*b*) the velocity of the bullet.

Also explain why the viscosity and compressibility of the air should not be regarded as additional independent variables.

10 Explain how the pressure rise across a particular type of centrifugal pump in a water pipeline can be influenced independently by (*a*) the size of the

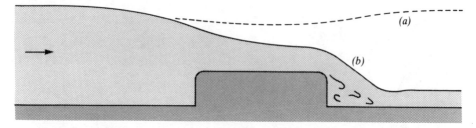

Figure 5-12 Free surface flow over a broad-crested weir. (a) Drowned weir. (b) Free overflow.

pump, (b) the speed of rotation of the impeller, (c) the rate of flow through the pump and (d) the density and viscosity of the water. Also explain why the pressure and temperature of the water should not be introduced as additional independent variables. In what way would cavitation invalidate the analysis?

[density varies]

11 A centrifugal pump delivers $0.9\,\text{m}^3/\text{s}$ water against a pressure head of 28 m when its shaft rotates at 450 rev/min. If the same pump is operated at 750 rev/min and the rate of water flow is $1.5\,\text{m}^3/\text{s}$, what will be the head difference? Neglect the influence of Reynolds number.

[77.8 m]

12 The drag force on a submarine is to be estimated from tests on a $\frac{1}{15}$-scale model in a wind tunnel. Show that the drag forces for the model and the prototype both satisfy

$$\frac{F}{\rho l^2 V^2} = f\left(\frac{\rho l V}{\mu}\right) \tag{5-76}$$

where F is the drag force, l is a typical length, V is the speed of flow relative to the submarine and ρ and μ are the density and viscosity of the fluid. Hence determine the required velocity in the wind tunnel to correspond to a speed of 10 m/s at full scale assuming that the wind tunnel operates at a pressure of ten atmospheres.

Also determine the force at full scale corresponding to a force of 1 kN in the model. The densities of water and air are $1000\,\text{kg/m}^3$ and $12\,\text{kg/m}^3$ and the corresponding viscosities are $1000\,\mu\text{Pa s}$ and $18\,\mu\text{Pa s}$.

[116 m/s, 37.0 kN]

13 Water in a pipe is heated by passing it over the outside of a second pipe containing hot steam moving in the opposite direction as illustrated in Fig. 5-13. Determine a functional relationship for the temperature rise in the water and hence show that this depends solely on other temperatures if the geometry of the system and the ratios of the densities and the velocities of the water and steam are held constant.

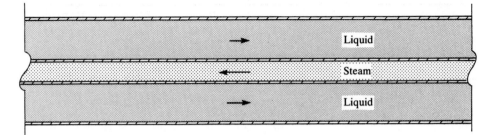

Figure 5-13 Heating water by a contra-flow of steam.

FURTHER READING

Barr, D.I.H. (1984) Consolidation of basics of dimensional analysis, *J. Engng Mech.*, **110** (9), 1357–1376.

Dunn, P.D. & Reay, D.A. (1982) *Heat Pipes*, 3rd edn, Pergamon.

Isaacson, E. de St Q. and Isaacson, M. de St. Q. (1975) *Dimensional Methods in Engineering*, Edward Arnold.

Kenn, M.J. and Garrod, A.D. (1981) Cavitation damage and the Tarbela tunnel collapse of 1974, *Proc. Instn Civ. Engrs*, Part 1, **70**, 65–89 and 779–810.

Novak, P. and Cabelka, J. (1981) *Models in Hydraulic Engineering*, Pitman.

Sharp, J.J. (1981) *Hydraulic Modelling*, Butterworths.

CHAPTER SIX

CONTINUITY

6-1 SYSTEMS AND CONTROL VOLUMES

When it is first encountered, the law of conservation of mass is usually applied to a *system*, that is to an *identifiable collection of matter*. The law states simply that the total mass of the system remains permanently unchanged irrespective of the conditions to which it is exposed. Thus, for example, the mass of a given sample of water remains permanently constant. This is true even after some evaporation has taken place because the definition of the system as 'a given sample of water' implies that the evaporated molecules must be considered along with the liquid when applying the law.

Although this form of the law can be useful in its own right, it is often preferable to use a somewhat different form which allows us to consider rates of change of mass. For example, if the above sample of water is imagined to be contained in a bowl, then it is natural to observe that the rate of change of mass in the bowl is equal to the rate of evaporation. In making this assertion, however, we are not concentrating our attention on the original sample of water, but rather on the container in which it is held. More precisely, we are focusing our thoughts on the region of three-dimensional space that happens to be occupied by the bowl of water for the time being. This region of space is an example of a *control volume*.

By allowing water to enter the bowl—from a tap, say—we can generalize the above discussion to include inflow into the region of space as well as outflow (evaporation in this case) from it. The law of conservation of mass, written for the control volume, is known as the *continuity equation*:

| The rate of increase of mass in a control volume | = | the rate of mass inflow | − | the rate of mass outflow | (6-1) |

Notice that the continuity equation tells us nothing at all about absolute quantities such as the total mass of water in the bowl. It merely describes their rates of change.

Use is made of the concept of control volumes in disciplines other than fluid mechanics, for example in the regulation of traffic densities in cities. Computerized traffic controls can be used to ensure that the number of vehicles entering a particular zone does not exceed the number leaving it. No information is gained about the distribution or total number of vehicles, but the total is prevented from increasing. Economists imagine whole countries as encompassed by control surfaces across which 'flow' imports and exports. By interpreting these terms as inclusive of all transactions involving finance, they try to determine the economic health of the country.

6-2 GENERAL FORM OF THE CONTINUITY EQUATION

Figure 6-1 depicts a fluid flow along a pipe. At the instant t_0, a certain sample of fluid occupies an arbitrarily chosen control volume ABCD and the region I (denoting inflow). At a later instant, $t_0 + \delta t$, the fluid sample has flowed so that it occupies the control volume and the region O (denoting outflow). The elemental mass of fluid within the region I at the time t_0, namely δm_I, crosses the surface AD of the control volume during the interval δt at an average *rate* of $\delta m_\mathrm{I}/\delta t$. Similarly, the average rate at which fluid flows across the surface *BC* of the control volume during the interval δt is $\delta m_0/\delta t$. In the limit, as δt tends to zero, we may define the *rate of mass flow* $\dot m$ across a surface as

$$\dot m \equiv \frac{dm}{dt} \tag{6-2}$$

The rate of mass flow is commonly referred to as the *mass flux*. For the simple case of incompressible flow along a rigid pipe, the total mass of fluid

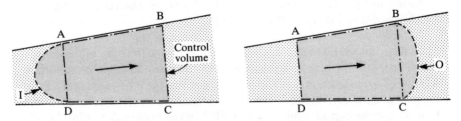

Figure 6-1 Flow along a tapered pipe. (a) System at $t = t_0$. (b) System at $t = t_0 + \delta t$.

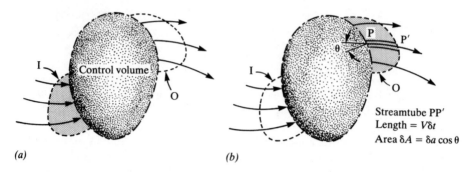

(a)   *(b)*

Streamtube PP'
Length = $V\delta t$
Area $\delta A = \delta a \cos \theta$

Figure 6-2 Flow through an arbitrary control volume. (a) System shaded at t_0. (b) System shaded at $t_0 + \delta t$.

within the control volume ABCD is the same at the time $t_0 + \delta t$ as it was at the instant t_0. It follows that the rate of mass inflow, $\dot{m}_I$, is equal to the rate of mass outflow, $\dot{m}_O$.

These ideas are now generalized to apply to flow through the arbitrarily shaped control volume depicted in Fig. 6-2. We wish to develop an analytical form of the continuity equation (6-1) and we begin by considering, as our *system*, the contents of the control volume and the elemental region I at the instant t_0. At the later time $t_0 + \delta t$, the *same* matter occupies the control volume and the elemental region O and its mass has not changed. That is,

$$(m_{cv} + \delta m_O)_{t_0 + \delta t} = (m_{cv} + \delta m_I)_{t_0} \tag{6-3}$$

After rearranging the terms and dividing throughout by δt, we obtain

$$\frac{m_{cv, t_0 + \delta t} - m_{cv, t_0}}{\delta t} = \frac{\delta m_{I, t_0}}{\delta t} - \frac{\delta m_{O, t_0 + \delta t}}{\delta t} \tag{6-4}$$

In the limit, as δt approaches zero, this becomes

$$\frac{\partial m_{cv}}{\partial t} = \dot{m}_I - \dot{m}_O \tag{6-5}$$

in which the suffix t_0 is discarded because it denotes an arbitrary instant. Equation (6-5) is the general mathematical form of the relationship (6-1) which was derived in a qualitative manner.

Mass flux It is useful to express the mass flow rate $\dot{m}$ in terms of the density and velocity. For example, we shall see that the mass flux across the surface AD in Fig. 6-1 satisfies $\dot{m} = \rho a V$ in which a denotes the cross-sectional area of the pipe and ρ and V are the density and velocity respectively. For the more general case, we consider the streamline PP' at the instant $t_0 + \delta t$ in Fig. 6-2. The length of this line is $V \delta t$ in which V denotes the local fluid velocity. The volume of an elemental streamtube around PP' is $V \, \delta t \, \delta A$ and the mass of fluid within it is $\rho V \, \delta t \, \delta A$, in which δA denotes the cross-sectional area of the streamtube perpendicular to its axis. If the angle between the velocity vector and the local surface normal is denoted by θ, then the area of the control surface in contact with the streamtube is δa, where $\delta A = \delta a \cos \theta$. The contribution of this streamtube to the *rate* of mass flow across the control surface is therefore

$$\delta \dot{m} = \rho V \cos \theta \, \delta a \qquad (6\text{-}6)$$

and the total mass flux is

$$\dot{m} = \int \rho V \cos \theta \, da \qquad (6\text{-}7)$$

Range of validity of the continuity equation Equations (6-5) and (6-7) are general expressions valid for any control volume. The *contents* of the control volume may include matter in solid, liquid or gaseous form. However, the use of (6-7) may be complex unless the matter entering and leaving the control volume is homogeneous.

There is no restriction on the size of the control volume. It can be elemental, but it can also be sufficiently large to encompass, say, an oil distribution network, a turbine or a harbour.

6-3 UNIAXIAL AND ONE-DIMENSIONAL FLOWS

In pipe and channel flows and in many other flows of practical importance, it is possible to choose the control volume so that the flow at an inlet or outlet section may be regarded as *uniaxial*, that is parallel to a single axis. In this case, the mass flux expression (6-7) can be simplified by choosing the control surface normal to the direction of flow. Then $\cos \theta = 1$ and

$$\dot{m} = \int \rho V \, da \qquad (6\text{-}8)$$

It is usual to neglect density variations across a cross-section (though not necessarily in the direction of flow). Alternatively, we may choose to define a *mean density* for the cross-section by

$$\bar{\rho} \equiv \frac{1}{a} \int \rho \, da \qquad (6\text{-}9)$$

In a similar manner, we may define a *mean velocity* by

$$\bar{V} \equiv \frac{1}{\bar{\rho} a} \int \rho V \, da \qquad (6\text{-}10)$$

Equations (6-8) and (6-10) may be combined to give

$$\dot{m} = \bar{\rho} a \bar{V}$$ (6-11)

and so the continuity equation for uniaxial flows is

$$\frac{\partial m_{cv}}{\partial t} = \bar{\rho}_1 a_1 \bar{V}_1 - \bar{\rho}_0 a_0 \bar{V}_0$$ (6-12)

In the remainder of this development, we shall write ρ instead of $\bar{\rho}$ and so the mass flux is $\dot{m} = \rho a \bar{V}$. Strictly, this is valid only for flows where density variations in any particular flow section are negligible. However, all subsequent equations may be used in the more general case if the density ρ is interpreted as $\bar{\rho}$. A similar procedure could be followed with the velocity V, but this would make less sense because significant variations in velocity at a flow section are commonplace.

Range of validity The only additional limitation of (6-12) in comparison with (6-4) is that the flows at each inlet and outlet section must be uniaxial. Note particularly that it is not necessary for the flows *within* the control volume to be uniaxial (see Fig. 6-4 for example). It is not even necessary for the flows at the inlets and outlets to be parallel to one another (Fig. 6-5).

When there is more than one inlet and/or outlet section, the product $\rho a \bar{V}$ must be evaluated separately for each. This implies that appropriate mean values must be defined for the velocity (and density) at each section.

In practice the uniaxial equation (6-12) is commonly used when the inflows and outflows are not quite uniaxial (e.g. Fig. 6-1). Strictly, in such circumstances, the implied interpretation of the velocity $\bar{V}$ is its component in the principal direction of flow.

One-dimensional flow It is often acceptable to neglect variations in parameters within a cross-section. In this case, there is no need to define mean values of the parameters and the flow is conveniently termed *one-dimensional*. In such flows, the velocity vectors at a flow section are equal in magnitude as well as direction. The mass flux is simply

$$\dot{m} = \rho a V$$ (6-13)

Some authors use the term *one-dimensional* to describe all uniaxial flows. In this book, however, flows such as those depicted at the inlet and outlet sections in Fig. 6-1 are regarded as uniaxial, but not as one-dimensional.

Steady flow In a *steady* flow the conditions at any position do not vary with time. In particular, there is no rate of change of mass within a control volume and so the uniaxial continuity equation is

$$\rho_1 a_1 \bar{V}_1 = \rho_0 a_0 \bar{V}_0$$ (6-14)

Incompressible flow When the fluid and its container may be regarded as incompressible, the uniaxial continuity equation reduces to

$$a_1 \bar{V}_1 = a_0 \bar{V}_0 \tag{6-15}$$

irrespective of whether the flow is also steady. The product aV is the *volumetric* flow rate; it is commonly denoted by the symbol Q.

6-4 APPLICATIONS OF THE CONTINUITY EQUATION

Example 6-1: Storage reservoir Discuss the factors influencing the rate of change of water quantity in the storage reservoir depicted in Fig. 6-3.

SOLUTION *Inflows* can occur by means of surface and ground water flows and by precipitation in the form of rain or snow. The major contributions are nearly always from rivers and streams, which are themselves a consequence of precipitation in upstream catchment areas.

The principal *outflows* are the discharge into the downstream river and the supply to the community for whose benefit the reservoir exists. Discharge into the river is usually necessary even when the reservoir is not full because downstream requirements will determine a minimum acceptable flow rate. Other forms of outflow are evaporation and seepage. Vast quantities of water are lost by evaporation, but that is another story. Seepage under the dam must be restricted, partly to prevent an undesirable loss of water, but also to protect the dam from the effects of erosion and from uplift due to pressure forces on its underside.

When all forms of inflow and outflow have been established, the rate of change of mass of water in the reservoir can be deduced from Eq. (6-1).

Comment on Example 6-1 The 'rates' of inflow and outflow need not be instantaneous values. For large reservoirs, the average monthly or annual rates are normally of greatest interest. A knowledge of rates of flow during shorter periods is rarely of practical significance except during flood conditions. The purpose of a storage reservoir is to ensure that sufficient quantities of water are retained during wet months or years to ensure adequate supplies during dry months or years.

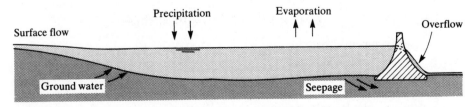

Figure 6-3 Reservoir inflows and outflows.

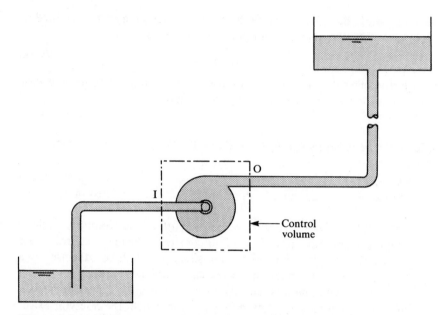

Figure 6-4 Pumping from a low-level reservoir to a high-level reservoir.

Example 6-2: Steady incompressible flow in a pipe A pump is used to deliver water to a high-level reservoir from a low-level reservoir. Estimate the mean velocity in its 50 mm bore delivery pipe when the mean velocity in its 75 mm bore suction pipe is 2 m/s. Also find the volumetric flux through the pump.

SOLUTION By choosing the control volume shown in Fig. 6-4 and noting that water may be regarded as incompressible in a steady flow, we obtain

$$a_1 \bar{V}_1 = a_O \bar{V}_O \qquad (6\text{-}16)$$

The ratio of the pipe cross-sectional areas a_1/a_O is equal to the square of the ratio of their diameters, namely $(75/50)^2 = 2.25$. Therefore $\bar{V}_O = 2.25 \times 2\,\text{m/s} = 4.5\,\text{m/s}$.

The volumetric flow rate Q may be evaluated from either $a_1 \bar{V}_1$ or $a_O \bar{V}_O$. Using the former,

$$Q = \tfrac{1}{4}\pi \times 0.75^2 \times 2\,\text{m}^3/\text{s} = 0.0088\,\text{cumec}$$

(see comment 2 below).

Comments on Example 6-2
1. It is quite common to have a suction pipe of larger diameter than the delivery pipe. It is also common to install the pump close to the upstream

reservoir. These precautions help to minimize the reduction in pressure along the suction pipe and thus the likelihood of cavitation within the pump.

2. The unit *cumec* is an abbreviation for m³/s, namely cubic metres per second. It is used quite frequently by water engineers.

Example 6-3: Steady flow at a junction Figure 6-5 shows a part of a water supply network where a 150 mm bore branch pipe leads off a 300 mm bore main. The steady volumetric flow rate in the main at the section A is 0.033 cumec and the mean velocity at B is 0.4 m/s. What is the rate of flow in the branch pipe?

SOLUTION Assume for now that the flow at C is outwards. If this assumption is incorrect then the error will become obvious later. Using Eq. (6-15) and allowing for all inlet and outlet sections we obtain

$$a_A\bar{V}_A = a_B\bar{V}_B + a_C\bar{V}_C \qquad (6\text{-}17)$$

Since $a_A\bar{V}_A = 0.033\,\text{m}^3/\text{s}$ and $a_B\bar{V}_B = \frac{1}{4}\pi \times 0.3^2 \times 0.4\,\text{m}^3/\text{s} = 0.0283\,\text{m}^3/\text{s}$, the volumetric flux at C is

$$a_C\bar{V}_C = (0.033 - 0.0283)\,\text{m}^3/\text{s} = 0.0047\,\text{m}^3/\text{s}$$

Since the result has a positive numerical value, the assumption that the flow at C is outwards is correct. A negative result would have indicated that the assumption was untrue, but the numerical value would still have been valid.

Example 6-4: Flow through a restriction
(a) Figure 6-6 depicts a liquid flowing through a convergent–divergent region in a pipe. The areas at the sections 1, 2 and 3 are 0.1 m², 0.05 m² and 0.1 m² respectively and the mean velocity at section 1 is $\bar{V} = 1$ m/s. Estimate the mean velocity at the sections 2 and 3.
(b) Repeat the calculation assuming that the fluid is air with densities of 1.5 kg/m³, 1.425 kg/m³ and 2.425 kg/m³ at the sections 1, 2 and 3 respectively and that $\bar{V}_1 = 50$ m/s.

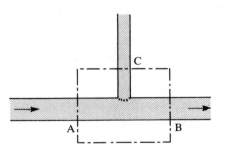

Figure 6-5 Steady flow at a pipe junction.

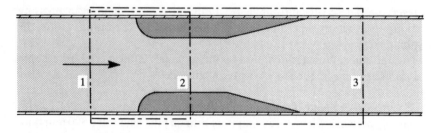

Figure 6-6 Flow through a venturimeter.

SOLUTION

(a) The steady flow-continuity equation (6-14) applied to the smaller of the control volumes in Fig. 6-6 yields

$$\rho_1 a_1 \bar{V}_1 = \rho_2 a_2 \bar{V}_2 \qquad (6\text{-}18)$$

Assuming $\rho_1 = \rho_2$ for a steady liquid flow and noting that $a_1 = 2a_2$, we obtain $\bar{V}_2 = 2 \bar{V}_1 = 2\,\text{m/s}$. Similarly, for the larger of the two control volumes,

$$\rho_1 a_1 \bar{V}_1 = \rho_3 a_3 \bar{V}_3 \qquad (6\text{-}19)$$

which shows that $\bar{V}_3 = \bar{V}_1 = 1\,\text{m/s}$.

(b) These two equations are equally valid for the gas flow. However, account must be taken of the changes in density as the gas flows along the duct. Numerically, therefore, we obtain $\bar{V}_2 = 175.4\,\text{m/s}$ and $\bar{V}_3 = 51.5\,\text{m/s}$.

Comments on Example 6-4

1. In both cases, the fluid travels more quickly at the narrow section than it does at the wider sections. This surprises some people, but only if they focus attention on the development of a restriction instead of the steady-state condition. At any instant, the velocity in the restriction is bigger than just upstream.

2. Since the area of flow in the restriction is half that in the pipe, the liquid velocity doubles. However, the gas velocity more than doubles and the reason is easy to understand. In both cases, the kinetic energy of the fluid in the restriction is bigger than that upstream. The total energy of the fluid cannot have increased so in practice the pressure must have decreased. For the gas (but not for the liquid) the pressure reduction causes a density reduction and the continuity equation tells the rest of the story.

3. Despite the compressibility of the gas, the behaviour of the two flows is similar. In both cases, the velocity increases where the cross-sectional area is reduced and it then decreases again. However, this reassuring state of affairs doesn't always exist, as the next example shows.

Example 6-5: de Laval nozzle The convergent–divergent nozzle shown in Fig. 6-7 is used to accelerate a stream of air ($R = 287 \, \text{J/kg K}$). Estimate the mean velocity at the sections 2 and 3 if the mean velocity at section 1 is $V_1 = 218 \, \text{m/s}$. The pressures, temperatures and cross-sectional areas at the sections 1, 2 and 3 are as follows:

flow section:	1	2	3
pressure, kPa:	843	528	96
temperature, °C:	203	144	−17
area, mm²:	7410	5530	10 920

SOLUTION Using the equation of state for a perfect gas we find that the densities at the three sections are $\rho_1 = p_1/RT_1 = 6.17 \, \text{kg/m}^3$, $\rho_2 = 4.41 \, \text{kg/m}^3$ and $\rho_3 = 1.31 \, \text{kg/m}^3$. The mass flux $\dot{m}$ is the same at all three sections. At section 1,

$$\dot{m} = \rho_1 a_1 \bar{V}_1 = 10.0 \, \text{kg/s} \tag{6-20}$$

and so $\bar{V}_2 = \dot{m}/\rho_2 a_2 = 409 \, \text{m/s}$ and $\bar{V}_3 = 697 \, \text{m/s}$.

Comments on Example 6-5
1. In this example the velocity increases continuously downstream. Within the convergent region upstream of the 'throat' of the restriction the conditions are qualitatively similar to those in the previous example. Downstream of the throat, however, the velocity continues to increase even though the area is also increasing. Since $\dot{m} = \rho a V$ is a constant, it follows that the density must decrease strongly.
2. The key to this behaviour lies at the throat, where in this example the velocity of flow is exactly equal to the local speed of sound. Upstream of the throat the velocity is less than the speed of sound, whereas downstream it is greater than the speed of sound. These conditions are usually termed subsonic and supersonic respectively.
3. If the nozzle converged again downstream then the velocity would decrease and the density would increase—the opposite of the behaviour in the subsonic approach from section 1 to section 2.

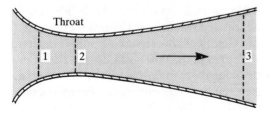

Figure 6-7 de Laval nozzle.

Example 6-6: Vehicular traffic flows Vehicular traffic moves steadily along a two-lane carriageway towards a restriction where the two lanes merge into one. In one of the approach lanes, the traffic speed is 60 km/h and the average spacing of the vehicles is 30 m. In the other the average speed and spacing are 90 km/h and 60 m respectively. What is the average spacing of the vehicles at the restriction where their average speed is 35 km/h?

SOLUTION In the slower of the two approaching lanes, the traffic 'density' is 1000/30 vehicles per kilometre, the speed is 60 km/h and so the traffic flux is $60 \times 1000/30$ vehicles/h $= 2000$ vehicles/h. In the faster lane, the flux is $90 \times 1000/60$ vehicles/h $= 1500$ vehicles/h and so the total influx is 3500 vehicles/h. Since the traffic flow is steady, the total efflux must also be 3500 vehicles/h, and at an average speed of 35 km/h this represents a traffic density of 100 vehicles per kilometre. Therefore the average vehicle spacing in the single lane at the restriction is 10 m.

Comments on Example 6-6

1. The behaviour of the traffic is analogous to a *supersonic* compressible flow. As it approaches the restriction, its density increases and its velocity decreases. With *subsonic* compressible flows the opposite behaviour occurs; that is, the density decreases and the velocity increases on approaching a restriction.
2. In practice, it is rare for the rate of traffic flow at a substantial restriction to be exactly equal to the rate at which traffic approaches. More commonly, a 'traffic jam' occurs and its length increases or decreases according to the relative magnitudes of traffic influx and efflux.

Example 6-7: Centrifugal blower Air flows through the rotating impeller of a small fan which raises its pressure. The impeller spins at 2000 rev/min and its radii at inlet and outlet are 50 mm and 100 mm respectively. The width of the impeller is 30 mm and the blade angle is $\alpha = 25°$. Estimate the mean air velocities at inlet and outlet if the rate of flow of air is $Q = 100$ l/s.

SOLUTION The flow through the fan is illustrated in Fig. 6-8. In (a) the inlet flow is purely radial ($\bar{V}_r \neq 0$; $\bar{V}_0 = 0$) and the outlet flow has a tangential component ($\bar{V}_r \neq 0$, $\bar{V}_0 \neq 0$). In (b) the impeller is shown in its housing to illustrate the method of collection and disposal of the air.

The circumference of the cylindrical flow section of radius r depicted in (a) and (b) is $2\pi r$, and so the area of the flow surface is $a = 2\pi rb$. Only radial components of velocity contribute to flow across this flow surface (tangential components merely imply movement around the axis) and so the volumetric rate of flow is

$$Q = a\bar{V}_r = 2\pi rb\bar{V}_r \tag{6-21}$$

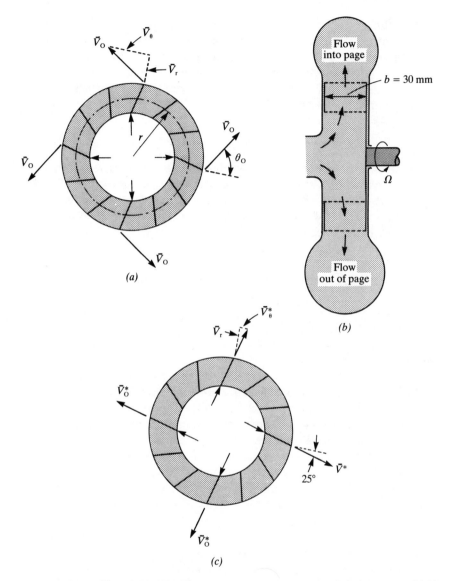

Figure 6-8 Airflows through a centrifugal blower. (a) Velocities relative to stationary axes. (b) Air flow in cross-section of blower. (c) Velocities relative to rotating axes.

For the specified geometry and rate of flow, the mean radial components of velocity at inlet and outlet are therefore

$$\bar{V}_{rI} = 8.84 \text{ m/s} \qquad \bar{V}_{rO} = 5.31 \text{ m/s}$$

Provided that nothing upstream has induced the air to spin, there will be no tangential velocity component at inlet and so the absolute velocity will be equal to the radial component. However, there will certainly be a tangential velocity component at outlet because this is the whole purpose of the impeller. To estimate its magnitude, we consider the flow relative to axes rotating with the impeller as shown in Fig. 6-8c.

Provided that there are sufficient blades, the air particles will be constrained to move parallel to them and so the approximate direction of flow is known at outlet. That is,

$$\bar{V}^*_{\theta O} = -\bar{V}_{rO} \tan 25° = -2.47 \text{ m/s} \qquad (6\text{-}22)$$

in which $\bar{V}^*_{\theta O}$ denotes the mean tangential component relative to the rotating axes and the minus sign indicates that this is clockwise. Now the axes are rotating with the impeller and so their tangential velocity at outlet is $r_O \Omega = 10.47$ m/s. Therefore, relative to the stationary axes in (a), the tangential velocity component is

$$\bar{V}_O = \bar{V}^*_\theta + r_O \Omega = 8.00 \text{ m/s} \qquad (6\text{-}23)$$

The magnitude and direction of the absolute velocity at outlet satisfy

$$\bar{V}^2 = \bar{V}_\theta^2 + \bar{V}_r^2 \qquad \text{and} \qquad \tan \theta = \bar{V}_r / \bar{V}_\theta \qquad (6\text{-}24)$$

and so $\bar{V}_O = 9.43$ m/s and $\theta_O = 58.0°$.

Comments on Example 6-7

1. It is important to choose axes carefully. In this instance a highly unsteady flow relative to conventional axes appears steady relative to rotating axes.
2. It is reasonable to assume that the flow will be approximately parallel to the blades at outlet. However, the conditions at inlet are determined before the air reaches the impeller. It is the designer's responsibility to ensure that the blade angle at inlet is approximately parallel to the oncoming flow.
3. After outlet from the impeller, the air enters the volute which leads to the final outlet. The area of the volute increases in the direction of flow along its axis because air is continually received laterally—hence the snail-like appearance of the overall unit.

Example 6-8: Rising water level in reservoir After a period of heavy rainfall, the volumetric flow rate down a river into a reservoir of surface area $A = 100$ hectares is measured at a gauging station and found to be 100 cumec. As shown in Fig. 6-9, discharge from the reservoir is over a spillway and records show that the volumetric rate of outflow Q_O (cumec) is related to the height H (metres) of the reservoir surface above the spillway crest by $Q_O = 20 H^{3/2}$ (see comment 2 below). Neglecting all other forms of inflow and outflow (reasonable after a period of heavy rainfall), estimate (a) the rate of rise of the water surface when $H = 2$ m and (b) the height of the water surface above the crest when the influx of 100 cumec causes no further change in the surface level.

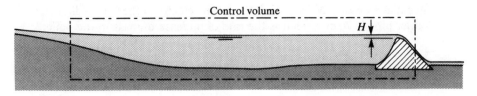

Figure 6-9 Discharge over a dam spillway.

SOLUTION

(a) When $H=2\,\text{m}$, the stipulated discharge relationship yields $Q_0 = 56.6\,\text{m}^3/\text{s}$. Since the influx is $100\,\text{m}^3/\text{s}$, the unsteady flow continuity equation (6-4) shows that the volumetric rate of increase of (incompressible) water in the reservoir is

$$\frac{\partial V_{cv}}{\partial t} = (100 - 56.6)\,\text{m}^3/\text{s} = 43.4\,\text{m}^3/\text{s} \qquad (6.25)$$

An increase of δH in the surface level in the reservoir represents a volumetric increase $A\,\delta H$ in the amount of water stored. Therefore $\partial V_{cv}/\partial t$ may be approximated by $A\,dH/dt$ in which the use of the *total* derivative d/dt indicates that the rise δH is assumed to be the same at all points on the surface. Since the surface area is a million square metres, the rate of rise of the water surface is

$$\frac{dH}{dt} = \frac{43.4}{10^6}\,\text{m/s} = 0.0434\,\text{mm/s} \qquad (6\text{-}26)$$

(b) When steady-state conditions prevail, $\partial V_{cv}/\partial t$ is zero and the rates of inflow and outflow are equal. Therefore the numerical value of H satisfies $100 = 20H^{3/2}$ and so $H = 2.92\,\text{m}$.

Comments on Example 6-8

1. For the given rate of inflow, the water surface will rise whenever H is less than $2.92\,\text{m}$ and will fall when it exceeds this value. In practice, of course, the rate of inflow does not remain constant during a flood.

2. Strictly speaking, the discharge equation is dimensionally inhomogeneous. It could be written more correctly as $Q_0 = CH^{3/2}$ in which $C = 20\,\text{m}^{3/2}/\text{s}$. Nevertheless, no numerical error will result from the use of the equation in the form $Q_0 = 20H^{3/2}$ provided that Q_0 and H are measured in cumec and metres respectively. Similar complications often arise when parameters such as the acceleration due to gravity are incorporated into coefficients.

Example 6-9: Flow through a boiler Water enters a boiler in liquid form $(\rho = 1000\,\text{kg/m}^3)$ at a rate of 1250 litres per minute. It is discharged as steam at a density of 3 kg/m^3. If the mean velocity in the 0.35 m diameter outlet pipe is 70 m/s, what is the rate of increase of mass in the boiler?

SOLUTION The volumetric influx is $Q = 1250 \text{ l/min} = 20.83 \times 10^{-3} \text{ m}^3/\text{s}$ and so the mass flux is

$$\dot{m}_1 = \rho_1 Q_1 = 20.83 \text{ kg/s} \tag{6-27}$$

The area of the outlet pipe is $a_0 = \pi D^2/4 = 0.0962 \text{ m}^2$ and so the mass flux at outlet is

$$\dot{m}_0 = \rho_0 a_0 V_0 = 20.20 \text{ kg/s} \tag{6-28}$$

If we use the continuity equation (6-5) then the rate of increase of the mass in the boiler is

$$\frac{\partial m}{\partial t} = \dot{m}_1 - \dot{m}_0 = 0.63 \text{ kg/s} \tag{6-29}$$

Comments on Example 6-9

1. As usual, the continuity equation tells us nothing at all about the total amount of water inside the boiler at any instant or about the relative proportions of liquid and vapour. Nevertheless, it enables us to determine the rate of increase of water in the boiler even though the fluid is a liquid at inlet and a gas at outlet.

2. In practice, the rate of increase of mass is most likely to occur when the heat supplied to the boiler is reduced because this leads to an increased proportion of liquid water. It also leads to a decrease in temperature and hence to an increase in the density of the steam—thus tending to bring the mass efflux into line with the influx.

Example 6-10: Arterial blood flow Figure 6-10 depicts a succession of pulses moving along a 12 mm diameter artery at a speed $c = 10 \text{ m/s}$, the dilation of the vessel being exaggerated for clarity. The blood is stationary between pulses and it moves at 0.75 m/s within a pulse. If 72 pulses arrive at any particular location every minute and each lasts for 0.25 s, what is the average rate of flow along the artery?

SOLUTION The velocity of flow at a typical flow section A, say, depends upon whether or not a pulse happens to be present. During periods between pulses (e.g. $t = t_1$ and $t = t_3$ in the figure) there is no flow. During a pulse, however, the mean velocity is 0.75 m/s and the instantaneous rate of flow is

$$Q = a\bar{V} = 0.0848 \text{ l/s} \tag{6.30}$$

The total volume of blood crossing the flow section during a pulse lasting

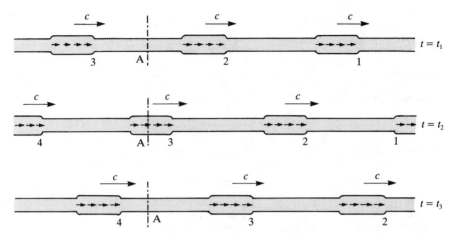

Figure 6-10 Idealized representation of arterial blood flow (the three diagrams represent successive instants).

$\Delta t_p = 0.25$ s is $Q \, \Delta t_p = 0.0212$ litres. Since 72 pulses cross the flow section in one minute, the average rate of flow is $72 \times 0.0212 \, 1/\text{min} = 1.53 \, 1/\text{min}$.

Comments on Example 6-10

1. The average rate of flow Q_{AV} in an interval Δt is simply the temporal average of the instantaneous rates of flow. It satisfies

$$Q_{AV} \Delta t = \int_0^{\Delta t} Q \, dt \qquad (6\text{-}31)$$

In this particular instance, the integration yields $Q \, \Delta t_p + 0(\Delta t - \Delta t_p)$ in which Δt_p denotes the portion of time during which a pulse is present. The temporal average need not be a constant in general.

2. Notice that the speed of travel of the pulses was not used explicitly in the solution. In practice, low wavespeeds imply supple artery walls and higher rates of flow for any particular pulse rate.

Example 6-11: Two-dimensional flow Figure 6-11 depicts a steady two-dimensional flow past a cylinder of radius R. The shape of each streamline can be represented in cylindrical polar coordinates by the equation

$$V_\infty \sin \theta \left(r - \frac{R^2}{r} \right) = \psi \qquad (6\text{-}32)$$

in which V_∞ is the uniform velocity remote from the cylinder and ψ is constant on any particular streamline. Different values of the constant ψ are applicable on different streamlines.

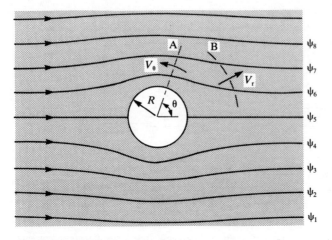

Figure 6-11 Inviscid flow past a cylinder.

(a) Verify that the radial and tangential velocity components at any position satisfy

$$V_r = \frac{1}{r}\frac{\partial \psi}{\partial \theta} \quad \text{and} \quad V_\theta = -\frac{\partial \psi}{\partial r} \tag{6-33}$$

and hence deduce a physical meaning for ψ.

(b) Show that the maximum velocity on the surface of the cylinder is equal to twice the free stream velocity.

SOLUTION

(a) The rate of flow across an elemental part of the flow section A shown in Fig. 6-11 is $\delta Q_A = V_\theta \, \delta r$. Therefore, if Eq. (6-33) is correct, the total rate of flow across the section between the streamlines 6 and 7 must be

$$\Delta Q_A = \int_6^7 \left(-\frac{\partial \psi}{\partial r} \right) dr = [-\psi]_6^7 = \psi_6 - \psi_7 \tag{6-34}$$

Similarly, for the flow section B, $\delta Q_B = V_r r \, \delta \theta$, and if (6-33) is correct then

$$\Delta Q_B = \int_6^7 \left(\frac{1}{r}\frac{\partial \psi}{\partial \theta} \right) r \, d\theta = [\psi]_6^7 = \psi_7 - \psi_6 \tag{6-35}$$

Now the positive directions of flow along the streamlines are different in these two cases. Hence, both results may be interpreted as a statement that the flow from left to right is equal to $\psi_7 - \psi_6$. It follows that we may interpret ψ as the rate of flow between any particular streamline and the streamline $\psi = 0$. In this case, it also follows that Eq. (6-33) correctly states the velocity components.

(b) At any point in the flow, Eq. (6-33) gives

$$V_r = V_\infty \cos\theta \left(1 - \frac{R^2}{r^2}\right) \quad \text{and} \quad V_\theta = -V_\infty \sin\theta \left(1 + \frac{R^2}{r^2}\right)$$

$$(6\text{-}36)$$

On the surface of the cylinder, $r = R$ and so

$$V_r = 0 \quad \text{and} \quad V_\theta = -2V_\infty \sin\theta \qquad (6\text{-}37)$$

Therefore the maximum velocity is $2V_\infty$.

Comments on Example 6-11

1. The parameter ψ is called a *stream function*. It can be defined for all two-dimensional and axi-symmetric flows, but simple analytical expressions such as Eq. (6-32) can be found for only a limited number. Nevertheless, ψ is a useful parameter, especially in certain types of numerical analysis.
2. In the special case of irrotational flows, it is also possible to define a parameter ϕ known as a *velocity potential*. In such flows, potential lines are always normal to streamlines and so the direction of flow is always normal to a potential surface.
3. Although a cylinder is not a very good approximation to the shape of a motor vehicle, it is plausible to deduce from this example that the occupants of a vehicle will detect high airspeeds just above the roof. In practice, this means that the pressure just above the roof must be lower than elsewhere—which is why the roof of a 'soft top' sports car billows outwards.

Example 6-12: Laminar flow between moving surfaces Figure 6-12 depicts a 100 mm wide rectangular block moving at a steady speed $V = 10\,\text{mm/s}$ over the flat, horizontal base of a rectangular tank of oil. The block causes oil to accumulate at the right-hand side of the tank, and so the surface level slowly increases. Estimate the rate of flow beneath the block (per unit width normal to the page) when the surface levels differ by 150 mm, expressing the result relative to (a) the tank and (b) the block. Also estimate the corresponding rate of flow if the block stops when the level difference reaches 150 mm. The density and viscosity of the oil are $875\,\text{kg/m}^3$ and $0.01\,\text{Pa s}$.

SOLUTION The velocity distribution in a laminar flow between two parallel surfaces is shown in section 4.3 to be

$$u = \frac{Vy}{H} + \frac{y(y - H)}{2\mu}\frac{\partial p}{\partial x} \qquad (6\text{-}38)$$

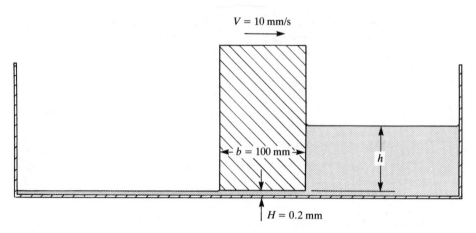

Figure 6-12 Rectangular block 'sweeping' an oil tank.

The volumetric flux through an elemental strip of height δy and width b normal to the page is $\delta Q = bu \, \delta y$ and the total flux between the plates is $Q = \int bu \, dy$. Therefore

$$Q = b \left[\frac{Vy^2}{2H} + \frac{1}{2\mu} \frac{\partial p}{\partial x} \left\{ \frac{y^3}{3} - \frac{Hy^2}{2} \right\} \right]_0^H \tag{6-39}$$

i.e.

$$Q = b \left[\frac{VH}{2} - \frac{H^3}{12\mu} \frac{\partial p}{\partial x} \right] \tag{6-40}$$

The pressure difference corresponding to an oil level difference of $h = 150$ mm is $\Delta p = \rho g h = 1.288$ kPa and so the pressure gradient beneath the block is $\partial p / \partial x = 12.88$ kPa/m. Relative to the tank, the rate of flow per unit width (normal to the page) is therefore $Q/b = 0.141$ ml/s. Relative to the block, the corresponding rate of flow is $Q/b - VH = -1.86$ ml/s.

When the block stops, $V = 0$ and Eq. (6-40) yields $Q/b = -0.859$ ml/s relative to either the tank or the block.

Comment on Example 6-12 Notice that the rate of leakage is strongly dependent upon the pressure gradient. When thin oil films are used for lubrication, it is usual for the gaps to be slightly tapered and so the movement of the block (bearing) induces large pressure gradients. These induce a high rate of leakage beneath the block, thus maintaining the oil film. The high pressures also tend to cause the block to maintain its separation from the surface over which it is sliding.

6-5 DIFFERENTIAL FORM OF THE UNIAXIAL CONTINUITY EQUATION

In all of the preceding examples, the continuity equation is used in its so-called *integral* form, that is in a form that can be applied to large control volumes. Sometimes, however, it is preferable to use a *differential* form of the equation. This is usually so, for example, when numerical solutions are sought to problems which are analytically intractable.

Before deriving the differential equation for three-dimensional flows, we consider the simpler but important case of uniaxial flows. Equation (6-5) may be applied to control volumes of elemental length such as those depicted in Fig. 6-13. Using ρ, a and $\bar{V}$ to denote values at the mid-section, the mass flow rates at inlet and outlet are

$$\dot{m}_1 = \rho a \bar{V} - \tfrac{1}{2} \frac{\partial}{\partial x}(\rho a \bar{V})\, \delta x \qquad (6\text{-}41)$$

and

$$\dot{m}_O = \rho a \bar{V} + \tfrac{1}{2} \frac{\partial}{\partial x}(\rho a \bar{V})\, \delta x \qquad (6\text{-}42)$$

and the rate of change of mass within the control volume is

$$\frac{\partial m_{cv}}{\partial t} = \frac{\partial}{\partial t}(\rho a\, \delta x) \qquad (6\text{-}43)$$

in all of which, second- and higher-order terms in δx are neglected. On substitution of these relationships into Eq. (6-5) we obtain

$$\frac{\partial}{\partial t}(\rho a\, \delta x) = -\frac{\partial}{\partial x}(\rho a \bar{V})\, \delta x \qquad (6\text{-}44)$$

or, since δx does not vary with time,

$$\boxed{\frac{\partial}{\partial t}(\rho a) = -\frac{\partial}{\partial x}(\rho a \bar{V})} \qquad (6\text{-}45)$$

Range of validity This equation is subject to the same restrictions as the integral formulation (6-12). However, since the inflow and outflow sections are only an elemental distance apart, uniaxial conditions are possible only if the flow is uniaxial throughout. Equation (6-45) cannot be used to analyse flows such as those depicted in Figs 6-4 and 6-5.

The differential formulation is very useful when a detailed knowledge of streamwise variations in flow parameters is required, for example gradually varied pipe and channel flows.

Numerical solutions of complex flows are often obtained by using computers to integrate differential equations. In this context, note that

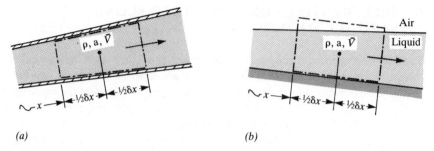

(a)   *(b)*

Figure 6-13 Elemental control volumes. (a) Pipe flow. (b) Channel flow.

Eq. (6-45) relates a single timewise derivative to a single spacewise derivative; it is known as the *conservation* form of the uniaxial differential continuity equation.

Considerable care must be exercised in the choice of control volumes, especially when the cross-sectional area of the flow is time-dependent. In unsteady free-surface flows, for instance, it is useful to ensure that the upper surface of the control volume is either wholly above or wholly below the free surface. In Fig. 6-13*b*, the control surface is chosen wholly within the air above the liquid surface. This simplifies the use of the continuity equation for the liquid because there are no inflows or outflows across the lateral surfaces of the control volume.

Steady flow When the flow is *steady*, the uniaxial continuity equation (6-45) becomes

$$\frac{\partial}{\partial x}(\rho a \bar{V})=0 \tag{6-46}$$

which may be integrated to give $\rho a \bar{V}=$ constant. This result was also obtained in section 6.3 in which it was shown to be true for the inlet and outlet sections even when there is a zone of three-dimensional flow between them.

Incompressible flow When the fluid may be regarded as incompressible, (6-45) reduces to

$$\frac{\partial a}{\partial t}=-\frac{\partial}{\partial x}(a \bar{V}) \tag{6-47}$$

When, additionally, the cross-sectional area of the flow does not vary with time,

$$\frac{\partial}{\partial x}(a \bar{V})=0 \tag{6-48}$$

which is noteworthy because it may be applied to unsteady flows and yet it does not contain a timewise derivative.

6-6 DIFFERENTIAL FORM OF THE GENERAL CONTINUITY EQUATION

Consider the elemental control volume shown in Fig. 6-14. At its centre, the fluid density is ρ and the velocity components parallel to the x, y and z axes are u, v and w respectively. The mass flux across the typical face ABCD is

$$\delta \dot{m}_{\text{ABCD}} = \left\{ \rho w + \tfrac{1}{2} \frac{\partial}{\partial z} (\rho w) \, \delta z \right\} \delta x \, \delta y \tag{6-49}$$

(NB: Only the velocity component w can contribute to the mass flux across a face perpendicular to the z-axis.) Using a similar expression for the opposite face, we find that the net *influx* across the two faces perpendicular to the z-axis is

$$\delta \dot{m}_z = - \frac{\partial}{\partial z} (\rho w) \, \delta x \, \delta y \, \delta z \tag{6-50}$$

Similar expressions may be written for the flows across the other two pairs of faces and so the overall net influx to the elemental cuboid is

$$\delta \dot{m}_1 = - \left\{ \frac{\partial}{\partial x} (\rho u) + \frac{\partial}{\partial y} (\rho v) + \frac{\partial}{\partial z} (\rho w) \right\} \delta x \, \delta y \, \delta z \tag{6-51}$$

The rate of increase of mass inside the control volume is

$$\frac{\partial m_{\text{cv}}}{\partial t} = \frac{\partial}{\partial t} (\rho \, \delta x \, \delta y \, \delta z) \tag{6-52}$$

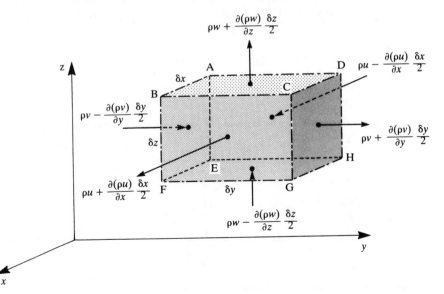

Figure 6-14 Flow through an elemental cuboid.

and so the general continuity equation (6-5) may be written in differential form as

$$\frac{\partial \rho}{\partial t} + \frac{\partial}{\partial x}(\rho u) + \frac{\partial}{\partial y}(\rho v) + \frac{\partial}{\partial z}(\rho w) = 0 \tag{6-53}$$

Incompressible flows For the special case of an incompressible flow, this becomes

$$\frac{\partial u}{\partial x} + \frac{\partial v}{\partial y} + \frac{\partial w}{\partial z} = 0 \tag{6-54}$$

which shows that the velocity cannot increase simultaneously in all three directions.

PROBLEMS

1 The rate of flow of water along a 100 mm bore hose is 0.01 m³/s. Determine the mean velocity in the hose and in its 25 mm outlet nozzle through which a fireman directs the flow at a fire.

[1.27 m/s; 20.4 m/s]

2 Steam exhausts from a 100 mm bore pipe through a 10 mm diameter nozzle. The mean velocity in the pipe is 2.4 m/s and the density of the steam at outlet is 55 per cent of its value in the pipe. What is the outlet velocity?

[436 m/s]

3 The rates of flow at various locations in the portion of a water distribution network shown in Fig. 6-15 are measured and found to be as follows:

location:	A	B	C	D	E	F	G
flow, m³/s:	1.4	1.2	0.8	0.7	0.6	0.2	0.7

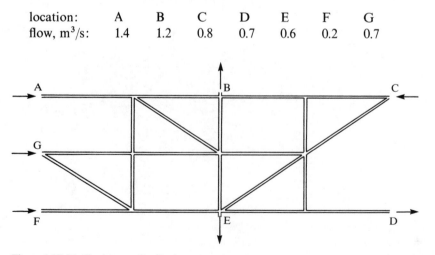

Figure 6-15 Idealized water distribution network.

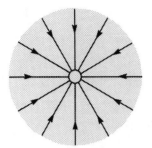 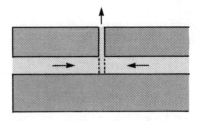

Figure 6-16 Radial flow towards a well in a confined aquifer.

Determine whether significant leakage occurs from the network between the measuring points.

[yes]

4 Natural gas flows isothermally in a long pipeline in which the temperature may be regarded as constant and equal to 12 °C. The mean velocity at an upstream section is 7 m/s and the absolute pressure is 750 kPa. Estimate the velocity 50 km downstream where the pressure is 100 kPa. Assume that the fluid is a perfect gas for which $R = 520$ J/kg K.

[52.5 m/s]

5 Water flows radially towards a well in a 3 m deep aquifer bounded by impermeable strata as shown in Fig. 6-16. If the steady rate of extraction from the well is 200 l/s, determine the velocity of the water at a radius of 80 m.

[0.133 mm/s]

6 In the laminar flow between stationary parallel plates shown in Fig. 6-17 the velocity distribution at each cross-section satisfies

$$u = ky(H - y) \qquad (6\text{-}55)$$

Show that the mean velocity $\bar{V}$ is two-thirds of the maximum velocity.

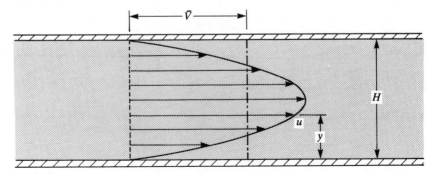

Figure 6-17 Laminar flow between stationary plates.

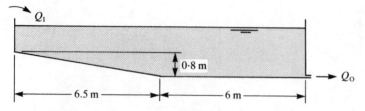

Figure 6-18 Longitudinal section of a swimming pool (distorted scale).

7 The swimming pool shown in Fig. 6-18 is 12.5 m × 5 m in plan. Water is supplied continuously at a rate of 1 l/s and drains unintentionally through an outlet at a rate Q_0 l/s which satisfies

$$Q_0 = 0.5H^{1/2} \tag{6-56}$$

where H is the height in metres of the water surface above the outlet. Estimate the time required for the water level to increase from 1 m to 2 m above the outlet. (*NB*: The integral of $1/(A - x^{1/2})$ is $2A - 2x^{1/2} - 2A \ln (A - x^{1/2})$.)

[45.5 h]

8 During exhalation, the velocity of air at a flow section in a 15 mm diameter trachea is found to vary as shown in Fig. 6-19. Determine the volume of air discharged by the lung.

[1.39 litres]

9 A displacement pump is used to deliver fresh concrete through a 100 mm bore pipe. The diameter of the piston is 200 mm and the stroke length is 400 mm. Estimate the average velocity of the concrete in the pipe if there is one stroke every 3 seconds.

[0.533 m/s]

10 Figure 6-20 depicts a line of wheeled capsules ($D_c = 360$ mm) being driven along a horizontal pipeline ($D = 400$ mm) by an airflow. If the rate of flow of air along the pipeline is 0.75 m³/s and the velocity of the capsules is 0.6 m/s, what is the mean velocity of air in the annulus around the capsules?

[28.9 m/s]

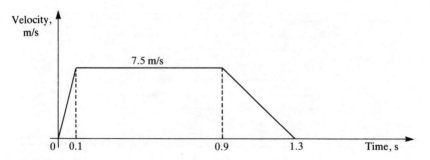

Figure 6-19 Velocity in a trachea during exhalation.

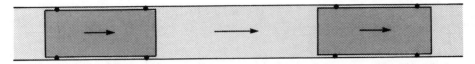

Figure 6-20 Capsular flow in a pipeline.

11 A long line of vehicles waits at a traffic signal, the distances between successive vehicles being 5 m. When the signal changes to green the first vehicle moves off and others follow at regular intervals. Assuming for simplicity that each vehicle instantly moves at a constant speed of 50 km/h from rest and that the moving vehicles are spaced at 20 m, determine (a) how many vehicles cross the signal in the 35 seconds before it changes to red and (b) how many other vehicles have begun to move by this time. Neglect the lengths of individual vehicles.

[25, 12]

12 Figure 6-21 shows the flow pattern around a cylinder moving from right to left at a steady speed V_0. The flow pattern can be represented by a stream function ψ satisfying

$$\psi = -\frac{R^2}{r} V_\infty \sin \theta \qquad (6\text{-}57)$$

By deriving an expression for the velocity distribution at any position in the flow-field and integrating it from the surface of the cylinder to infinity along $\theta = \frac{1}{2}\pi$, show that the total rate of flow from left to right exactly counterbalances the volume displaced by the cylinder. Also show that 50

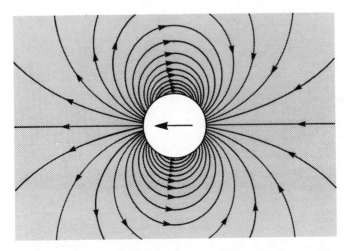

Figure 6-21 Instantaneous streamline pattern when a cylinder moves from right to left.

per cent of the flow at $\theta = \frac{1}{2}\pi$ is contained within a distance R from the cylinder.

Notes

(*a*) This flow is unsteady. Individual particle pathlines do not coincide with the streamlines. As the cylinder moves, the whole streamline pattern moves with it.

(*b*) The same flow is shown relative to the cylinder in Fig. 6-11.

FURTHER READING

Sellin, R.H.J. (1970) *Flow in Channels*, Macmillan.

Van Dyke, M. (1982) *An Album of Fluid Motion*, Parabolic Press.

SEVEN

LINEAR MOMENTUM

7-1 NEWTON'S SECOND LAW OF MOTION

In Chapter Six attention is focused on fluid motion and on rates of flow. We now turn our attention to the forces that generate these movements and we begin with Newton's second law of motion. For a system composed of elements that may be moving at different velocities, the law may be written as

$$\boxed{\text{Force} = \text{rate of change of momentum}} \qquad (7\text{-}1)$$

in which the force and the rate of change of momentum are in the same direction.

Forces The forces that act on fluids are described in Chapters Two and Four, and only a brief summary is given here. They are conveniently classified into two types, namely *body* and *surface* forces. Body forces may be due to gravitational, electrical and magnetic fields or to the inertia of the fluid relative to an accelerating frame of reference. Surface forces are principally due to normal and shear stresses, but can also include capillarity forces in the special case of a free liquid surface. In the absence of shear stresses, the normal stress acts equally in all directions and is called the fluid pressure. When shear stresses are very small, little error is introduced by continuing to regard the components of normal stress as being equal to the pressure.

For most of the requirements of this chapter the precise identity of the applied forces is not important. It is therefore convenient to describe them by

179

F_x, F_y and F_z in the x-, y- and z-directions respectively. When the body forces are considered independently, they are written for an elemental volume δV as

$$\text{Body force} = (\rho X \, \delta V, \, \rho Y \, \delta V, \, \rho Z \, \delta V) \tag{7-2}$$

in which X, Y and Z denote the body forces per unit mass in the x, y and z directions. In the usual case where the only body force is that due to gravity and the z-axis is chosen vertically upwards, $X = Y = 0$ and $Z = -g$, where g is the gravitational force per unit mass (otherwise called the gravitational acceleration).

Impulse equation Equation (7-1) relates the *instantaneous rate of change* of momentum to the instantaneous applied force. Often, however, we are more interested in the *total change* of momentum caused by an application of forces to a body for a short, but finite, time. For example, when a bat strikes a ball or when a fluid jet impinges on a vane, the net effect from an observer's point of view is that the velocity and momentum of the ball (jet) suddenly change in both magnitude and direction. To obtain the total change of momentum in the x-direction, say, we write Eq. (7-1) as

$$F_x = \frac{\mathrm{d}}{\mathrm{d}t} \, (\text{change of } x\text{-momentum}) \tag{7-3}$$

and integrate for the duration of the impact to give

$$\int F_x \, \mathrm{d}t = \text{change of } x\text{-momentum} \tag{7-4}$$

The integral $\int F \, \mathrm{d}t$ is usually called an *impulse*.

Reference axes Newton's second law is direction dependent. We must apply it only in definite directions—although we are free to choose these to suit ourselves. Often a single application in the principal direction of flow is sufficient, but in general the law yields independent relationships in three independent directions. Usually we choose these to be mutually perpendicular.

The reference axes need not be stationary, but it is usually advantageous to ensure that they do not rotate or accelerate. The development in this chapter is valid for stationary axes and for axes moving with a uniform velocity. In either case, all conditions must be specified relative to the axes, so that for instance the occupants of an aeroplane may interpret their vehicle as stationary. From their standpoint, the air far ahead is moving towards the aircraft at, say, 600 km/h. Notice that for these axes (stationary relative to the aeroplane, but not relative to the earth) the flow is steady. This simplifies the analysis enormously.

When axes rotate or accelerate, inertial terms must be included in the equation of motion in addition to anything discussed in this chapter. Their inclusion is straightforward—being analogous to the inclusion of gravitational acceleration—but it is unnecessary for present purposes.

7-2 GENERAL FORM OF THE LINEAR MOMENTUM EQUATION

Figure 7-1 depicts two flows for which the momentum principle can give useful information. In one, known as a *hydraulic jump*, the inlet and outlet flows are uniaxial and parallel to the channel bed. In the other, a liquid jet deflected through an angle α by a curved vane, the inlet and outlet flows are uniaxial but their directions are different.

Equation (7-1) can be applied to typical fluid samples such as those shown shaded at the instant t_0 in Fig. 7-1a and at $t_0 + \delta t$ in Figure 7-1b. However, the flows are so complex that we are unlikely to be able to proceed very far with the analysis. Instead, we follow the method used in the development of the continuity equation (6-5) and develop a form of Newton's second law of motion that is applicable to events inside a control volume.

Consider the flow through the arbitrarily shaped control volume depicted in Fig. 7-1. We shall apply Eq. (7-1) to the system shown shaded at the instant t_0, namely the contents of the control volume and the elemental region I. At any instant the total mass within the control volume is denoted by m_{cv} and its x-momentum is denoted by $M_{x,cv}$. The mass and x-momentum of the contents of the region I are δm_I and δM_{xI} respectively. During the interval δt the mass δm_I flows across the control surface into the control at an average rate $\delta m_I/\delta t$. In the limit, as δt approaches zero, this is the mass flux, $\dot m_I$. Similarly we define the *momentum flux* $\dot M$ across a flow section by, typically,

$$\dot M_x \equiv \frac{dM_x}{dt} \qquad (7\text{-}5)$$

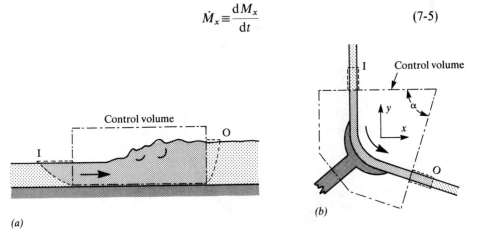

(a)

(b)

Figure 7-1 Examples of uniaxial inlet and outlet flows. (a) Hydraulic jump (liquid highlighted at $t = t_0$). (b) Jet deflection (liquid highlighted at $t = t_0 + \delta t$).

Equation (7-1) relates the force F_x acting on the system to the rate of change of momentum of the system, yielding

$$F_x = \frac{(M_{x,\text{cv}} + \delta M_{x,\text{O}})_{t_0 + \delta t} - (M_{x,\text{cv}} + \delta M_{x,\text{I}})_{t_0}}{\delta t} \tag{7-6}$$

which may be rearranged as

$$F_x = \frac{(M_{x,t_0 + \delta t} - M_{x,t_0})_{\text{cv}}}{\delta t} + \frac{\delta M_{x,\text{O},t_0 + \delta t}}{\delta t} - \frac{\delta M_{x,\text{I},t_0}}{\delta t} \tag{7-7}$$

In the limit, as δt approaches zero, this becomes the *momentum equation*:

$$\boxed{F_x = \frac{\partial M_{x,\text{cv}}}{\partial t} + \dot{M}_{x,\text{O}} - \dot{M}_{x,\text{I}}} \tag{7-8}$$

in which the suffix t_0 denoting an arbitrary instant is discarded. Independent equations can be derived similarly for the y- and z-directions.

Momentum flux It is useful to express the momentum flux $\dot{M}_x$ as a function of the density and velocity. For example, at a one-dimensional flow section of area a, the momentum flux can be shown to be equal to $\rho a V^2$. As a more general case, consider the streamline PP' at the instant $t_0 + \delta t$ in Fig. 7-2. Following the method used in section 6.2, we see that the mass within an elemental streamtube around PP' is $\delta m = \rho V \cos \theta \, \delta a \, \delta t$, in which δa is an elemental area of the outflow surface and θ is the angle between the velocity vector and the local surface normal. The x-momentum is the product of the

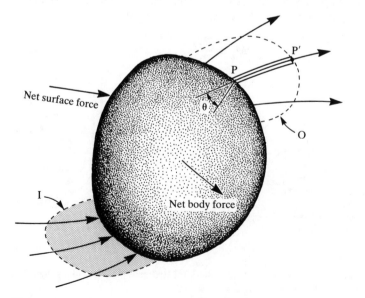

Figure 7-2 General control volume (system shaded at t_0).

mass δm and its velocity component u in the x-direction. Therefore the contribution of this streamtube to the *rate* at which x-momentum is convected across the surface is

$$\delta \dot{M}_x = u \, \delta \dot{m} = \rho u V \cos \theta \, \delta a \qquad (7\text{-}9)$$

and the total x-momentum flux is

$$\dot{M}_x = \int u \, d\dot{m} = \int \rho u V \cos \theta \, da \qquad (7\text{-}10)$$

Simplified forms of this expression are developed in section 7-3 for uniaxial and one-dimensional flow sections.

Force The force F_x is the net force acting on the contents of the control volume in the x-direction at the instant t_0. It is composed of body and surface forces, and the latter comprises both normal and shear forces over the whole of the control surface. In the particular case of a flow section, however, it is usually sufficiently accurate to approximate the surface force by regarding it as being due to the pressure alone. In this case, the force on an elemental area of the flow section is $p \, \delta a$ and the component of this in the x-direction is the required contribution to F_x.

When the control surface at a flow section is *flat*, it is useful to define a *mean pressure* and a *mean elevation* for it by

$$\bar{p} \equiv \frac{1}{a} \int p \, da \qquad \text{and} \qquad \bar{z} \equiv \frac{1}{\bar{\rho} a} \int \rho z \, da \qquad (7\text{-}11)$$

The pressure force is $\bar{p} a$ and, in the special case when the pressure distribution over the section is linear, $\bar{p}$ is the pressure at the centroid (section 3-4-1). The mean elevation $\bar{z}$ is the elevation of the centre of mass of a fluid lamina of elemental thickness at the section. When density variations in the lamina are negligible (as is usually the case), $\bar{z}$ is the elevation of the centroid.

Range of validity of the linear momentum equation Equations (7-8) and (7-10) are general expressions which are valid for any control volume. However, in common with the corresponding continuity relationships (6-5) and (6-7), their use can be somewhat cumbersome unless it is possible to make use of one of the following simplifications.

7-3 UNIAXIAL AND ONE-DIMENSIONAL FLOWS

At a uniaxial flow section where the control surface is normal to the direction of flow, $\cos \theta = 1$ and (7-10) becomes

$$\dot{M}_x = \int \rho u V \, da \qquad (7\text{-}12)$$

In the special case where the direction of flow coincides with the positive x-direction, $u = V$ (and $v = w = 0$). In general, however, $u = V \cos \phi$ where ϕ is the angle between the velocity vector and the positive x-direction. For example, in Fig. 7-1b, $\phi_1 = \frac{1}{2}\pi$ and $\phi_0 = \frac{1}{2}\pi - \alpha$ provided that the jet has no velocity component normal to the page. Since ϕ is a constant for any section, the momentum flux is

$$\dot{M}_x = \cos \phi \int \rho V^2 \, da \qquad (7\text{-}13)$$

It would be convenient to be able to replace the integral $\int \rho V^2 \, da$ by the product $\bar{\rho} a \bar{V}^2$ in which $\bar{\rho}$ and $\bar{V}$ are the mean density and velocity defined by Eqs (6-9) and (6-10). However, the substitution is not permissible in general because $\int \rho V^2 \, da$ is equal to $\bar{\rho} a \bar{V}^2$ only in the special case of one-dimensional flow. This is verified in Example 7-1; it is a consequence of the arithmetical fact that the mean of the squares of any set of unequal positive numbers exceeds the square of their mean. To allow for the difference, a *momentum flux coefficient* β is introduced, defined by

$$\beta \equiv \frac{1}{\bar{\rho} a \bar{V}^2} \int \rho V^2 \, da \qquad (7\text{-}14)$$

Using this definition and noting that $\dot{m} = \bar{\rho} a \bar{V}$, we may write the momentum flux as

$$\dot{M}_x = \beta \bar{\rho} a \bar{V}^2 \cos \phi = \beta \dot{m} \bar{V} \cos \phi \qquad (7\text{-}15)$$

and so the momentum equation for uniaxial flows is

$$F_x = \frac{\partial M_{x,cv}}{\partial t} + (\beta \dot{m} \bar{V} \cos \phi)_0 - (\beta \dot{m} \bar{V} \cos \phi)_1 \qquad (7\text{-}16)$$

Range of validity The only additional limitation of Eq. (7-16) in comparison with (7-8) is that the flows at each inlet and outlet section must be uniaxial. When there is more than one inlet and/or outlet section, an appropriate value must be defined for β at each of them.

When dealing with flows that are 'nearly' uniaxial, it is acceptable (but not precisely correct) to interpret V as the velocity component normal to the control surface. Errors due to this approximation will almost certainly be smaller than those introduced in the choice of an assumed velocity distribution for use with Eq. (7-14). It is rarely possible to deduce an accurate value for β and we commonly approximate it to unity even though this is merely the lower limit of its possible values. Further information is given in Example 7-1.

One-dimensional flow When the flow at a section may be assumed to be one dimensional, the x-momentum flux is

$$\dot{M}_x = \dot{m} \, V \cos \phi = \rho a V^2 \cos \phi \qquad (7\text{-}17)$$

in which $\cos \phi = 1$ if the flow is in the positive x-direction.

Steady flow In a steady flow, there is no rate of change of momentum within the control volume and so the uniaxial momentum equation is

$$F_x = (\beta \dot{m} \bar{V} \cos \phi)_0 - (\beta \dot{m} \bar{V} \cos \phi)_1 \qquad (7\text{-}18)$$

When there is only one inlet and one outlet, $\dot{m}_0 = \dot{m}_1$. If the flows at these sections are parallel to each other and the x-axis is chosen in this direction, the momentum equation becomes

$$F_x = \dot{m}(\beta_0 \bar{V}_0 - \beta_1 \bar{V}_1) \qquad (7\text{-}19)$$

When the flow may also be regarded as one-dimensional, this simplifies even further to

$$F_x = \dot{m}(V_0 - V_1) \qquad (7\text{-}20)$$

7-4 APPLICATIONS OF LINEAR MOMENTUM

Example 7-1: Momentum flux coefficient It can be shown from Eqs (4-43) and (4-44) that the velocity distribution in a steady laminar flow of an incompressible fluid along a pipe (Fig. 7-3) satifies

$$u = K(R^2 - r^2) \qquad (7\text{-}21)$$

in which K is a constant, R is the pipe radius and u is the axial velocity at the radius r. Evaluate the momentum flux coefficient for this flow.

SOLUTION We first deduce an expression for the mean velocity $\bar{V}$. The cross-sectional area of an elemental annulus of radius r and width δr is $\delta a = 2\pi r \, \delta r$ and the volumetric flux through it is $2\pi u r \, \delta r$. The total volumetric flux along the pipe is

$$Q = \int_0^R 2\pi u r \, dr \qquad (7\text{-}22)$$

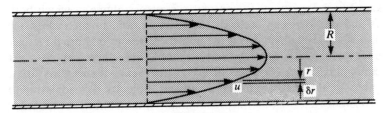

Figure 7-3 Velocity distribution in a Poiseuille flow.

For the particular velocity distribution given by Eq. (7-21) we obtain

$$Q = 2\pi K \int_0^R (R^2 r - r^3)\, dr \tag{7-23}$$

and so

$$Q = \tfrac{1}{2}\pi K R^4 \tag{7-24}$$

Therefore the mean velocity $\bar V = Q/a = Q/\pi R^2$ is

$$\bar V = \tfrac{1}{2} K R^2 \tag{7-25}$$

For an incompressible fluid, the momentum flux coefficient defined by (7-14) satisfies

$$\beta a \bar V^2 = \int_0^R u^2\, da \tag{7-26}$$

Writing πR^2 for a and $2\pi r\ \delta r$ for δa as before and using (7-25) we obtain

$$\tfrac{1}{4}\beta \pi K^2 R^6 = 2\pi \int_0^R u^2 r\, dr \tag{7-27}$$

which, on substituting (7-21) for u, gives

$$\tfrac{1}{8}\beta K^2 R^6 = K^2 \int_0^R (R^4 r - 2R^2 r^3 + r^5)\, dr \tag{7-28}$$

Therefore

$$\tfrac{1}{8}\beta R^6 = \frac{R^6}{2} - \frac{R^6}{2} + \frac{R^6}{6} \tag{7-29}$$

and so

$$\beta = 4/3 = 1.333 \tag{7-30}$$

Comments on Example 7-1

1. Since $\beta = 4/3$ in this example, it is clear that the practice of approximating it to unity is not universally satisfactory. However, this is a particularly severe case in which the mean velocity $\tfrac{1}{2}K R^2$ is only half the maximum velocity (obtained by substituting $r = 0$ into Eq. (7-21)). In a steady *turbulent* flow along a circular section pipe, the mean velocity is typically about 80 per cent of the maximum velocity and β is typically about 1.02.

2. It is sometimes convenient to consider a control volume in which both inflows and outflows can occur at the same section. In such cases, the definition (7-14) is still valid, but the numerical values of β are no longer restrained to be approximately unity. Indeed, since there can be a net momentum flux across a flow section even when there is no net mass flux, it is possible for β to be infinite. To avoid this difficulty it is advisable to choose reference axes so that the predominant flow direction at each 'inlet' and 'outlet' section is clearly defined.

3. In order to develop a feel for the extent to which β will differ from unity in typical cases it is useful to choose particular sets of numbers—e.g. 2, 3, 4, 4.5, 4, 3, 2—and to compare the mean of their squares $\frac{1}{7}(4+9+16+20.25+16+9+4)=11.18$ with the square of their mean $(\frac{1}{7}(2+3+4+4.5+4+3+2))^2 = 10.33$. The former invariably exceeds the latter.

Example 7-2: Pelton wheel turbine Figure 7-4a depicts a jet of water striking the bucket of a Pelton wheel turbine and separating into two parts, both of which are deflected through an angle α. Estimate the force exerted on the bucket by the water if the incident jet is 150 mm in diameter, the inlet and outlet velocities are 40 m/s and 38 m/s respectively, and the angle α is 165°. The density of the water is 1000 kg/m³.

SOLUTION The cross-sectional area of the incident jet is $a = \frac{1}{4}\pi \times 0.15^2$ m² $= 0.0177$ m² and so the mass flux is $\dot{m} = \rho a V = 706.9$ kg/s. The force F shown in Fig. 7-4a is the external force exerted on the bucket to prevent it

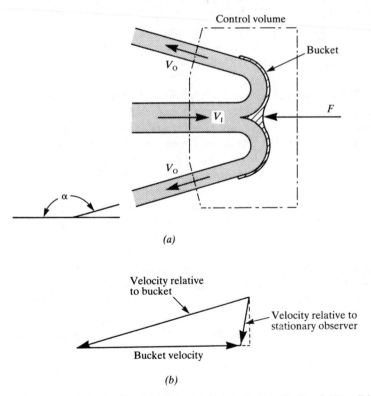

(a)

(b)

Figure 7-4 Jet deflection in a Pelton wheel turbine. (a) Pelton wheel bucket (conditions relative to the bucket). (b) Outlet velocity triangle.

accelerating. Therefore the steady flow momentum equation (7-18) may be written in the direction of the incident jet as

$$-F = \dot{m}(V_0 \cos \alpha - V_1) \tag{7-31}$$

in which the momentum flux correction coefficient is approximated to unity at the inlet and outlet sections. On substituting the known numerical values we obtain $F = 54.2$ kN. The water exerts a force of 54.2 kN on the bucket in the opposite direction to the force shown in the figure.

Comments on Example 7-2

1. The velocity at exit is less than the velocity at inlet because of friction on the bucket surface. A reduction of about 5 per cent is typical.
2. When designing a Pelton wheel it is necessary to take account of the motion of the bucket because this is fixed to the periphery of a large rotating disc. A stationary observer might see the bucket moving from left to right in Fig. 7-4a at a velocity of, say, 35 m/s. Relative to this observer, the incident velocity is 75 m/s and the component of the outlet velocity in this direction is $-V_0 \cos \alpha + 35$ m/s $= -1.71$ m/s. The outlet velocity component perpendicular to the incident jet direction is $V_0 \sin \alpha = 9.84$ m/s and so this observer sees the angle through which the jet has turned as $\cos^{-1}(-1.71/9.84) = 100°$. This result is illustrated in Fig. 7-4b, which shows the velocity vectors for both observers.

Example 7-3: Jet thrust Figure 7-5 depicts air discharging horizontally to the atmosphere through an orifice in the side of a large container in which the absolute pressure is 150 kPa and the temperature is 40 °C. Just downstream of the orifice the jet converges rapidly until it is nearly uniaxial at the section 2. It subsequently gradually expands again. This behaviour is typical of a flow through an orifice and the short, nearly parallel section is called a *vena contracta*. If the velocity at the 50 mm diameter vena contracta is 262 m/s and the atmospheric pressure is 100 kPa, estimate the force that must be applied to the container to hold it stationary. For the air use $R = 287$ J/kg K and $\gamma = 1.40$.

SOLUTION Consider the control volume shown in Fig. 7-5. Although there is a significant rate of decrease of *mass* within the control volume, the rate of change of *momentum* is very small in comparison with the momentum flux through the vena contracta. The steady-flow momentum equation (7-20) may therefore be applied in the horizontal direction to give

$$F = \dot{m}_2 V_2 = \rho_2 a_2 V_2^2 \tag{7-32}$$

The density of the stationary air within the container can be deduced from the absolute pressure of 150 kPa and the absolute temperature of

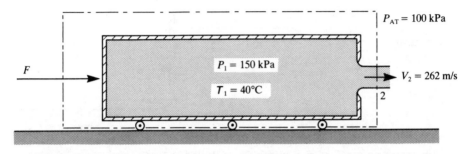

Figure 7-5 Thrust from a jet.

$(273+40)\,\text{K} = 313\,\text{K}$ using the equation of state (1-11). Thus $\rho_1 = p_1/RT_1 = 1.67\,\text{kg/m}^3$. The acceleration of the fluid particles from rest to $262\,\text{m/s}$ takes place with negligible heat transfer and negligible shear resistance. For reasons presented in Chapter Twelve, it follows that the pressures and densities at the positions 1 and 2 are related by the isentropic expression (2-31), that is

$$\rho_2 = \rho_1 \left(\frac{p_2}{p_1}\right)^{1/\gamma} \tag{7-33}$$

and so $\rho_2 = 1.25\,\text{kg/m}^3$. Therefore the applied force is

$$F = \rho_2 a_2 V_2{}^2 = 1.25 \times \tfrac{1}{4}\pi \times 0.05^2 \times 2.62^2\ N = 168\ N \tag{7-34}$$

Comments on Example 7-3
1. If the restraining force was not applied, the container would accelerate in the opposite direction to the jet velocity. This is the principle used in rocket propulsion, the high-pressure gases being produced by burning fuel. A similar principle is used in jet engines, but in this case most of the mass flow is air which passes completely through the engine; there is an inflow section as well as an outflow section.
2. The expression $F = \dot{m}_2 V_2$ indicates that the thrust of a rocket can be increased by increasing either $\dot{m}_2$ or V_2. For practical reasons concerning the size and weight of a rocket, it is important that the maximum possible thrust is obtained from each element of fuel, and so the velocity V_2 must be maximized. Unfortunately, there is a well-defined upper limit to the attainable velocity for any particular mass flux through a simple orifice. Rockets are always fitted with convergent–divergent nozzles (Fig. 6-7) to overcome this difficulty.

Example 7-4: Force on a sluice gate Figure 7-6 depicts water $(\rho = 1000\,\text{kg/m}^3)$ flowing steadily along a horizontal, rectangular channel of width $b = 3m$ and passing under a sluice gate. The depths a short

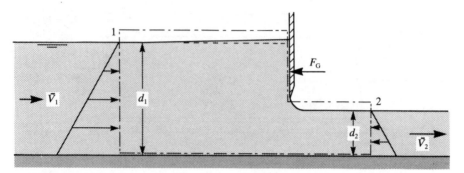

Figure 7-6 Free surface flow under a sluice gate.

distance upstream and downstream are $d_1 = 2$ m and $d_2 = 0.8$ m respectively and the upstream mean velocity is $\bar{V}_1 = 2$ m/s. Estimate the force exerted by the water on the gate assuming that the momentum flux coefficients are $\beta_1 = 1.02$ and $\beta_2 = 1.04$.

SOLUTION The flow is steady and is parallel to the bed at the flow sections 1 and 2. Using Eq. (7-19), the net force experienced by the water is

$$F_x = \dot{m}(\beta_2 \bar{V}_2 - \beta_1 \bar{V}_1) \qquad (7\text{-}35)$$

The mass flux is $\dot{m} = \rho b d_1 \bar{V}_1 = 12\,000$ kg/s and the downstream velocity can be deduced from continuity to be $\bar{V}_2 = d_1 \bar{V}_1 / d_2 = 5$ m/s. Therefore Eq. (7-35) shows the net force to be $F_x = 37.9$ kN.

There are three major contributions to the net force, namely the force exerted by the gate and the pressure forces on the upstream and downstream flow sections. Each of the latter can be regarded as the product of the area of the flow section bd and the pressure at its centroid $\frac{1}{2}\rho gd$ and so

$$F_x = \frac{1}{2}\rho gbd_1{}^2 - \frac{1}{2}\rho gbd_2{}^2 - F_G \qquad (7\text{-}36)$$

in which F_G denotes the force exerted on the water by the gate. For the given values, $F_G = 11.5$ kN. The water exerts an equal force on the gate in the opposite direction (i.e. in the direction of flow).

Comments on Example 7-4
1. It is notionally possible to determine the force by integrating the pressure distribution on the surface of the gate. However, the distribution can be predicted only from a two-dimensional study of the flow. If the approximate shape in Example 3-6 (p. 73) is assumed and the height of the gate is known, Eq. (3-55) can be used to determine the value of the exponent N. The line of action of the resultant then follows from Eq. (3-57).

2. The influence of skin friction on the bed of the channel has been neglected in this analysis. This is nearly always acceptable when considering local flow features in a pipe or channel. Shear stresses are usually much smaller than pressures and so their cumulative effect is important only in quite long lengths of channel.

Example 7-5: Surge wave Figure 7-7a depicts a surge wave travelling down a rectangular channel of small slope and width $b = 6$ m, causing the depth of flow to increase suddenly from a constant value of 1.2 m to 2.0 m. If the mean flow velocity ahead of the surge is 1.6 m/s, what is the speed of the surge along the channel?

SOLUTION Figure 7-7a depicts the unsteady-flow conditions that would be seen by a stationary observer. It is possible to use the unsteady-flow momentum equation to analyse these conditions, but it is easier to adopt an alternative approach illustrated in Fig. 7-7b. This shows the surge as seen by an observer moving down the channel alongside the surge. The figure could be recorded on a television screen if the observer continually directed a television camera at the surge. The conditions are *steady* and we can use the steady-flow momentum equation (7-20) to represent them. Thus

$$F_x = \dot{m}\{(V_2 - V_w) - (V_1 - V_w)\} \tag{7-37}$$

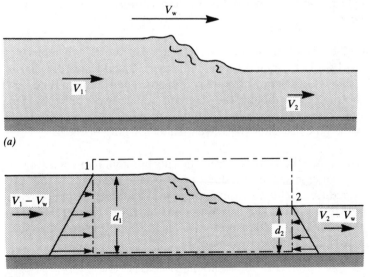

(a)

(b)

Figure 7-7 Propagation of a flood surge wave. (a) Relative to stationary axes. (b) Relative to axes moving with the surge.

in which $\qquad \dot{m} = \rho b d_1 (V_1 - V_w) = \rho b d_2 (V_2 - V_w)$ $\qquad$ (7.38)

Neglecting forces due to gravity and shear, the only contributions to F_x are the pressure forces on the flow sections 1 and 2. That is

$$F_x = \tfrac{1}{2}\rho g b d_1{}^2 - \tfrac{1}{2}\rho g b d_2{}^2 \qquad (7\text{-}39)$$

and, for the given values of breadth and depth, $F_x = 75.3\,\text{kN}$. On substituting this into Eq. (7-37) and using (7-38) to eliminate $\dot{m}$ and $(V_1 - V_w)$ we obtain $(V_2 - V_w) = \pm 5.11\,\text{m/s}$. Thus V_w is either $6.71\,\text{m/s}$ or $-3.51\,\text{m/s}$ and since the wave is moving downstream only the former of these is valid.

Comments on Example 7-5

1. It is not always possible to make an unsteady flow appear steady by viewing it relative to a second set of axes moving at a constant speed. However, analyses of wave-like motions can often be simplified considerably in this way.
2. The influences of skin friction and gravity have been neglected in this analysis. However, the prescription of a steady flow at constant depth before the arrival of the surge wave implies that their effects counterbalance in this region. Since skin friction forces are invariably small in a short length of channel, gravitational forces must also be small in this instance.
3. Common sense was used above to discount the possibility of a negative surge speed. It is shown in Chapter Twelve that this possibility could be discounted rigorously by considering the implications of the second law of thermodynamics.

Example 7-6: Acceleration from rest Figure 7-8 shows two reservoirs connected by a horizontal pipeline of diameter D and length $L = 200\,\text{m}$. Initially, a valve at the downstream end is closed and so there is no flow along the pipe even though the reservoir water levels differ by $5\,\text{m}$. Neglecting compressibility effects, estimate the initial acceleration of the water in the pipe just after the valve is suddenly fully opened.

SOLUTION While the valve is closed, the only horizontal forces acting on the control volume shown are due to hydrostatic pressures. Using Eq. (3-22), we deduce that the horizontal gauge pressure forces on the upstream and downstream control surfaces of area $\tfrac{1}{4}\pi D^2$ are $F_1 = \tfrac{1}{4}\pi D^2 \rho g(H_1 - \tfrac{1}{2}D)$ and $F_2 = \tfrac{1}{4}\pi d^2 \rho g(H_2 - \tfrac{1}{2}D)$ respectively. The force on the upstream face exceeds that on the downstream face by

$$F_1 - F_2 = \tfrac{1}{4}\pi D^2 \rho g(H_1 - H_2) \qquad (7\text{-}40)$$

but no flow occurs because the valve provides a balancing force.

When the valve is suddenly opened, the balancing force disappears and the fluid accelerates. Since the pipe is of constant cross-section, the

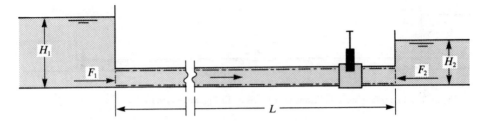

Figure 7-8 Initiation of flow between two reservoirs.

incompressible flow continuity equation (6-15) shows that $V_1 = V_2$ during the early stages of the flow when one-dimensional conditions may be assumed. It follows that the momentum flux $\rho a V^2$ is the same at both sections and so the momentum equation (7-16) gives

$$\tfrac{1}{4}\pi D^2 \rho g(H_1 - H_2) = \frac{\partial M_{cv}}{\partial t} \tag{7-41}$$

The mass of water within the control volume is $\rho a L$ and its momentum at any instant is $\rho a L V$. The rate of change of momentum is therefore $\rho a L \, dV/dt$ in which the total derivative d/dt is used because none of ρ, a, L and V varies with distance. Since $a = \tfrac{1}{4}\pi D^2$, the momentum equation becomes

$$\tfrac{1}{4}\pi D^2 \rho g(H_1 - H_2) = \tfrac{1}{4}\pi D^2 \rho L \frac{dV}{dt} \tag{7-42}$$

and so $\qquad \dfrac{dV}{dt} = \dfrac{g}{L}(H_1 - H_2) = 0.245 \, \text{m/s}^2 \tag{7-43}$

Comments on Example 7-6
1. It is emphasized that the analysis may be used only in the first few instants after the opening of the valve. As the velocity increases, the acceleration will decrease because shear stresses on the wall increase and because of local effects at the pipe inlet and outlet.
2. Notice that the result (7-43) is independent of the pipe diameter. (Why?) Readers might find it useful to rework the analysis for a *sloping* pipe of the same length. The same result should be obtained provided that the same difference in reservoir levels is assumed.
3. The above type of analysis yields a useful approximation to the behaviour of unsteady flows in pipes when valves are opened or closed sufficiently slowly. However, account must be taken of compressibility effects when rapid valve movements (or pump trips, etc.) are considered. It is easy to imagine, for example, that the first fluid particles to respond to the movement of a valve will be those close to the valve itself. The time delay

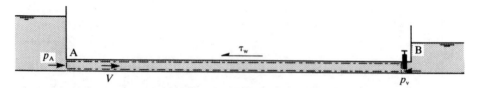

Figure 7-9 Instantaneous conditions in a pipeline during valve closure.

before other particles 'realize' what is happening is a function of the speed at which pressure waves travel through the fluid.

Example 7-7: Slow valve closure A valve at the downstream end of a 100 m long, 0.3 m bore horizontal pipeline connecting two reservoirs is closed in such a manner that the pressure just upstream of it rises linearly during a computer-controlled automatic shutdown lasting 8 seconds. Before closure, oil with a density of 850 kg/m^3 flowed along the pipeline at a steady velocity of 1.6 m/s. By making simplifying assumptions, estimate the maximum pressure rise in the pipeline during closure. The skin friction coefficient of the pipeline is 0.005.

SOLUTION For the control volume shown in Fig. 7-9 the unsteady flow momentum equation can be written in its one-dimensional form as

$$(p_A - p_v)a - \tau_w \pi D L = \rho a L \frac{dV}{dt} \tag{7-44}$$

in which p_v is the pressure just upstream of the valve. Account is taken of pressure forces on the upstream and downstream faces of the control volume and also of skin friction on its sides. The momentum flux terms at the inlet and outlet sections are omitted because they are equal (i.e. the velocity at any instant is the same at all sections in the pipeline upstream of the valve).

The stipulated linear rise in pressure upstream of the valve, starting from $p_v = p_B$ at $t = 0$ s, may be written as

$$p_v - p_B = C_1 \frac{t}{T} \tag{7-45}$$

in which C_1 is a constant of unknown value and $T = 8$ s is the closure time. By eliminating p_v from Eqs (7-44) and (7-45) we obtain

$$(p_A - p_B) - \tau_w \frac{\pi D L}{a} = \rho L \frac{dV}{dt} + C_1 \frac{t}{T} \tag{7-46}$$

which must be integrated over the period of value closure to yield the velocity history and hence the pressure history. This can be done only approximately because the value of the wall shear stress τ_w is unknown.

In order to derive an approximate relationship suitable for integration, we note that the left-hand side of (7-46) is identically zero during the steady flow preceding the valve closure. Also, it is equal to $(p_A - p_B)$ at the instant of closure when the velocity (and hence the shear stress) is zero. As a first approximation, we may assume that the left-hand side of (7-46) increases linearly during the closure—i.e. it is equal to $(p_A - p_B)(t/T)$. With this substitution, Eq. (7-46) may be written as

$$\rho L \frac{dV}{dt} \simeq (p_A - p_B - C_1)\frac{t}{T} \qquad (7\text{-}47)$$

and integrated to give

$$\rho L V \simeq (p_A - p_B - C_1)\frac{t^2}{2T} + C_2 \qquad (7\text{-}48)$$

in which C_2 is another constant.

The constants C_1 and C_2 may be determined from the boundary conditions, namely $V = V_0$ at $t = 0$ and $V = 0$ at $t = T$. Hence the velocity history (7-48) satisfies

$$V \simeq V_0 \left(1 - \frac{t^2}{T^2}\right) \qquad (7\text{-}49)$$

and the pressure history (7-45) satisfies

$$p_v - p_B \simeq \left(p_A - p_B + \frac{2\rho L V_0}{T}\right)\frac{t}{T} \qquad (7\text{-}50)$$

To evaluate these expressions, we note that the left-hand side of (7-46) is zero in the initial steady-state flow. Thus

$$p_A - p_B = \tau_w \frac{\pi D L}{a} = 4f \frac{L}{D} \tfrac{1}{2}\rho V_0^2 = 7.25 \text{ kPa} \qquad (7\text{-}51)$$

Also $2\rho L V_0/T = 34.0 \text{ kPa}$ and so the maximum pressure rise, at $t = T$, is $p_v - p_B = 41.3 \text{ kPa}$.

Comments on Example 7-7

1. Equation (7-49) can be used to plot a graph of V^2 against time as shown in Fig. 7-10. Since this is approximately linear and the wall shear stress τ_w is roughly proportional to V^2, the assumed relationship (7-47) is approximately valid.

2. The analysis is valid only until the instant of closure, at which time there is a pressure gradient in the pipeline (p_A at the reservoir rising to p_v at the valve) but the flow is stationary. Further flow disturbances will occur while the pressures in the pipeline subsequently equalize.

3. In the derivation, the pressure inside the pipeline close to reservoir A has been approximated to p_A, the hydrostatic pressure at the same depth in the

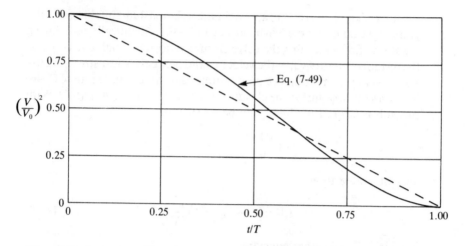

Figure 7-10 Comparison of Eq. (7-49) with the assumed linear distribution.

reservoir. Strictly, however, account should be taken of local inertial and resistance effects that cause reductions in the pressure. Nevertheless (a) these reductions are often small in comparison with pressure variations due to skin friction and/or accelerations and (b) account can be taken of their influence by introducing additional terms that do not complicate the method of solution. These effects are discussed in Chapter Nine.

4. In all cases of unsteady flow in pipelines, it is necessary to be aware of the possible occurrence of *water-hammer*. This is the name given to the propagation of pressure waves analogous to surge waves in channels. Water-hammer waves travel along a pipeline at the speed of sound and they can be highly destructive. The conditions described in the present example were chosen to ensure that the influence of pressure wave action could be neglected. This would not have been acceptable, however, if the assumed valve closure had occurred significantly more quickly (Example 7-10, p. 201).

Example 7-8: Pressure recovery in a diffuser Derive the following relationship for an inviscid, incompressible flow along a tapered pipe assuming one-dimensional flow at each location:

$$(p + \tfrac{1}{2}\rho V^2 + \rho gz)_1 - (p + \tfrac{1}{2}\rho V^2 + \rho gz)_2 = \rho \int_1^2 \frac{\partial V}{\partial t} dx \qquad (7\text{-}52)$$

Hence estimate the piezometric head increase in a water turbine diffuser with inlet and outlet diameters of 1 m and 2 m when the steady rate of flow is 15 cumec.

SOLUTION Consider the elemental control volume shown in Fig. 7-11. The x-direction components of the forces acting on it are

(a) pressure forces on the flow sections:

$$\left\{ pa - \frac{1}{2}\frac{\partial(pa)}{\partial x}\,\delta x \right\} \quad \text{and} \quad -\left\{ pa + \frac{1}{2}\frac{\partial(pa)}{\partial x}\,\delta x \right\}$$

(b) pressure forces on the sides: $\quad p\dfrac{\partial a}{\partial x}\,\delta x$

(c) the weight $(-W\sin\theta)$: $\quad -\rho g a\,\delta x\dfrac{\partial z}{\partial x}$

in which p, ρ and a denote values at the mid-section. Thus the total force acting on the element in this direction is

$$\delta F_x = -\frac{\partial(pa)}{\partial x}\,\delta x + p\frac{\partial a}{\partial x}\,\delta x - \rho g a\,\delta x\frac{\partial z}{\partial x} \tag{7-53}$$

i.e.

$$\delta F_x = -\left\{ \frac{\partial p}{\partial x} + \rho g\frac{\partial z}{\partial x} \right\} a\,\delta x \tag{7-54}$$

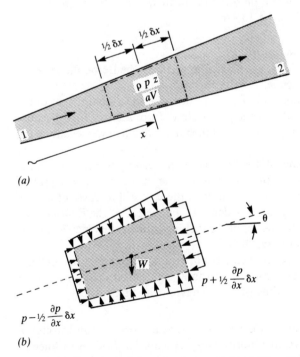

(a)

(b)

Figure 7-11 Flow through an inclined, tapered pipe. (a) Location of typical control volume. (b) Forces acting on the control volume.

The rate of change of x-momentum within the control volume is $(\partial/\partial t)/(\rho a V\,\delta x)$ and the excess of the momentum efflux over the influx is $\dot{m}(\partial V/\partial x)\,\delta x$. Therefore the momentum equation (7-16) gives

$$-\left\{\frac{\partial p}{\partial x}+\rho g\frac{\partial z}{\partial x}\right\}a\,\delta x=\rho a\,\delta x\frac{\partial V}{\partial t}+\rho a V\frac{\partial V}{\partial x}\,\delta x \qquad (7\text{-}55)$$

After dividing throughout by $a\,\delta x$ and noting that $V\,\partial V/\partial x=\tfrac{1}{2}\partial(V^2)/\partial x$, we may write this equation in the form

$$\frac{\partial}{\partial x}\{p+\tfrac{1}{2}\rho V^2+\rho gz\}+\rho\frac{\partial V}{\partial t}=0 \qquad (7\text{-}56)$$

which on integration with respect to x yields (7-52). For the special case of steady flow, the result may be written as

$$\left(\frac{p}{\rho g}+\frac{V^2}{2g}+z\right)_1=\left(\frac{p}{\rho g}+\frac{V^2}{2g}+z\right)_2 \qquad (7\text{-}57)$$

For the turbine diffuser, the inlet and outlet diameters are $1\,\text{m}$ and $2\,\text{m}$ respectively and the corresponding areas are $0.785\,\text{m}^2$ and $3.14\,\text{m}^2$. The corresponding velocities at a flow rate of $15\,\text{m}^3/\text{s}$ are $19.1\,\text{m/s}$ and $4.77\,\text{m/s}$ and so the increase in the piezometric head $(p/\rho g+z)_2-(p/\rho g+z)_1$ is found from Eq (7-57) to be $13.9\,\text{m}$.

Comments on Example 7-8
1. Equations (7-52) and (7-57) are particular forms of the Bernoulli equation developed in Chapter Nine.
2. The use of the expression $p\,(\partial a/\partial x)\,\delta x$ to describe the x-component of the pressure force on the sides of the pipe follows from section 3-4-1 because $\delta a=(\partial a/\partial x)\,\delta x$ is the projected area of the pipe wall when viewed along the x-axis.
3. The purpose of a turbine diffuser is to 'convert' velocity to pressure. In the absence of the diffuser, the turbine would waste the high kinetic energy of the water. In the presence of the diffuser, there is still a waste of kinetic energy, but it is not so great. Incidentally, one effect of the diffuser is to reduce the pressure at the outlet of the turbine runner, thus increasing the head difference across the machine.
4. Diffusers are never 100 per cent efficient in practice. Boundary layer effects such as skin friction and, more important, separation cause significant variations from the one-dimensional behaviour assumed above.

Example 7-9: Lubrication Figure 7-12 shows a cross-section through a rectangular body of width b (normal to the page). The body is inclined at an angle θ to a flat plate moving at a steady velocity V in the negative x-direction. Skin friction on the surface of the plate causes fluid to be dragged through the narrow gap below the body.

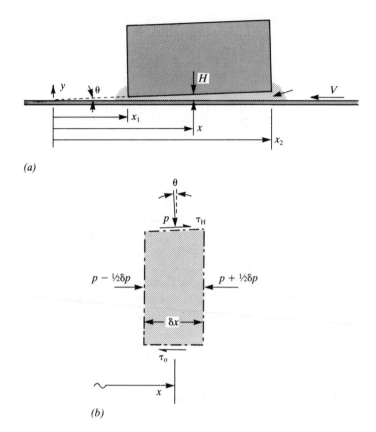

(a)

(b)

Figure 7-12 Flow through a tapered gap. (a) Overall geometry. (b) Elemental control volume.

Derive an expression describing the pressure distribution along the gap and hence estimate the uplift on the body. Assume that the flow is laminar and incompressible, and that the velocity distribution at any section is parabolic. The pressure on the other surfaces of the block is equal to p_0.

SOLUTION We begin by considering the conditions at any particular cross-section. It is easily verified that the only parabolic velocity distribution which satisfies $u = -V$ at $y = 0$ and $u = 0$ at $y = H$ is

$$u = -V\left(1 - \frac{y}{H}\right) + K(y^2 - Hy) \qquad (7\text{-}58)$$

in which K is a constant. The volumetric flux Q along the gap can be determined by integrating this expression, and is found to be

$$Q = \int_0^H bu\,dy = -bH(\tfrac{1}{2}V + \tfrac{1}{6}KH^2) \qquad (7\text{-}59)$$

By using this relationship to eliminate the constant K from Eq. (7-58), the parabolic velocity distribution may be written as

$$u = -V + \left(\frac{6Q}{bH} + 4V\right)\frac{y}{H} - \left(\frac{6Q}{bH} + 3V\right)\left(\frac{y}{H}\right)^2 \tag{7-60}$$

in which Q/bH is the mean velocity in the cross-section. A second integration yields the momentum flux $\dot{M}$, which after a little manipulation can be shown to be

$$\dot{M} = \int_0^H \rho b u^2 \, dy = \frac{\rho b H}{15}\left\{2V^2 + \frac{3VQ}{bH} + 18\left(\frac{Q}{bH}\right)^2\right\} \tag{7-61}$$

The axial pressure gradient at any section can be estimated by applying the momentum equation to the elemental control volume shown in Fig. 7-12b, in which the shear stresses τ_H and τ_0 are drawn in their mathematically positive directions. The pressure force on the left-hand face is $pbH - \frac{1}{2}(\partial/\partial x)(pbH)\,\delta x$ and the corresponding force on the right-hand face is $pbH + \frac{1}{2}(\partial/\partial x)(pbH)\,\delta x$. The axial component of the pressure force on the inclined face is $p(\partial/\partial x)(bH)\,\delta x$ and the net pressure force is the sum of these three contributions. Therefore the steady-flow momentum equation yields

$$-bH\frac{\partial p}{\partial x}\,\delta x + (\tau_H - \tau_0)b\,\delta x = \frac{\partial \dot{M}}{\partial x}\,\delta x \tag{7-62}$$

in which $(\partial \dot{M}/\partial x)\,\delta x$ represents the increase in the x-momentum flux between the sections x and $x + \delta x$.

Since the flow is laminar, the shear stresses can be deduced from Eq. 7-60 using $\tau = \mu(\partial u/\partial y)$. This leads to

$$\tau_H - \tau_0 = -\mu\left(\frac{12Q}{bH^2} + \frac{6V}{H}\right) \tag{7-63}$$

In principle, the momentum flux term in Eq. (7-62) could be deduced by differentiating Eq. (7-61), but we shall assume that it is negligible (see comment 4 below). In this case Eq. (7-62) becomes

$$\frac{\partial p}{\partial x} = -6\mu\left(\frac{2Q}{bx^3 \tan^3 \theta} + \frac{V}{x^2 \tan^2 \theta}\right) \tag{7-64}$$

in which the gap size H has been replaced by $x \tan \theta$. On integrating with respect to x, the pressure distribution along the gap is found to be

$$p = \frac{6\mu}{\tan^2 \theta}\left(\frac{Q}{bx^2 \tan \theta} + \frac{V}{x}\right) + C \tag{7-65}$$

The constant of integration C and the rate of flow Q can both be determined by substituting $p = p_0$ at $x = x_1$ and at $x = x_2$, giving

$$Q = -bV \tan \theta \left(\frac{x_1 x_2}{x_1 + x_2} \right) \qquad (7\text{-}66)$$

and
$$p - p_0 = \frac{6\mu V}{\tan^2 \theta} \left\{ \frac{1}{x} - \frac{(x_1 x_2 / x^2 - 1)}{(x_1 + x_2)} \right\} \qquad (7\text{-}67)$$

which is the required pressure distribution. The lift force on the block is

$$F_y = \int_{x_1}^{x_1} (p - p_0) b \, dx = \frac{6b\mu V}{\tan^2 \theta} \left\{ \frac{2(x_1 - x_2)}{(x_1 + x_2)} + \ln \left(\frac{x_2}{x_1} \right) \right\} \qquad (7\text{-}68)$$

Comments on Example 7-9

1. It is instructive to consider a typical numerical example. Suppose that the body length $(x_2 - x_1) = 0.2$ m, the gaps at the ends of the taper are 0.05 mm and 0.10 mm, the fluid viscosity is 0.05 Pa s and the plate velocity is 3 m/s. Geometrical considerations show that $\tan \theta = 0.00025$, $x_1 = 0.2$ m and $x_2 = 0.4$ m, and so the uplift per metre width given by Eq. (7-68) is $F_Y = 19.6$ MN. Thus very high loads can be supported by this mechanism.
2. This is the general principle underlying lubrication theory. The geometrical configuration in Fig. 7-12 resembles a slipper bearing. Alternatively, by imagining the plate to be rolled up into a circle around an eccentrically positioned cylindrical shaft, we see that the essential features of a journal bearing are also illustrated.
3. The above conditions can be steady only if a force is applied to the plate to resist the skin friction force on its surface. The magnitude of this force (per unit width normal to the page) may be determined from $F_x = \int \tau_w \, dx$ in which $\tau_w = \mu \, du/dy$ is evaluated at the plate surface. The power absorbed is $\dot{W} = F_x V$.
4. It is assumed in the analysis that the momentum flux term in Eq. (7-62) may be neglected. To investigate the validity of this assumption, we can differentiate Eq. (7-61) with respect to x and compare the result with $bH(\partial p/\partial x)$. Readers may verify that, for the numerical values used in comment 1, $\partial \dot{M}/\partial x$ is much smaller than $bH(\partial p/\partial x)$. (Assume any realistic value of the density.)

Example 7-10: Water-hammer Figure 7-13a depicts water ($\rho = 1000$ kg/m³) flowing steadily along a rigid, horizontal pipeline in which a control valve is fully open. Figure 7-13b shows the conditions in the same section of pipe a few moments after the control valve has been closed suddenly. The flow is stationary on both sides of the valve itself and pressure wavefronts are rushing upstream and downstream (at the speed of sound, $c = 1350$ m/s) to communicate news of the closure to the remote fluid. Estimate the pressure force on the valve if the initial steady velocity in the 0.3 m diameter pipeline was $V_0 = 1.4$ m/s. Assume one-dimensional conditions and neglect skin friction.

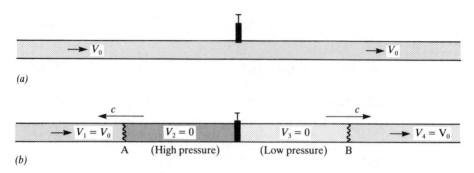

(a)

(b)

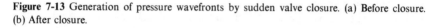

A   (High pressure)   (Low pressure)   B

Figure 7-13 Generation of pressure wavefronts by sudden valve closure. (a) Before closure.
(b) After closure.

SOLUTION Following the method adopted in Example 7-5, we observe
each pressure wavefront in turn relative to axes moving with it. These are
shown in Fig. 7-14a and b and they are steady flows in both cases. That is,
in front of each wavefront the conditions are the same as they were before
valve closure. Behind each wavefront the conditions are the same as at the
valve. All arrows in Fig. 7-14 are drawn in the positive x-direction, but the
fluid is moving from right to left in case (b). The net forces on the control
volumes are $(p_1 - p_2)a$ and $(p_3 - p_4)a$ respectively and we may write
$p_1 = p_4 = p_0$, the original pressure. Therefore the steady-flow momentum
equation (7-20) gives

$$(p_0 - p_2)a = \dot{m}_A\{c - (V_0 + c)\} = -\dot{m}_A V_0 \qquad (7\text{-}69)$$

and
$$(p_3 - p_0)a = \dot{m}_B\{(V_0 - c) - (-c)\} = \dot{m}_B V_0 \qquad (7\text{-}70)$$

Since the mass fluxes are approximately $\dot{m}_A \approx \rho a c$ and $\dot{m}_B \approx -\rho a c$
respectively, we obtain

$$(p_2 - p_0)a \simeq \rho a c V_0 \qquad (7\text{-}71)$$

and
$$(p_3 - p_0)a \simeq -\rho a c V_0 \qquad (7\text{-}72)$$

For the particular numerical values in this example, $a = 0.707\,\text{m}^2$ and so
$(p_2 - p_0)a \approx (p_0 - p_3)a \approx 134\,\text{MN}$. The force on the valve is $(p_2 - p_3)a$
$= 2.67\,\text{MN}$.

Comments on Example 7-10
1. Equations (7-71) and (7-72) show that the pressure changes across the
   wavefronts are independent of the cross-sectional area of the pipe. They
   depend upon the fluid density, the speed of sound and the velocity changes
   and are equal to 1.89 MPa in this example.
2. The predicted force on the valve is very large. In practice, the valve will

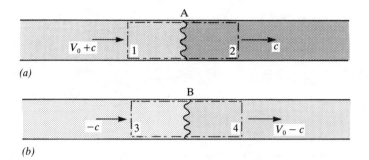

Figure 7-14 Velocities relative to axes moving with the wavefronts. (a) Upstream wavefront. (b) Downstream wavefront.

move in response to the force (unless it is remarkably rigidly supported) and stress waves will propagate axially along the pipe walls.

3. The pressure rise predicted by Eq. (7-71) is a good approximation. The pressure drop predicted by Eq. (7-72) may also be a good approximation, but only if the pressures are not so low that cavitation occurs. The minimum possible pressure p_3 is zero (or slightly positive because of vapour).

4. Because of the very large pressure changes associated with water-hammer, it is good practice to avoid sudden valve closures whenever possible. It can be especially important to avoid the development of sub-atmospheric pressures inside pipes with little buckling resistance.

7-5 DIFFERENTIAL FORM OF THE UNIAXIAL MOMENTUM EQUATION

The uniaxial momentum equation is now developed for control volumes of elemental length such as those depicted in Fig. 7-15. Using β, $\dot{m}$, ρ, a and $\bar{V}$ to denote values at the mid-section, the momentum fluxes at the inlet and outlet sections are

$$\dot{M}_{x,1} = \beta\dot{m}\bar{V} - \tfrac{1}{2}\frac{\partial}{\partial x}(\beta\dot{m}\bar{V})\,\delta x \tag{7-73}$$

$$\dot{M}_{x,0} = \beta\dot{m}\bar{V} + \tfrac{1}{2}\frac{\partial}{\partial x}(\beta\dot{m}\bar{V})\,\delta x \tag{7-74}$$

and the rate of change of momentum within the control volume is

$$\frac{\partial M_{x,\text{cv}}}{\partial t} = \frac{\partial}{\partial t}(\rho a\bar{V}\,\delta x) \tag{7-75}$$

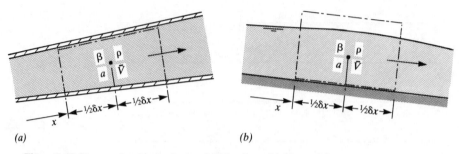

Figure 7-15 Elemental control volumes. (a) Pipe flow. (b) Channel flow.

in all of which, second- and higher-order terms in δx are neglected. On substitution of these relationships into Eq. (7-8) and writing δF_x for the net force on the elemental control volume, we obtain

$$\delta F_x = \frac{\partial}{\partial t}(\rho a \bar{V}\, \delta x) + \frac{\partial}{\partial x}(\beta \dot{m} \bar{V})\, \delta x \tag{7-76}$$

In the limit, after dividing by δx and allowing it to approach zero, we obtain

$$\boxed{\frac{\partial F_x}{\partial x} = \frac{\partial}{\partial t}(\rho a \bar{V}) + \frac{\partial}{\partial x}(\beta \dot{m} \bar{V})} \tag{7-77}$$

This form of the momentum equation is particularly useful in numerical computations because it can be written in the *conservation* form:

$$\frac{\partial}{\partial t}(\rho a \bar{V}) = \frac{\partial}{\partial x}(F_x - \beta \dot{m} \bar{V}) \tag{7-78}$$

However, it is not the form in which the momentum equation is most commonly used. The preferred form is obtained by regarding $\rho a \bar{V}$ as the product of ρa and $\bar{V}$ and treating $\beta \dot{m} \bar{V}$ as the product of $\dot{m}$ and $\beta \bar{V}$ to give

$$\frac{\partial F_x}{\partial x} = \rho a \frac{\partial \bar{V}}{\partial t} + \bar{V} \frac{\partial(\rho a)}{\partial t} + \beta \bar{V} \frac{\partial \dot{m}}{\partial x} + \dot{m} \frac{\partial(\beta \bar{V})}{\partial x} \tag{7-79}$$

Since $\dot{m} = \rho a \bar{V}$, multiplication of the continuity equation (6-45) by $\bar{V}$ yields

$$\bar{V} \frac{\partial(\rho a)}{\partial t} + \bar{V} \frac{\partial \dot{m}}{\partial x} = 0 \tag{7-80}$$

On subtracting this from (7-79) we obtain

$$\frac{\partial F_x}{\partial x} = \rho a \frac{\partial \bar{V}}{\partial t} + (\beta - 1)\bar{V} \frac{\partial \dot{m}}{\partial x} + \dot{m} \frac{\partial(\beta \bar{V})}{\partial x} \tag{7-81}$$

which is as general as Eq. (7-77) but has the advantage that it contains no timewise derivatives of ρ or a.

Range of validity Equations (7-77) and (7-81) are less general than the corresponding integral formulation (7-16) because they are valid only when the whole of the flow is uniaxial and the x-axis is chosen in the direction of flow. Also, strictly, the density ρ should be interpreted throughout as the mean density defined by Eq. (6-9).

One-dimensional flow When the flow is one-dimensional, $\beta = 1$ and so the substitution of $\rho a V$ for $\dot{m}$ in (7-81) yields

$$\frac{\partial F_x}{\partial x} = \rho a \left(\frac{\partial V}{\partial t} + V \frac{\partial V}{\partial x} \right) \tag{7-82}$$

Along a particle path, $dx/dt = V$ and so the expression in parentheses may be written as dV/dt. This equation could therefore be obtained directly by applying Newton's second law of motion (7-1) to an infinitesimal fluid element of area a.

7-6 DIFFERENTIAL FORM OF THE GENERAL MOMENTUM EQUATION

Consider the elemental control volume shown in Fig. 7-16a in which the surface shear stresses acting in the z-direction are denoted by σ_z, τ_{yz}, and τ_{xz} at the centre. For generality, the normal stresses σ_x, σ_y and σ_z are not assumed to be equal to the pressure p. The force acting on the lower face of the cuboid in the z-direction is the product of the normal stress $\{\sigma_z - \frac{1}{2}(\partial\sigma_z/\partial z)\,\delta z\}$ and the area of the face $\delta x\,\delta y$. Similarly, the normal force acting in the opposite direction on the upper face of the cuboid is $\{\sigma_z + \frac{1}{2}(\partial\sigma_z/\partial z)\,\delta z\}\,\delta x\,\delta y$. The *net* force on the two faces in the positive z-direction is therefore $-(\partial\sigma_z/\partial z)\,\delta x\,\delta y\,\delta z$. The net forces due to shear stresses on the lateral faces of the cuboid are found similarly and the body force (see section 7-1) is denoted by $\rho Z\,\delta x\,\delta y\,\delta z$. The total force acting on the fluid in the control volume in the z-direction is

$$F_z = \left(\rho Z + \frac{\partial \tau_{xz}}{\partial x} + \frac{\partial \tau_{yz}}{\partial y} - \frac{\partial \sigma_z}{\partial z} \right) \delta x\,\delta y\,\delta z \tag{7-83}$$

The mass flux per unit area across each face of the control volume is shown in Fig. 7-16b in which the arrows denote the positive direction of the flux. The associated z-momentum fluxes are deduced as follows. The velocity components at the centre of the cuboid are u, v and w in the x-, y- and z-directions respectively. The *mass* flux across a central plane normal to, say, the x-axis (i.e. across a y-z plane) is $\delta\dot{m}_x = \rho u\,\delta y\,\delta z$ and the x-, y- and z-*momentum* fluxes associated with the velocity components are $u\,\delta\dot{m}_x, v\,\delta\dot{m}_x$ and $w\,\delta\dot{m}_x$, that is $\rho u^2\,\delta y\,\delta z$, $\rho uv\,\delta y\,\delta z$ and $\rho uw\,\delta y\,\delta z$ respectively. The z-momentum efflux

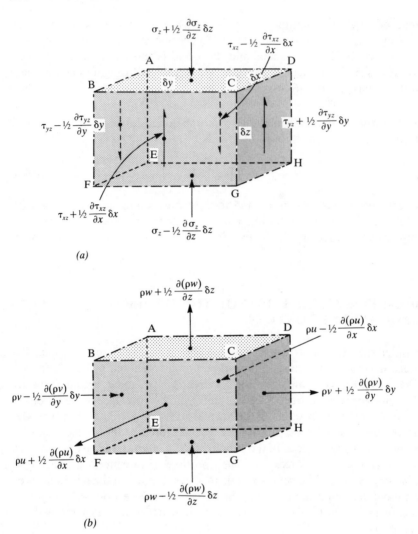

Figure 7-16 Flow through an elemental cuboid. (a) Surface stresses. (b) Mass fluxes per unit area.

across the face CDHG is $\{\rho uw + \frac{1}{2}(\partial/\partial x)(\rho uw)\,\delta x\}\,\delta y\,\delta z$ and the excess of this over the influx across the opposite face is $(\partial/\partial x)(\rho uw)\,\delta x\,\delta y\,\delta z$. For all six faces, the net z-momentum efflux is

$$\dot{M}_{z,0} - \dot{M}_{z,1} = \left\{\frac{\partial}{\partial x}(\rho uw) + \frac{\partial}{\partial y}(\rho vw) + \frac{\partial}{\partial z}(\rho w^2)\right\}\delta x\,\delta y\,\delta z \qquad (7\text{-}84)$$

Since the rate of change of z-momentum within the control volume is $(\partial/\partial t)(\rho w)\,\delta x\,\delta y\,\delta z$, the general z-momentum equation is

$$\rho Z + \frac{\partial \tau_{xz}}{\partial x} + \frac{\partial \tau_{yz}}{\partial y} - \frac{\partial \sigma_z}{\partial z} = \frac{\partial}{\partial t}(\rho w) + \frac{\partial}{\partial x}(\rho u w) + \frac{\partial}{\partial y}(\rho v w) + \frac{\partial}{\partial z}(\rho w^2) \quad (7\text{-}85)$$

The terms on the right-hand side of this equation can be regarded as the differentials of the products of w and ρ, ρu, ρv and ρw respectively. They can therefore be expanded as

$$w \left\{ \frac{\partial \rho}{\partial t} + \frac{\partial}{\partial x}(\rho u) + \frac{\partial}{\partial y}(\rho v) + \frac{\partial}{\partial z}(\rho w) \right\}$$

$$+ \rho \left\{ \frac{\partial w}{\partial t} + u \frac{\partial w}{\partial x} + v \frac{\partial w}{\partial y} + w \frac{\partial w}{\partial z} \right\} \quad (7\text{-}86)$$

in which the first expression in brackets is identically zero because of the continuity equation (6-53). Thus the general form of the z-momentum equation, together with similar expressions for x- and y-momentum, is

$$\rho X - \frac{\partial \sigma_x}{\partial x} + \frac{\partial \tau_{yx}}{\partial y} + \frac{\partial \tau_{zx}}{\partial x} = \rho \left\{ \frac{\partial u}{\partial t} + u \frac{\partial u}{\partial x} + v \frac{\partial u}{\partial y} + w \frac{\partial u}{\partial z} \right\}$$

$$\rho Y + \frac{\partial \tau_{xy}}{\partial x} - \frac{\partial \sigma_y}{\partial y} + \frac{\partial \tau_{zy}}{\partial z} = \rho \left\{ \frac{\partial v}{\partial t} + u \frac{\partial v}{\partial x} + v \frac{\partial v}{\partial y} + w \frac{\partial v}{\partial z} \right\} \quad (7\text{-}87)$$

$$\rho Z + \frac{\partial \tau_{xz}}{\partial x} + \frac{\partial \tau_{yz}}{\partial y} - \frac{\partial \sigma_z}{\partial z} = \rho \left\{ \frac{\partial w}{\partial t} + u \frac{\partial w}{\partial x} + v \frac{\partial w}{\partial y} + w \frac{\partial w}{\partial z} \right\}$$

Range of validity These equations are valid for all fluid flows. However, they cannot be integrated unless additional relationships are known for the stresses.

Inviscid flow When the fluid is assumed to be inviscid, the shear stresses vanish and the normal stresses are equal to the pressure. In this case, the Eqs (7-87) reduce to the *Euler* equations,

$$X - \frac{1}{\rho} \frac{\partial p}{\partial x} = \frac{\partial u}{\partial t} + u \frac{\partial u}{\partial x} + v \frac{\partial u}{\partial y} + w \frac{\partial u}{\partial z}$$

$$Y - \frac{1}{\rho} \frac{\partial p}{\partial y} = \frac{\partial v}{\partial t} + u \frac{\partial v}{\partial x} + v \frac{\partial v}{\partial y} + w \frac{\partial v}{\partial z} \quad (7\text{-}88)$$

$$Z - \frac{1}{\rho} \frac{\partial p}{\partial z} = \frac{\partial w}{\partial t} + u \frac{\partial w}{\partial x} + v \frac{\partial w}{\partial y} + w \frac{\partial w}{\partial z}$$

Viscous incompressible flow For the laminar flow of an incompressible Newtonian fluid, the normal and shear stresses may be replaced by (4-79) and (4-82). After making use of the continuity equation (6-53), we obtain

$$X - \frac{1}{\rho}\frac{\partial p}{\partial x} + v\left(\frac{\partial^2 u}{\partial x^2} + \frac{\partial^2 u}{\partial y^2} + \frac{\partial^2 u}{\partial z^2}\right) = \frac{\partial u}{\partial t} + u\frac{\partial u}{\partial x} + v\frac{\partial u}{\partial y} + w\frac{\partial u}{\partial z}$$

$$Y - \frac{1}{\rho}\frac{\partial p}{\partial y} + v\left(\frac{\partial^2 v}{\partial x^2} + \frac{\partial^2 v}{\partial y^2} + \frac{\partial^2 v}{\partial z^2}\right) = \frac{\partial v}{\partial t} + u\frac{\partial v}{\partial x} + v\frac{\partial v}{\partial y} + w\frac{\partial v}{\partial z} \qquad (7\text{-}89)$$

$$Z - \frac{1}{\rho}\frac{\partial p}{\partial z} + v\left(\frac{\partial^2 w}{\partial x^2} + \frac{\partial^2 w}{\partial y^2} + \frac{\partial^2 w}{\partial z^2}\right) = \frac{\partial w}{\partial t} + u\frac{\partial w}{\partial x} + v\frac{\partial w}{\partial y} + w\frac{\partial w}{\partial z}$$

which are known as the *Navier–Stokes* equations. For completeness, it should be pointed out that the 'pressure' p used in Eq. (7-89) is *defined* as the mean of the three normal stresses; that is

$$p \equiv \tfrac{1}{3}(\sigma_x + \sigma_y + \sigma_z) \qquad (7\text{-}90)$$

Nevertheless it may be regarded as being equal to the thermodynamic pressure described in section 2-4 for a stationary fluid—see, for example, Schlichting (1979) or Batchelor (1973).

PROBLEMS

1 A horizontal jet of water ($\rho = 998 \text{ kg/m}^3$) with a diameter of 3 mm and a velocity of 150 m/s impacts on a stationary vertical surface as shown in Fig. 7-17. Estimate the horizontal force exerted on the wall.

[159 N]

2 A jet of water ($\rho = 998 \text{ kg/m}^3$) approaches a stationary vane tangentially and is deflected horizontally through an angle of 60° as shown in Fig. 7-18. The velocity of the jet at inlet is 20 m/s and its area is 0.001 m². The velocity

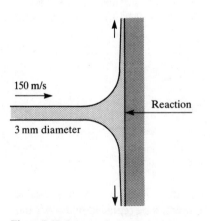

150 m/s

Reaction

3 mm diameter

Figure 7-17 Impact of a water jet on a rigid surface.

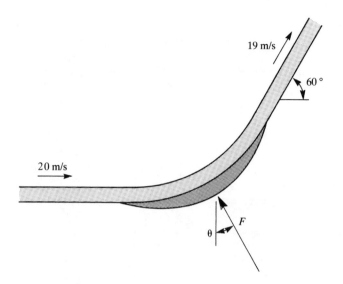

Figure 7-18 Deflection of a water jet by a stationary vane.

at outlet is 19 m/s. Determine the magnitude and direction of the external force F needed to hold the vane stationary.

[390 N, $\theta = 32.5°$]

3 In Example 7-2 a jet of water moving at 75 m/s relative to a stationary observer impacts on a Pelton wheel bucket that is moving at 35 m/s in the same direction. Determine the force exerted by the same jet on the same bucket when the speed of the bucket is (*a*) 25 m/s and (*b*) 45 m/s. Assume that $|V_0| = 0.95|V_1|$ in both cases.

[84.7 kN, 30.5 kN]

4 Water ($\rho = 1000$ kg/m³) enters a 50 mm diameter horizontal pipe at a mean velocity of $\bar{V}_1 = 2$ m/s and exits to the atmosphere in the opposite direction as shown in Fig. 7-19. The gauge pressure at the section A is 650 kPa. Estimate the axial force in the walls of the pipe at A if the diameter of the jet at outlet is 12 mm.

[1.42 kN]

5 Figure 7-20 illustrates an elementary jet pump in which a 6 mm diameter jet of air ($\rho = 1.17$ kg/m³) induces airflow along a 25 mm bore pipe. Estimate the pressure rise $(p_3 - p_2)$ when $V_1 = 5$ m/s and $V_J = 80$ m/s. Neglect skin friction and assume that the velocity distribution at the flow section 3 is uniform.

[354 Pa]

6 A valve in a 50 mm diameter pipe conveying water ($\rho = 1000$ kg/m³) has become jammed in an almost closed position. The rate of flow that can be

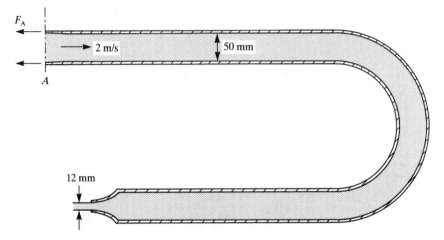

Figure 7-19 Flow-induced stress in a pipe wall.

induced along the pipe by a 2 m difference in levels in tanks 10 m apart is found to be 0.75 l/s. If the corresponding rate with a fully open valve used to be 3.5 l/s, estimate the force imposed by the water on the valve. Assume that all resistances are proportional to V^2.

[36.8 N]

7 Water flows steadily over a bed hump weir in a 1 m wide rectangular channel. The depths of flow at the points 1 and 2 in Fig. 7-21 are 2.1 m and 0.8 m respectively and the upstream velocity is 1.1 m/s. Estimate the magnitude and direction of the force exerted by the water on the weir in the direction of flow.

[14.4 kN, downstream]

8 The depth and velocity of water at a particular flow section in a wide channel are $d_1 = 1.6$ m and $\bar{V}_1 = 2.1$ m/s. The channel slopes downhill at

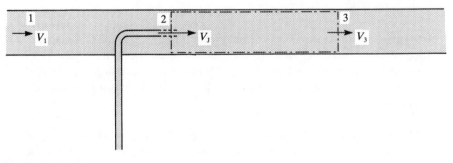

Figure 7-20 Elementary jet pump.

$d_1 = 2.1$ m

$V_1 = 1.1$ m/s

$d_2 = 0.8$ m

Figure 7-21 Steady flow over a bed hump weir.

1 in 300. Estimate the average skin friction coefficient of the channel bed if the depth 10 m downstream is (a) 1.56 m and (b) 1.64 m.

[0.0423, 0.0031]

9 The velocity distribution in a turbulent flow in a pipe of radius R is found to satisfy

$$\left(\frac{u}{V_0}\right) = \left(\frac{y}{R}\right)^{0.15} \tag{7-91}$$

where u is the velocity at a distance y from the wall and V_0 is the velocity on the axis. Estimate the values of the ratio $\bar{V}/V_0$ and the momentum coefficient β.

[0.809, 1.022]

10 The velocity distributions upstream and downstream of an object in a uniform stream of air ($\rho = 1.1$ kg/m^3) are as shown in Fig. 7-22. Neglecting any difference in pressure between the two locations, estimate the drag force on the object per unit width (normal to the page). Use the control volume shown and assume the axial velocity component of any flow leaving its sides to be 10 m/s.

[3.67 N]

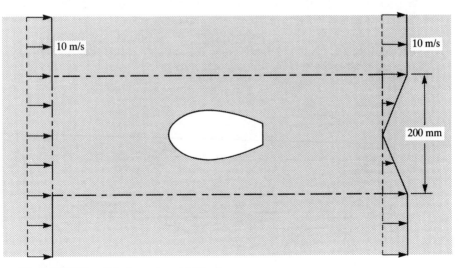

Figure 7-22 Flow disturbance by a bluff body.

11 The velocity distribution at a section in the boundary layer alongside a flat plate at zero incidence to a uniform stream of air ($\rho = 1.2 \, \text{kg/m}^3$) satisfies

$$\left(\frac{u}{V}\right) = \left(\frac{u}{H}\right)^{0.13} \qquad (7\text{-}92)$$

Estimate the skin friction force on the plate between the leading edge and this section if $H = 12 \, \text{mm}$ and the velocity of the uniform stream is 60 m/s (see Fig. 7-23). Choose a control volume with its lower edge along the surface of the plate and its upper edge in the free stream.

[4.73 N per metre width]

12 Estimate the axial pressure gradient in the air in a railway carriage when the train decelerates at a constant rate of 0.75 m/s². Neglect leakage into or out of the carriage and assume $\rho = 1.2 \, \text{kg/m}^3$.

[0.90 Pa/m]

13 (a) A football is held at rest on the floor of a railway carriage which is decelerating at 0.75 m/s². The average density of the ball is 70 kg/m³. What will be its initial acceleration if it is suddenly released? In which direction will it move along the carriage? The density of the air is 1.2 kg/m³. (*Hint*: Think 'Archimedes'.)

[−0.013 m/s², forwards]

(b) A helium balloon is held at rest on the ceiling of the same carriage. Its average density is 0.9 kg/m³. What will be its initial acceleration if it is released at the same instant as the football? In which direction will it move along the carriage?

[1.00 m/s², backwards]

14 A valve at the downstream end of a 30 m long pipeline (Fig. 7-24) is closed in such a way that the velocity of flow in the pipe reduces linearly to zero. Neglecting skin friction and other sources of resistance, estimate the pressure head difference between the two ends of the pipe during decelerations of 0.1 m/s², 10 m/s² and 1000 m/s². Assume that the fluid is incompressible.

[0.306 m, 30.6 m, 3058 m]

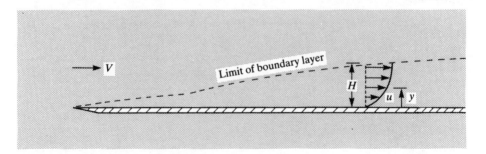

Figure 7-23 Momentum flux deficit in a boundary layer.

Figure 7-24 Deceleration of flow by a controlled valve closure.

15 Because water is slightly compressible, the maximum head difference that could actually exist during the valve closure described in the preceding problem is cV_0/g where c is the speed of propagation of a pressure wavefront. Compare this result with the incompressible analysis and decide which gives the more realistic upper limit in the various cases. Assume that $c = 1250$ m/s (typical for water in a pipeline).

[255 m]

16 Following the sudden closure of a valve in a pipeline, a pressure wavefront propagates upstream towards a reservoir. The conditions shortly before the wavefront arrives at the reservoir are depicted in Fig. 7-25 together with the corresponding conditions a few moments later. By considering the conditions relative to observers moving with the wavefronts, determine the magnitude and direction of the velocity V_2. Assume that the pressure in the pipeline to the left-hand side of the wavefronts is always equal to the reservoir pressure.

[2 m/s into the reservoir]

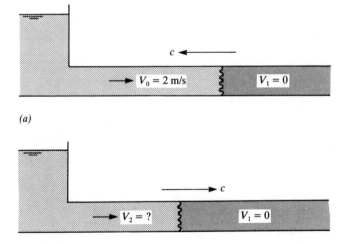

(a)

(b)

Figure 7-25 Reflection of a wavefront at a reservoir. (a) Incident wavefront. (b) Reflected wavefront.

FURTHER READING

Abramovich, G.N. (1963) *The Theory of Turbulent Jets*, MIT Press.
Batchelor, G.K. (1973) *An Introduction to Fluid Dynamics*, Cambridge University Press.
Henderson, F.M. (1966) *Open Channel Flow*, Macmillan.
Iversen, H.W. (1975) An analysis of the hydraulic ram, *J. Fluids Engng, Trans. ASME*, **97**, 191–196.
Schlichting, H. (1979) *Boundary Layer Theory*, 7th edn, McGraw-Hill.
Zucrow, M.J. and Hoffman, J.D. (1977) *Gas Dynamics* (2 vols), Wiley.

EIGHT

MOMENT OF MOMENTUM

8-1 ANGULAR MOMENTUM

Throughout Chapter Seven we dealt with *linear* momentum, that is momentum in a particular direction. Newton's second law of motion can also be used to describe changes in the *moment of momentum* of a system. The relationship is

> Torque = rate of change of moment of momentum $\qquad$ (8-1)

in which the *torque* is the net moment of all of the applied forces about an axis. It causes a rate of change of the moment of momentum (i.e. of the moment of the linear momentum) about the same axis.

Torque All forces are vectors, that is they have a definite magnitude and they act in a definite direction. In practical applications each particular force also acts at a particular position and this makes it possible to quantify its *torque* about any particular axis—that is, the moment of the force about the axis. For example, the torque associated with the force F_1 about the axis O normal to the page in Fig. 8-1a is $F_1 r_1$ in which r_1 is the perpendicular distance from the axis to the line of action of the force. The total torque T about the chosen axis for all the various forces can be written in general as

$$T = \sum F_i r_i \qquad (8-2)$$

Any component of a force in the direction normal to the page is discounted when evaluating the torque about an axis normal to the page.

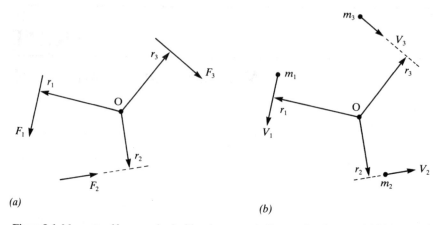

Figure 8-1 Moments of forces and velocities about an axis O normal to the page. (a) Moments of forces. (b) Moments of momentum.

Angular momentum The moment of momentum of a non-rotating object about an arbitrarily chosen axis is the product of its linear momentum and the perpendicular distance from the axis to the instantaneous velocity vector of its centre of mass. In Fig. 8-1b for example, the moment of momentum of the particle of mass m_1 at the instant shown is $m_1 V_1 r_1$. The total moment of momentum J about the chosen axis for all the various masses is

$$J = \sum m_i V_i r_i \qquad (8\text{-}3)$$

In general, the masses may be moving in three-dimensional space. However, only the components of their velocities in a plane normal to an axis contribute to the moment of momentum about the axis.

The moment of momentum is often called the *angular momentum*. The reason for this becomes clear when the discussion is generalized to include rotating objects. For example, consider the uniform rotation of a body such as the pump impeller shown in Fig. 8-2. It is natural to choose the centre of the shaft as the axis O_1 about which the moment of momentum will be evaluated. For this axis, each elemental mass δm_i contributes $V_i r_i \, \delta m_i$ to the total moment of momentum, which is therefore

$$J = \int V r \, dm \qquad (8\text{-}4)$$

For the special case of a body rotating at a uniform angular velocity Ω, the tangential velocity at a radius r is $V = r\Omega$. In this case, the total moment of momentum about the axis O_1 can be written as

$$J_1 = \Omega \int r^2 \, dm \qquad (8\text{-}5)$$

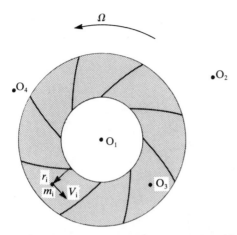

Figure 8-2 Moment of momentum of a body with uniform angular velocity.

in which $\int r^2 \, dm$ is the polar moment of inertia of the object about the axis of rotation.

Consider now the moment of momentum of the same body about a different axis O_2 parallel to the first. It can be shown fairly easily by following the same procedure as that adopted for the axis O_1 that the same value is obtained in both cases, that is $J_2 = J_1$. The moment of momentum exists about the axis O_2 even though the centre of mass of the body has no tangential velocity relative to O_2.

In general, the moment of momentum of any body about any axis can be regarded as the sum of (a) the moment of momentum of the body about its centre of mass and (b) the moment of the linear momentum of the body about the chosen axis. It is not *essential* to evaluate the total moment of momentum in this way because (8-4) is generally valid, but it is often *convenient* to do so.

8-2 GENERAL FORM OF THE MOMENT OF MOMENTUM EQUATION

In developing an analytical form of Eq. (8-1) it is convenient to use cylindrical polar coordinates (r, θ, x) instead of Cartesian coordinates. The set (r, θ, x) is used in preference to (r, θ, z) because z is used subsequently to denote elevation.

Figure 8-3 shows an arbitrarily shaped control volume viewed along the x-axis which appears as a point at the origin in the r-θ plane. Following the method used in section 7.2, we begin to develop the control volume form of the equation by considering a *system* composed of the combined contents of the control volume and the elemental region I at an arbitrary instant t_0. At the later time $t_0 + \delta t$, the same system occupies the control volume and the elemental region O.

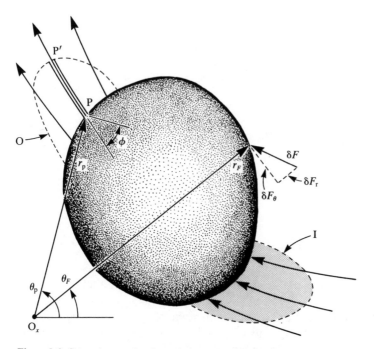

Figure 8-3 General control volume (system shaded at t_0).

In general, the system may experience body forces as well as normal and tangential surface forces, and all of these may contribute to the torque about the x-axis. For example, the elemental force δF shown acting on the surface of the control volume in Fig. 8-3 may be resolved into the components δF_r, δF_θ and δF_x. Of these, only δF_θ has a moment about the x-axis and so the contribution of δF to the torque is $r_F \, \delta F_\theta$. The net torque applied to the system is the sum of all such elemental quantities.

At any instant, the total mass of the contents of the control volume is m_{cv} and the counterclockwise angular momentum about the x-axis is J_{cv}. The mass and angular momentum of the contents of the elemental region I are δm_1 and δJ_1 respectively. During the interval δt, this mass flows into the control volume at an average rate $\delta m_1/\delta t$ which is used to define a mass flux $\dot m$ in the limit as δt tends to zero. Similarly we define the *angular momentum flux* across a flow section as

$$\dot{J} \equiv \frac{dJ}{dt} \tag{8-6}$$

Equation (8-1) relates the total torque T acting on the chosen system to the rate of change of angular momentum, giving

$$T = \frac{(J_{cv} + \delta J_0)_{t_0 + \delta t} - (J_{cv} + \delta J_1)_{t_0}}{\delta t} \tag{8-7}$$

which may be rearranged as

$$T = \frac{(J_{t_0+\delta t} - J_{t_0})_{cv}}{\delta t} + \frac{\delta J_{O, t_0+\delta t}}{\delta t} - \frac{\delta J_{1, t_0}}{\delta t} \tag{8-8}$$

In the limit, as δt approaches zero, this becomes the *moment of momentum equation*

$$\boxed{T = \frac{\partial J_{cv}}{\partial t} + \dot{J}_O - \dot{J}_1} \tag{8-9}$$

in which the suffix t_0 is discarded because it denotes an arbitrary instant. Independent equations can be derived about the y- and z-axes.

Angular momentum flux The mass flux $\dot{m}$ across the control surface at a typical inflow or outflow section may be regarded as the sum of the elemental fluxes $\delta \dot{m}$ through elemental areas δa of the surface. If the velocity components at a typical position are denoted by V_r, V_θ and V_x, the θ-momentum flux through an elemental area is $\delta \dot{M}_\theta = V_\theta \, \delta \dot{m}$ and the angular momentum flux about the x-axis is $\delta \dot{J}_x = rV_\theta \, \delta \dot{m}$. The total angular momentum flux for the section is

$$\dot{J}_x = \int rV_\theta \, d\dot{m} \tag{8-10}$$

Range of validity of the moment of momentum equation Equations (8-9) and (8-10) are general expressions which are valid for any control volume. However, they describe the angular momentum about the x-axis only. Independent relationships can be developed for the angular momentum about any two mutually perpendicular axes in the r-θ plane. However, their use can be complex unless the flow sections are uniaxial. This special case is now considered.

8-3 UNIAXIAL AND ONE-DIMENSIONAL FLOWS

At a uniaxial flow section where the control surface is chosen normal to the direction of flow, $\delta \dot{m} = \rho V \, \delta a$ and the angular momentum flux (8-10) can be written as

$$\dot{J}_x = \int \rho r V_\theta V \, da \tag{8-11}$$

In the special case where the direction of flow coincides with the positive θ-direction, $V_\theta = V$. In general, however, $V_\theta = V \cos \phi$ where ϕ is the angle between the velocity vector and the positive θ-direction (Fig. 8-3). When ϕ is approximately constant for the section, the angular momentum flux is

$$\dot{J}_x = \cos \phi \int \rho r V^2 \, \mathrm{d}a \tag{8-12}$$

In order to eliminate the integral before using this expression in the angular momentum equation, it would be natural to follow the procedure used for linear momentum flux (section 7.3) by defining an angular momentum flux coefficient as the ratio of the integral $\int \rho r V^2 \, \mathrm{d}a$ and the product $\bar{\rho} \bar{r} a \bar{V}^2$ in which $\bar{r}$ may conveniently be measured to the centre of area of the flow section. In practice, however, it will nearly always be sufficiently accurate to approximate Eq. (8-12) by

$$\dot{J}_x \simeq \bar{r} \cos \phi \int \rho V^2 \, \mathrm{d}a \tag{8-13}$$

In this case, direct use can be made of the momentum flux coefficient β defined by Eq. (7-14) to give

$$\dot{J}_x \simeq \beta \bar{\rho} \bar{r} a \bar{V}^2 \cos \phi = \beta \bar{r} \dot{m} \bar{V} \cos \phi \tag{8-14}$$

The angular momentum equation for uniaxial flows is therefore

$$T_x = \frac{\partial J_{x,\text{cv}}}{\partial t} + (\beta \bar{r} \dot{m} \bar{V} \cos \phi)_\text{o} - (\beta \bar{r} \dot{m} \bar{V} \cos \phi)_\text{I} \tag{8-15}$$

Range of validity The main additional limitation of this equation in comparison with Eq. (8-9) is that the flows at each inlet and outlet section must be uniaxial. Strictly, however, the approximation of Eq. (8-12) by the expression (8-13) is a further limitation.

One-dimensional flow When the flow at a section may be regarded as one-dimensional, the angular momentum flux for non-rotating fluid particles becomes

$$\dot{J}_x = \bar{r} \dot{m} V \cos \phi \tag{8-16}$$

in which $\cos \phi = 1$ if the flow is in the positive θ-direction. Notice that the distance $\bar{r}$ must be measured to the centroid of the flow section.

Steady flow In a steady flow, there is no rate of change of angular momentum within the control volume and so the moment of momentum equation is

$$T_x = (\beta \bar{r} \dot{m} \bar{V} \cos \phi)_\text{o} - (\beta \bar{r} \dot{m} \bar{V} \cos \phi)_\text{I} \tag{8-17}$$

When there is only one inlet and one outlet and these flow sections are parallel to the θ-direction,

$$T_x = \dot{m}(\beta_\text{o} \bar{r}_\text{o} \bar{V}_\text{o} - \beta_\text{I} \bar{r}_\text{I} \bar{V}_\text{I}) \tag{8-18}$$

and in the special case of one-dimensional inflows and outflows,

$$T_x = \dot{m}(\bar{r}_0 V_0 - \bar{r}_1 V_1)$$ (8-19)

8-4 APPLICATIONS OF ANGULAR MOMENTUM

Example 8-1: Resultant force The intensity of the normal stresses acting on the plane rectangular surface AB in Fig. 8-4 varies linearly (i.e. $\sigma = Kx$). Show that their overall influence is equivalent to a single force acting normal to the surface at a distance $0.556L$ from A.

SOLUTION The force acting on an element of the surface of length δx and width b normal to the paper is $\sigma b \, \delta x$ and the total force on the surface is

$$F = \int_L^{2L} \sigma b \, dx = Kb \int_L^{2L} x \, dx = 1.5KbL^2$$ (8-20)

The moment of the typical elemental force about the axis O is $\delta T = -\sigma b x \, \delta x$ and the total torque is

$$T = -\int_L^{2L} \sigma b x \, dx = -Kb \int_L^{2L} x^2 \, dx = -2.33KbL^3$$ (8-21)

The same torque could alternatively be induced by a force F acting at a distance T/F, namely $1.556L$, from the origin. Hence the stresses are equivalent to a single force of $1.5KbL^2$ acting at a distance of $0.556L$ from A.

Comments on Example 8-1
1. This is a slightly simplified form of the analysis used to derive Eqs (3-22) and (3-24). The hydrostatic pressure field is an example of a linear stress distribution.
2. It is often possible to deduce results such as this geometrically. For example, the stress block shown in Fig. 8-4 can be regarded as being comprised of two stress blocks ABCE and CDE. The total force is the area of the blocks, namely KL^2 and $\frac{1}{2}KL^2$ respectively (per unit width normal to the page). The lines of action of the resultants are at the centroids of the blocks and the distance of the overall resultant from A is

$$x' = \frac{(KL^2 \times \frac{1}{2}L) + (\frac{1}{2}KL \times \frac{2}{3}L)}{(KL^2 + \frac{1}{2}KL^2)} = 0.556L$$ (8-22)

Example 8-2: Deflection of a jet Estimate the bending moment at the cross-section A of the vane support shown in plan in Fig. 8-5. A water jet approaches the vane horizontally in the negative y-direction with a uniform velocity of 30 m/s and is deflected in a horizontal plane through

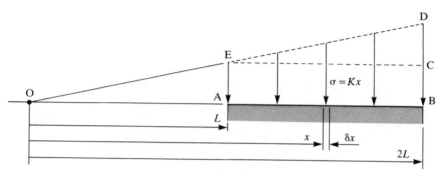

Figure 8-4 Linear distribution of normal stress on a flat surface.

an angle of 65°. The vertical thickness of the jet is 95 mm and its width is 40 mm at inlet and 45 mm at outlet. The positions of the centre lines of the jet at inlet and outlet are indicated in the figure in which the origin of the axis is chosen at the centroid of the section A. Neglect any movement of the air.

SOLUTION The normal stresses on the surface of the control volume shown in the figure are equal to atmospheric pressure at all points except in the support where they constitute a net axial force and the required bending moment. Shear and surface tension forces acting normal to the control surface at the inflow and outflow sections are assumed to be negligible. The momentum flux at inlet is $\rho a V^2 = 1000 \times (0.095 \times 0.04) \times 30^2$ N $= 3.42$ kN. By continuity, the velocity at outlet is $30 \times 40/45$ m/s and so the momentum flux is $\rho a V^2 = 3.04$ kN.

The clockwise moments of these momentum fluxes about the origin are 3.42×100 kNm and 3.04×250 kNm respectively and their difference, namely 418 kNm, is equal to the clockwise bending moment in the support at the section A.

Comments on Example 8-2

1. Notice that the use of the angular momentum equation is not restricted to applications involving circular motion. The above result *could* be obtained by using the linear x-momentum and y-momentum equations to determine the magnitude and line of action of the fluid force on the vane and then considering the equilibrium of the vane and its support. However, the above method is more direct.

2. Also notice that the angular momentum equation (like all other control volume equations) applies to *all* of the contents of the control volume, including solid as well as fluid material. Strictly, the above analysis should be modified to take account of the angular momentum of air set in motion by the jet.

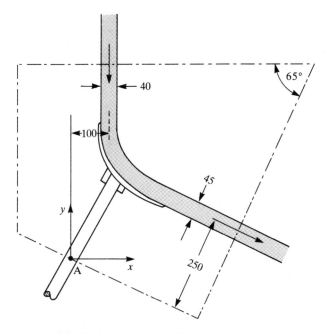

Figure 8-5 Deflection of a water jet by a vane.

Example 8-3: Pelton wheel The Pelton wheel water turbine shown schematically in Fig. 8-6 has two 150 mm diameter jets which are tangential to a 1.75 m diameter circle at inlet. Under ideal conditions, the water leaves the turbine parallel to the shaft (i.e. normal to the page). Estimate the power supplied to the wheel which rotates at $N = 450$ rev/min if the mean velocity in the jets is 75 m/s. The density of the water is $1000 \, \text{kg/m}^3$.

SOLUTION Consider the control volume shown in Fig. 8-6. The flow is steady and there is no angular momentum about the shaft at outlet. Therefore Eq. (8-18) shows that the torque exerted by the wheel on the water is $-\beta \dot{m} R V_1$. The water exerts an equal and opposite torque on the wheel.

The cross-sectional area of each jet is $a = \frac{1}{4} \pi D^2 = 0.0177 \, \text{m}^2$ and the total mass flux through the two jets is $\dot{m} = 2 \rho a \bar{V} = 2651 \, \text{kg/s}$. Assuming a typical value of 1.02 for the momentum flux coefficient in a turbulent jet, the torque exerted on the wheel is

$$T_x = \beta_1 \dot{m} R V_1 = 1.02 \times 2651 \times 0.875 \times 75 \, \text{Nm} = 177 \, \text{kN m} \quad (8\text{-}23)$$

The angular velocity of the wheel is $2 \pi N/60 = 47.1 \, \text{rad/s}$ and so the power supplied by the water is $177 \times 47.1 \, \text{kW} = 8.36 \, \text{MW}$.

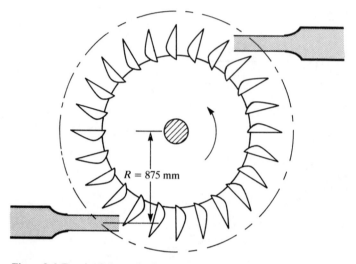

Figure 8-6 Two-jet Pelton wheel.

Comments on Example 8-3

1. The Pelton wheel turbine is an example of an *impulse* turbine. It is used for power generation when water is available at very high pressure. After passing through nozzles, the water is supplied to the machine at atmospheric pressure in the form of high-velocity jets which are deflected by the buckets in the manner described in Example 7-2 (p. 187).

2. Other types of turbine can operate efficiently with lower-pressure supplies. Francis turbines operate at medium pressure but need larger rates of flow than Pelton wheels. Kaplan turbines need very low pressures but huge rates of flow.

Example 8-4: Centrifugal pump Water enters a 0.75 m diameter pump impeller with no initial swirl and leaves with an absolute velocity of 26.5 m/s at an angle of 35° to the tangent (Fig. 8-7). The width of the impeller at outlet is 85 mm. Estimate the power supplied to the water and the manometric head difference across the complete pump if its hydraulic efficiency is 80 per cent and the impeller rotates at 850 rev/min.

SOLUTION At the impeller outlet, $D=0.75$ m and $b=0.085$ m and so the cross-sectional area of flow is $\pi Db = 0.200$ m². The radial velocity component is $26.5 \sin 35°$ m/s $= 15.2$ m/s and so the volumetric flux through the impeller is $Q = 0.200 \times 15.2$ m³/s $= 3.04$ m³/s. Assuming the density of water to be 1000 kg/m³, the mass flux is 3040 kg/s.

The angular momentum equation (8-9) written for the steady flow through the control volume shown in Fig. 8-7 is

$$T_x = \dot{J}_0 - \dot{J}_1 \tag{8-24}$$

There is no swirl at inlet and so $\dot{J}_i$ is zero. At outlet, the tangential velocity is $26.5 \cos 35°$ m/s $= 21.7$ m/s and so the angular momentum flux is $\dot{J}_o = r V_\theta \dot{m} = 0.375 \times 21.7 \times 3040$ Nm $= 24.75$ kNm. It follows from Eq. (8-24) that the torque T_x is also 24.75 kNm. The impeller rotates at 850 rev/min, i.e. 89 rad/s, and so the power supplied to the water is $\dot{W} = T_x \Omega = 24.75 \times 89$ kW $= 2.2$ MW.

If the hydraulic efficiency of the pump was 100 per cent, the manometric head difference ΔH across it would satisfy $\dot{m}g \, \Delta H = 2.2 \, MW$ and so ΔH would be 73.8 m. However, the hydraulic efficiency is 80 per cent and so the actual head difference is 59.0 m.

Comments on Example 8-4

1. The true area of flow at the impeller outlet is typically about 10 per cent less than πDb because of the thickness of the vanes. Therefore the true mass flux through the pump is less than the calculated value.
2. The power supplied to the water is not all converted into useful 'mechanical' energy. The residual 20 per cent is converted into internal energy and so the temperature of the water increases. In practice, account must also be taken of other sources of inefficiencies in pumps. For example, electrical and mechanical inefficiencies on the supply side typically waste about 5 per cent of the power.

Example 8-5: Polar moment of inertia Determine the polar moment of inertia of the Francis turbine runner shown in Fig. (8-8). Assume that the average thickness of the steel including an allowance for the blades is 12 mm where $r < 125$ mm and is 25 mm where $r > 125$ mm. Also determine the moment of inertia of the water in the blade region, assuming the

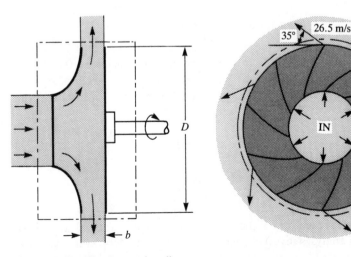

Figure 8-7 Centrifugal pump impeller.

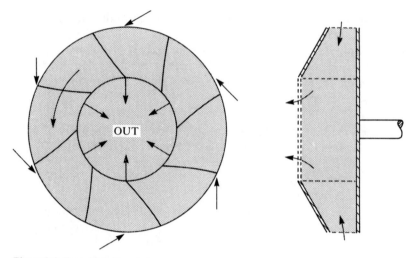

Figure 8-8 Francis turbine runner.

effective width of flow to vary linearly from $b = 100$ mm at $r = 250$ mm to $b = 200$ mm at $r = 125$ mm. The densities of the water and steel are 1000 kg/m³ and 7850 kg/m³ respectively.

SOLUTION The polar moment of inertia of an object about any particular axis is $\int r^2 \, dm$. For the turbine runner, the volume of an elemental annulus of radius r, width δr and thickness b is $2\pi r b \, \delta r$ and so its mass is $2\pi b \rho r \, \delta r$. Therefore the polar moment of inertia of the runner of density ρ_s is

$$I_x = 2\pi \rho_s b_1 \int_0^{R_1} r^3 \, dr + 2\pi \rho_s b_2 \int_{R_1}^{R_2} r^3 \, dr \qquad (8.25)$$

i.e.

$$I_x = \tfrac{1}{2} \pi \rho_s [b_1 R_1{}^4 + b_2 (R_2{}^4 - R_1{}^4)] \qquad (8\text{-}26)$$

For the particular values in this example, this gives $I_x = 1.165$ kg m². The effective width of flow in the blade region satisfies $b = 0.8(R_0 - r)$ where $R_0 = 375$ mm. At a typical radius r, the mass of an elemental annular strip is $2\pi \rho_w r \, \delta r \{0.8(R_0 - r)\}$ and so the polar moment of inertia of the water is

$$I_{xw} = 1.6\pi \rho_w \int_{R_1}^{R_2} (R_0 r^3 - r^4) \, dr = 1.6\pi \rho_w \left[R_0 \frac{r^4}{4} - \frac{r^5}{5} \right]_{R_1}^{R_2} \qquad (8\text{-}27)$$

which for this particular example is $I_{xw} = 0.775$ kg m².

Comment on Example 8-5

The contribution of the water in the blade region is significant and should not be neglected in analyses involving accelerating runners. However, the above result may be misleading in practice because the tangential velocity of the water is not the same as that of the runner. Further details are presented in Chapter Ten.

Example 8-6: Turbine runaway The power delivered by a generator connected to a water turbine when its shaft rotates at 100 rev/min is 50 MW. The polar moment of inertia of the shaft, armature, runner and water in the blade passages is 300 000 kg m². Estimate the initial rate of acceleration of the shaft if a power line failure suddenly releases the load on the generator.

SOLUTION In the normal operation, the power delivered by the generator is $\dot{W} = 50$ MW. For the particular values given above, the restraining torque provided by the electrical field is $T_G = \dot{W}/\Omega = 4.77$ MNm. The removal of the electrical load of the generator has no immediate impact on the water flow through the turbine. This continues as before, but all of the available torque is now used to accelerate the shaft. Using the control volume shown schematically in Fig. 8-9, Eq. (8-9) may be written for the two states as

$$\text{Before power loss:} \quad -(T_L + T_G) = \dot{J}_0 - \dot{J}_1 \qquad (8\text{-}28)$$

$$\text{After power loss:} \quad -T_L = \frac{\partial J_{cv}}{\partial t} + \dot{J}_0 - \dot{J}_1 \qquad (8\text{-}29)$$

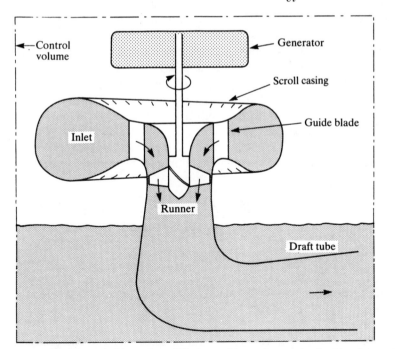

Figure 8-9 Schematic illustration of Kaplan turbine and generator (water enters scroll casing normal to page).

where T_L denotes torque due to resistances other than the generator. By subtraction, $T_G = \partial J_{cv}/\partial t$. Now $J_{cv} = I_x \Omega$ where I_x is the polar moment of inertia of the shaft and its attachments. Therefore $T_G = I_x \partial \Omega/\partial t$ and so the acceleration is

$$\frac{\partial \Omega}{\partial t} = \frac{T_G}{I_x} = 15.9 \text{ rad/s}^2 \qquad (8\text{-}30)$$

Comments on Example 8-6

1. Notice that all the contents of the control volume are considered when evaluating J_{cv}. In the derivation of the angular momentum equations, no restriction has been placed on these contents.
2. The rate of increase of the rotational speed of the shaft is alarming. The rate of increase will reduce as the turbine moves away from its design condition, but the acceleration will nevertheless continue until high stresses in the components cause the machine to disintegrate in a highly dangerous manner.
3. In practice, some means of stopping the flow through the turbine in the event of power failure is always provided. One option is to interrupt the flow supplying the turbine. In high pressure systems, this leads to water-hammer in the pipeline, which is also highly undesirable. Usually, a surge tank is provided at a suitably high point not far upstream to protect the remainder of the pipeline from the water-hammer.

Example 8-7: Hydraulic jump The depth and velocity upstream of the hydraulic jump shown in Fig. 8-10 are 0.6 m and 3.5 m/s respectively. Assuming that $\beta_1 = 1.02$ and $\beta_2 = 1.06$, estimate the depth and velocity downstream of the jump. Hence show that the pressure distribution at a typical flow section within the jump is not hydrostatic. The channel section is rectangular.

SOLUTION The linear momentum equation may be written for the jump as

$$\bar{p}_1 a_1 - \bar{p}_2 a_2 = \dot{m}(\beta_2 \bar{V}_2 - \beta_1 \bar{V}_1) \qquad (8\text{-}31)$$

Since $a = bd$, $\dot{m} = \rho a \bar{V}$ and $\bar{p} = \frac{1}{2}\rho gd$, this equation may be expressed as

$$\tfrac{1}{2}\rho gb(d_1{}^2 - d_2{}^2) = \dot{m}\left(\frac{\beta_2 \dot{m}}{\rho bd_2} - \beta_1 \bar{V}_1\right) \qquad (8\text{-}32)$$

in which the only unknown parameter is d_2. For the specified values, the solution is $d_2 = 0.928$ m and the corresponding velocity is $\bar{V}_2 = \dot{m}/\rho bd_2$ $= 2.26$ m/s.

The resultant pressure force on a flow section acts at the centre of pressure, namely at a depth of $\frac{2}{3}d$ in a rectangular section channel. The

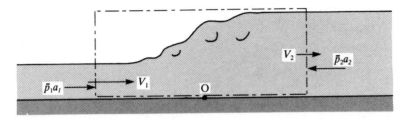

Figure 8-10 Hydraulic jump.

resultant momentum flux acts approximately at the centroid, namely at a depth of $\frac{1}{2}d$. Therefore the angular momentum equation for the axis O aligned normal to the page in Fig. 8-10 is (using clockwise = positive for convenience)

$$\tfrac{1}{3}d_1\bar{p}_1a_1 - \tfrac{1}{3}d_2\bar{p}_2a_2 + \Sigma C = \dot{m}\{\tfrac{1}{2}d_2\beta_2\bar{V}_2 - \tfrac{1}{2}d_1\beta_1\bar{V}_1\} \qquad (8\text{-}33)$$

in which ΣC denotes the couple due to (a) the weight of all fluid particles in the control volume and (b) the pressure force on the bed of the channel. In this example, we obtain $\Sigma C = -760\,\text{Nm}$. Since this is not equal to zero, the contributions of the weight and the bed pressure force cannot be equal—i.e. the pressure distribution is not hydrostatic.

Comments on Example 8-7

1. It is not possible to use this analysis to determine the pressure distribution on the bed. Vertical equilibrium requires that the total pressure force on the bed is equal to the weight of the water in the control volume, but we cannot deduce the local distribution in the absence of information about vertical components of velocity.
2. On entering the jump, a typical fluid particle will gradually acquire an upwards component of velocity which it will subsequently lose because the mean direction of outflow is parallel to the bed. The accelerations associated with the increasing and decreasing velocity components imply increased and decreased piezometric pressures which together induce a clockwise couple at the bed. Chapter Ten deals extensively with radial pressure gradients in curved flows.
3. The predicted depth of flow downstream of the jump is relatively insensitive to the assumed value of β_2 when this is close to unity, but it is highly sensitive at larger values. The preceding solution may be compared with predicted depths of $d_2 = 0.972\,\text{m}$ and $0.857\,\text{m}$ that are obtained with assumed values of $\beta_2 = 1.02$ and 1.10 respectively.

Example 8-8: General flow in a curved path Show that the rates of change of the flow parameters along a curved streamline in a three-dimensional,

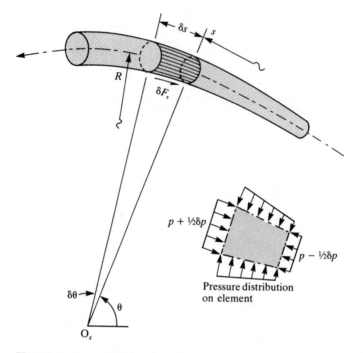

Figure 8-11 Forces on a streamtube element.

unsteady, inviscid flow satisfy the Euler equation:

$$\frac{1}{\rho}\frac{\partial p}{\partial s}+V\frac{\partial V}{\partial s}+g\frac{\partial z}{\partial s}=-\frac{\partial V}{\partial t} \tag{8-34}$$

SOLUTION Consider the elemental length of streamtube depicted in Fig. 8-11 in which the origin is chosen at the centre of curvature of the element. The streamwise component of the pressure force on the sides of the element is $p\,\delta a$ and the difference between the pressure forces on the two flow sections is $\delta(pa)$. The net streamwise force on the element is therefore

$$F_\theta=p\,\delta a-\delta(pa)-\rho g\frac{\partial z}{\partial s}a\,\delta s \tag{8-35}$$

in which z denotes the elevation and the only body force is due to gravity. The couple exerted by this force about the axis O_x is

$$C_x=RF_\theta=-aR\,\delta p-\rho gaR\frac{\partial z}{\partial s}\,\delta s \tag{8-36}$$

The rate of change of angular momentum within the control volume is

$$\frac{\partial J_{cv}}{\partial t} = \frac{\partial}{\partial t}(\rho a\, \delta s R V) = \left\{\rho a R \frac{\partial V}{\partial t} + R V \frac{\partial(\rho a)}{\partial t}\right\} \delta s \qquad (8\text{-}37)$$

and the excess of the angular momentum efflux over the angular momentum influx is

$$\dot{J}_0 - \dot{J}_1 = \frac{\partial}{\partial s}(\dot{m} R V)\, \delta s = \left\{\dot{m} R \frac{\partial V}{\partial s} + R V \frac{\partial \dot{m}}{\partial s}\right\} \delta s \qquad (8\text{-}38)$$

in which $\dot{m} = \rho a V$.

The angular momentum equation relates the expression (8-36) to the sum of the expressions (8-37) and (8-38). Since $\delta p = (\partial p / \partial s)\, \delta s$, we obtain

$$-aR \frac{\partial p}{\partial s} - \rho g a R \frac{\partial z}{\partial s} = \rho a R \frac{\partial V}{\partial t} + \rho a R V \frac{\partial V}{\partial s} \qquad (8.39)$$

in which use has been made of the continuity equation (6-45) to eliminate $\partial(\rho a)/\partial t + \partial \dot{m}/\partial s$. On division by $-\rho a R$, this yields Eq. (8-34) as required.

Comments on Example 8-8

1. It is not strictly necessary to use angular momentum concepts in the preceding derivation. Indeed, the same result could have been obtained by applying the linear momentum equation in the direction of the local tangent to a streamline. This may be verified by comparing the incompressible form of Eq. (8-34), namely

$$\frac{\partial}{\partial s}\left(\frac{p}{\rho} + \tfrac{1}{2}V^2 + gz\right) + \frac{\partial V}{\partial t} = 0 \qquad (8\text{-}40)$$

with Eq. (7-56) obtained for flow in a straight diffuser. Nevertheless, it is intuitively more satisfying to use the principle of angular momentum because this takes explicit account of the changing direction of flow.

2. The analysis can be extended to include flows involving shear stresses. Using δF_τ to denote the net shear force on the elemental length of streamtube shown in Fig. 8-11 we find the resultant force to be

$$F_\theta = -a\, \delta p - \delta F_\tau - \rho g a \frac{\partial z}{\partial s}\, \delta s \qquad (8\text{-}41)$$

and so an extended form of Eq. (8-34) is

$$\frac{1}{\rho}\frac{\partial p}{\partial s} + V \frac{\partial V}{\partial s} + g \frac{\partial z}{\partial s} + \frac{\partial V}{\partial t} + \frac{1}{\rho a}\frac{\partial F_\tau}{\partial s} = 0 \qquad (8\text{-}42)$$

3. In some flows—notably through pumps and turbines—lateral accelerations give rise to pressure changes that do not satisfy Eq. (8-42). Using δF_p to denote the equivalent axial force on an elemental length of streamtube,

we obtain the even more general expression

$$\frac{1}{\rho}\frac{\partial p}{\partial s} + V\frac{\partial V}{\partial s} + g\frac{\partial z}{\partial s} + \frac{\partial V}{\partial t} + \frac{1}{\rho a}\frac{\partial F_\tau}{\partial s} + \frac{1}{\rho a}\frac{\partial F_p}{\partial s} = 0 \qquad (8\text{-}43)$$

On integration between the arbitrary points 1 and 2 on a particular streamline, this yields:

$$\left[\int \frac{dp}{\rho} + \tfrac{1}{2}V^2 + gz\right]_2^1 = \int_1^2 \frac{\partial V}{\partial t}\,ds + \int_1^2 \frac{dF_\tau}{\rho a} + \int_1^2 \frac{dF_p}{\rho a} = 0 \qquad (8\text{-}44)$$

In practice, this equation can be integrated exactly only for simple classes of flow. The most important cases are considered in Chapter Nine.

PROBLEMS

1 Figure 8-12*a* depicts a jet of water being deflected through an angle of 65° by a stationary vane. The rate of flow is 20 l/s and the velocities at inlet and outlet are both equal to 16 m/s. Estimate the force exerted by the vane in the *y*-direction.

$$[-290\,\text{N}]$$

(*NB*: Problems 1, 2 and 3 should be solved using the linear momentum equation. Their relevance to this chapter is made clear in Problems 4 and 5.)

2 Figure 8-12*b* depicts a uniform stream of water being deflected through an angle of 65° by a line of stationary vanes. The rate of flow between adjacent

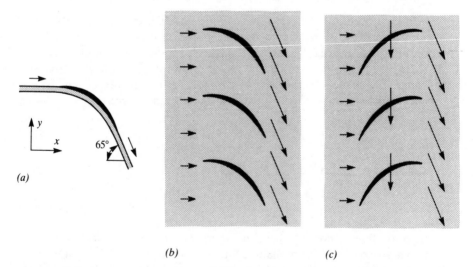

(a)

(b)   (c)

Figure 8-12 Deflection of a jet and streams by curved vanes. (a) Single vane. (b) Cascade. (c) Moving cascade.

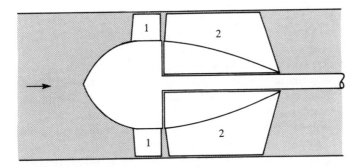

Figure 8-13 Axial-flow pump.

vanes is 20 l/s and the velocity at inlet is 16 m/s. Determine the velocity components in the x- and y-directions at outlet and estimate the force exerted by each vane in the y-direction.

[16 m/s, 34.3 m/s, −686 N]

3 Figure 8-12c depicts a uniform stream of water being deflected through an angle of 65° by a line of moving vanes. The inlet and outlet conditions are the same as in Problem 2. Estimate the force exerted by each vane in the y-direction and the power required to maintain its velocity at 35 m/s. (*Hint*: Consider the flow relative to the vanes.)

[−686 N, 24.0 kW]

4 Figure 8-13 depicts an axial flow pump in which the radius at the mid-height of the rotor blades is 350 mm. Water approaches the blades with no swirl and leaves with a swirl (circumferential) velocity of 25 m/s. Estimate the torque exerted by the rotor if the rate of flow is 1.6 cumec. Also estimate the power required to drive the rotor if the speed of rotation is 955 rev/min.

[14.0 kNm, 140 kW]

5 Steam at a temperature of 560°C and a density of 45 kg/m³ approaches the inlet guide vanes in a high-pressure steam turbine axially at a velocity of 75 m/s (Figure 8.14). The vanes deflect the flow so that its circumferential

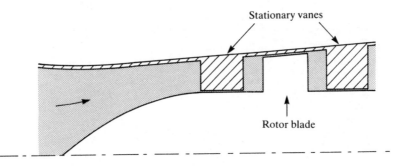

Figure 8-14 Idealized steam turbine.

velocity component at inlet to the first rotor is equal to the local velocity of the moving vanes, namely 150 m/s. On exit from the rotor, the flow is again axial (ready for entry to the next set of guide vanes).

Steam with the same initial state enters a second turbine with the same velocity and leaves the inlet guide vanes with a circumferential velocity component of 300 m/s, double the rotor blade velocity. On exit from the rotor, the flow is axial.

Neglecting non-uniformities, estimate the powers transferred to the rotors per kg of steam. In the first machine the diameter of the rotor (measured at the mid-height of the blades) is 600 mm. In the second it is 300 mm.

[22.5 kW, 45.0 kW]

6 Estimate the bending moment at the section A in the horizontal pipe shown in Fig. 7-19 if the distance between the axes of the straight lengths of pipe is 650 mm. The flow conditions are described in Problem 7-4 at the end of Chapter Seven.

[88.6 Nm]

7 A rectangular tank is supported on a hinged bearing and a roller bearing as shown in Fig. 8-15. The tank is being filled by a jet of water and emptied through an orifice at the same level as the hinge. The rates of inflow and outflow are both equal to 125 l/s and the respective velocities are (a) 20 m/s at 45° to the horizontal and (b) 5 m/s horizontally. Neglecting air resistance and water surface disturbances, estimate the magnitude of the vertical reaction at the roller support if it is known to be 50 kN with the same water level, but no inflow or outflow.

[52.5 kN]

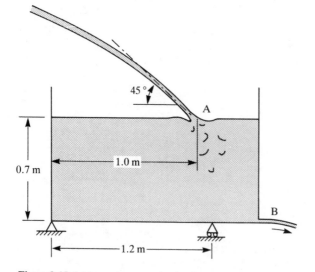

Figure 8-15 Jet impact on a tank of water.

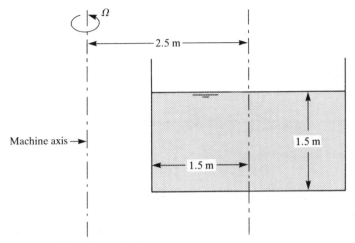

Figure 8-16 Rotating tank of water.

8 At the inlet and outlet of a centrifugal pump impeller, the tangential components of velocity of the water are zero and 30 m/s respectively. The mean velocity in the 125 mm bore inlet pipe is 7.5 m/s and the outer diameter of the impeller is 300 mm. Estimate the torque exerted by the shaft and hence deduce the power required if the shaft rotates at 1000 rev/min.

[414 Nm, 43.4 kW]

9 A Pelton wheel is to be designed to deliver a power of 10 MW when rotating at 750 rev/min. The velocity of the water jets on exit from the nozzles will be 65 m/s and there is to be no circumferential component of velocity after the jets leave the buckets. If the maximum rate of flow that can be provided at this velocity is 2 m³/s, what is the minimum acceptable diameter of the wheel? (See Fig. 8-6.)

[2.04 m]

10 The polar moment of inertia of a Pelton wheel unit is 50 000 kg m² and the power supplied at a rotational speed of 600 rev/min is 6 MW. If the water flow suddenly ceases but the electrical load remains, what will be the initial deceleration of the wheel?

[1.91 rads/s²]

11 The cylindrical tank of water shown in Fig. 8-16 is rigidly supported with its axis vertical at a radial distance of 2.5 m from the vertical axis of a fairground machine. The internal diameter of the cylinder is 3 m and the depth of water is 1.5 m. Neglecting changes in surface level, estimate the polar moment of inertia of the water about the axis of the machine assuming (*a*) that the water does not rotate (approximately true during acceleration) and (*b*) that the water rotates like a rigid body (approximately true if the tank has sufficient internal vanes).

[66.3 Mg m², 78.2 Mg m²]

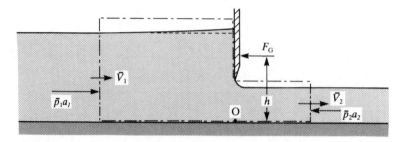

Figure 8-17 Force on a sluice gate.

12 Using an analysis similar to that in Example 8-7, prove that the pressure distribution does not vary hydrostatically in the flow under the sluice gate depicted in Fig. 8-17. The depths and velocities are described in Example 7-4. (*Hint*: Show that the assumption $\Sigma C = 0$ leads to an impossible line of action of the resultant force on the gate.)

FURTHER READING

Addison, H. (1966) *Centrifugal and Other Rotodynamic Pumps*, 3rd edn, Chapman & Hall.
Dixon, S.L. (1978) *Fluid Mechanics, Thermodynamics of Turbomachinery*, 3rd edn, Pergamon.
Stepanoff, A.J. (1957) *Centrifugal and Axial Flow Pumps*, 2nd edn, Wiley.
Whittle, Sir Frank (1981) *Gas Turbine Aerodynamics with Special Reference to Aircraft Propulsion*, Pergamon.

NINE

BERNOULLI EQUATIONS

9-1 EULER EQUATION

In many of the examples considered in previous chapters, the conditions at particular flow sections can be approximated to one-dimensional or at least to uniaxial. In this case, the equations of motion adopt a comparatively simple form amenable to analysis. However, there are many flows of interest in which uniaxial conditions do not exist over complete flow sections—as a glance at some of the figures in this chapter and in Chapter Ten will show. We need a method of analysing these in a simple manner, and this is provided by *Bernoulli equations*.

Differential forms of Bernoulli equations were developed in Examples 7-8 and 8-8. In the latter case, attention was focused on a streamtube of elemental cross-sectional area around an arbitrarily chosen streamline. For an axis s in the direction of flow at any point on the streamline, the momentum equation can be written as

$$\frac{1}{\rho}\frac{\partial p}{\partial s} + V\frac{\partial V}{\partial s} + g\frac{\partial z}{\partial s} = -\frac{\partial V}{\partial t} \qquad (9\text{-}1)$$

when the influence of shear stresses may be neglected. Equation (9-1) is known as an *Euler equation* because it can alternatively be derived directly from any of Eqs (7-88) by choosing the appropriate axis parallel to the velocity vector and introducing a gravitational body force.

9-1-1 Natural coordinates

It is desirable to use integral rather than differential forms of equations whenever possible. In this particular case, the equation we wish to integrate, namely (9-1), is valid in the direction of a streamline, and so our axis s must be chosen along a streamline. This choice is illustrated in Fig. 9-1 together with an axis n chosen normal to streamlines. The axes s and n are known as *natural coordinates* or as *intrinsic coordinates* because their directions are determined by the flowfield, not chosen independently by us in the manner of previous chapters.

In some texts, the axis n has a more precise meaning than is implied by Fig. 9-1, namely along the principal normal to the streamline. There is then no curvature locally in the direction of a third axis mutually perpendicular to s and n. In this book, the axis n typifies any direction normal to a streamline, but it coincides with the principal normal in most examples.

Notice that no restriction is placed on the allowable shape of the streamline. The results obtained in this chapter are valid for highly three-dimensional flows. However, it must always be remembered that the coordinate s is chosen along a *streamline*. As explained in section 1-2, this is not necessarily the same as a pathline or a streakline unless the flow is steady.

9-2 VARIATION OF THE BERNOULLI SUM ALONG A STREAMLINE

No general solution exists for Eq. (9-1), but it can be integrated exactly for certain important classes of flow. We shall consider it in turn for (a) steady incompressible flows, (b) steady compressible flows and (c) unsteady flows.

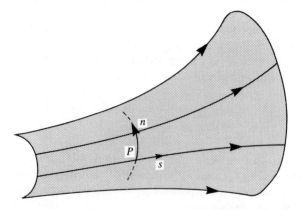

Figure 9-1 Natural coordinates (s = along a streamline; n = normal to streamlines).

9-2-1 Steady incompressible flows

In a steady incompressible flow, $\partial V/\partial t = 0$ and the density ρ is constant. The integration of Eq. (9-1) along a streamline is therefore straightforward. Since $V \partial V/\partial s$ is equivalent to $(\partial/\partial s)(\frac{1}{2}V^2)$, the integration yields

$$\frac{p}{\rho} + \tfrac{1}{2}V^2 + gz = \text{constant} \tag{9-2}$$

which is the basic form of the celebrated *Bernoulli equation* for steady incompressible inviscid flows. It may be written for any two points 1 and 2 *on the same streamline* as

$$\frac{p_1}{\rho} + \tfrac{1}{2}V_1^2 + gz_1 = \frac{p_2}{\rho} + \tfrac{1}{2}V_2^2 + gz_2 \tag{9-3}$$

Example 9-1 illustrates the use of this equation.

Bernoulli sum The three terms p/ρ, $\frac{1}{2}V^2$ and gz commonly occur as a sum in the analysis of fluid flows. The sum is conveniently called the *Bernoulli sum B* defined as

$$B \equiv \frac{p}{\rho} + \tfrac{1}{2}V^2 + gz \tag{9-4}$$

and the Bernoulli equation (9-3) becomes $B_1 = B_2$ or

$$B_1 - B_2 = 0 \tag{9-5}$$

Some authors refer to the sum as the Bernoulli *constant*. In view of the above equations, the reason for this practice is obvious. Nevertheless, it is misleading because the sum is commonly used in the description of flows where its value is not constant. The sum often has different values on different streamlines—as we shall see in Chapter Ten—and it can also vary along individual streamlines (see section 9-3).

Total head On dividing the Bernoulli sum by g, we obtain another widely used parameter, namely the *total head* h_0. This is

$$h_0 \equiv \frac{p}{\rho g} + \frac{V^2}{2g} + z \tag{9-6}$$

It may be regarded as the sum of the piezometric head $h^* = p/\rho g + z$ and the kinetic head $V^2/2g$.

Range of validity of the Bernoulli equation Equation (9-2) is valid along any streamline in any steady, inviscid, incompressible flow. There are no restrictions on the shape of the streamline or on the geometry of the overall flow. The equation is valid for flows in one, two or three dimensions and it is especially useful in the latter cases.

More general expressions are developed in the following sections for unsteady and compressible flows, but the requirement that fluid shear must be negligible is not relaxed. This is the most important limitation in the usefulness of the Bernoulli equation because shear stresses exist in most flows of practical interest. An empirical method of dealing with this complication is developed in section 9-3.

Line integrals The integration of Eq. (9-1) along a streamline is an example of a line integral. The procedure, devoid of details, is as follows. Integration over the elemental distance between the points 0 and 1, say, in Fig. 9-2 yields $B_0 = B_1$. Similarly, integration from the point 1 to the point 2 yields $B_1 = B_2$ and successive integrations give $B_2 = B_3 = B_4$, etc. The Bernoulli equation is obtained by repeating this process indefinitely along the streamline. *Vortex equations* are developed in a similar manner in Chapter Ten by integrating Euler equations along paths that are everywhere normal to the local direction of flow, for instance the path 0ABC in Fig. 9-2.

9-2-2 Steady Compressible Flows

When the density of the fluid may not be regarded as constant, the first term in the Euler equation (9-1) can be integrated along a streamline only if an analytical relationship between p and ρ is known. Fortunately, most compressible flows of practical interest involve fluids that may be treated as perfect gases and so the equation of state (1-11) is applicable. The Euler equation becomes

$$\frac{RT}{p}\frac{\partial p}{\partial s} + \frac{\partial}{\partial s}(\tfrac{1}{2}V^2) + g\frac{\partial z}{\partial s} = 0 \qquad (9\text{-}7)$$

in which R is the gas constant and T is the absolute temperature.

The most important cases where this equation can be integrated are isothermal flows and polytropic flows.

Isothermal flows There are relatively few instances where isothermal flows are also inviscid. Nevertheless, when the temperature of every particle on a particular streamline can be regarded as constant—equal to T_0, say—Eq. (9-7) can be integrated to give

$$RT_0 \ln(p) + \tfrac{1}{2}V^2 + gz = \text{constant} \qquad (9\text{-}8)$$

By analogy with Eq. (9-2), the expression on the left-hand side may be regarded as the Bernoulli sum for an isothermal flow.

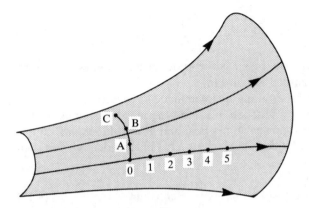

Figure 9-2 Integration along natural coordinates.

Polytropic flows In some flows, the pressure and density may be assumed to be related by the polytropic expression

$$p = k\rho^N \tag{9-9}$$

in which the coefficient k and the polytropic exponent N may be regarded as constants. In this case, the integration of the term $(1/\rho)(\partial p/\partial s)$ along a streamline may be carried out as follows

$$\int \frac{1}{\rho} \frac{\partial p}{\partial s} ds = \int \frac{1}{\rho} dp = k^{1/N} \int p^{-1/N} dp = \frac{N}{N-1} k^{1/N} p^{1-1/N} = \frac{N}{N-1} \frac{p}{\rho} \tag{9-10}$$

and the integration of Eq. (9-1) for a steady flow yields

$$\frac{N}{N-1} \frac{p}{\rho} + \tfrac{1}{2} V^2 + gz = \text{constant} \tag{9-11}$$

The expression on the left-hand side of this equation is the Bernoulli sum for a polytropic flow.

It is shown in Chapter Twelve that when an inviscid fluid particle undergoes an adiabatic (no heat) process, its specific entropy remains constant. The process is termed *isentropic*. When all particles undergo isentropic processes and they all have the same value of specific entropy then the flow is termed *homentropic*. In this case the polytropic index N is equal to the ratio of the specific heat capacities γ (see section 2-9). Example 9-3 (p. 257) illustrates the use of Eq. (9-11) in homentropic flows.

9-2-3 Unsteady Flows

When the flow is unsteady, the integral form of Eq. (9-1) is

$$\int \frac{1}{\rho} dp + \tfrac{1}{2} V^2 + gz + \int \frac{\partial V}{\partial t} ds = f(t) \qquad (9\text{-}12)$$

At any particular instant, the time t is the same for all particles and so the function $f(t)$ may be regarded as constant.

The integration of $\partial V/\partial t$ along a streamline can be carried out only if $\partial V/\partial t$ can be expressed as a function of s. No general expression exists, but appropriate relationships can sometimes be found. For instance, in Example 7-7 (p. 194) we considered a flow in which $\partial V/\partial t$ is independent of s. In such cases Eq. (9-12) becomes

$$\int \frac{1}{\rho} dp + \tfrac{1}{2} V^2 + gz + s \frac{dV}{dt} = \text{constant} \qquad (9\text{-}13)$$

or $$B_1 - B_2 = \Delta s \frac{dV}{dt} \qquad (9\text{-}14)$$

in which $\Delta s = s_2 - s_1$ is the distance along the streamline from the point 1 to the point 2. Note that this expression is valid for flows along curved streamlines as well as straight ones.

One-dimensional, incompressible flow In the special case of one-dimensional incompressible flow along a rigid, tapered duct, the volumetric flux $Q = aV$ is independent of the distance s. The acceleration at any position along the duct may be written as

$$\frac{\partial V}{\partial t} = \frac{1}{a} \frac{\partial Q}{\partial t} \qquad (9\text{-}15)$$

and the unsteady-flow Bernoulli equation becomes

$$B_1 - B_2 = \frac{\partial Q}{\partial t} \int_1^2 \frac{ds}{a} \qquad (9\text{-}16)$$

This is a convenient form of the equation because the integral $\int ds/a$ is dependent only upon the geometry of the duct.

9-3 CHANGES IN THE BERNOULLI SUM

In many flows of practical interest it is not reasonable to neglect fluid shear and the Bernoulli sum does not remain constant along a streamline. Neither does it remain constant when it flows through pumps and turbines because work is done on or by the fluid. In principle we can take account of these effects in the manner indicated in the comments following Example 8-8. However, the integration of the various terms is rarely practicable so we usually abandon the

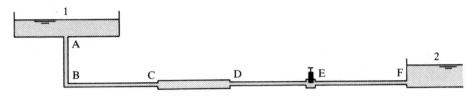

Figure 9-3 Sources of local losses in a pipeline.

purely theoretical approach and adopt an empirical method of estimating changes in the Bernoulli sum (but see Example 10-4).

Suppose that we use suitable instruments to measure the pressure, velocity and elevation at various positions along the pipeline shown in Fig. 9-3. Having done so, we can use, say, Eq. (9-4) to deduce the corresponding values of the Bernoulli sum. Experience shows that in a steady flow

1. there are sudden reductions in the Bernoulli sum at geometrical discontinuities A, B, C, etc. and
2. there is a gradual reduction in the Bernoulli sum along each length of pipe.

In addition, there may be increases or decreases in the Bernoulli sum because of pumps (or compressors) and turbines.

9-3-1 Local losses

The origins of typical abrupt reductions in the Bernoulli sum are illustrated by Fig. 9-4. In all cases there is a region of recirculating flow, downstream of which the streamlines diverge. Within the recirculating region and for several diameters downstream the flow is highly disturbed. Usually there is considerable small scale turbulence.

If our measuring instruments are located at the points 1 and 2, a few diameters upstream and downstream of the discontinuities, they will detect a reduction ΔB in the Bernoulli sum which is conventionally related to the local value of $\frac{1}{2}\bar{V}^2$ by a so-called *energy loss coefficient* k. By definition, k satisfies

$$k \equiv \frac{\Delta B}{\frac{1}{2}\bar{V}^2} \tag{9-17}$$

and so the local reduction in the Bernoulli sum is

$$B_1 - B_2 = \frac{1}{2}k\bar{V}^2 \tag{9-18}$$

The energy 'loss' is not really a loss, of course. It is a conversion from mechanical energy to internal energy. Notice that we use the mean velocity $\bar{V}$ in this definition. In practice it is not possible to distinguish individual

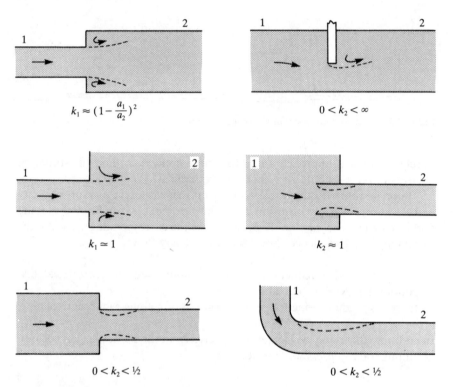

Figure 9-4 Typical local loss coefficients.

streamlines in turbulent flows and it is rarely practicable to do so in laminar flows.

Typical values of k are indicated in Fig. 9-4 and an approximate analysis is given in Example 9-8 (p. 263) for the special case of a sudden geometrical expansion.

Some authors use the term *minor losses* instead of local losses. This practice developed in applications where these losses are small in comparison with the other common cause of reductions in the Bernoulli sum, namely skin friction. However, the term is disliked by the author because it is grossly misleading when local losses are the major cause of changes in the Bernoulli sum. This is always the case when pipes are sufficiently short.

9-3-2 Skin friction

The influence of skin friction has been discussed in previous chapters. For present purposes it is sufficient to note that our measuring instruments will detect reductions in the Bernoulli sum that can be related to the shear stresses on the walls of the pipe (or channel, etc.).

In a pipe of length L and perimeter l (e.g. $l = \pi D$ for a circular section pipe of diameter D and $l = 4D$ for a square section pipe of side D), the skin friction force due to a wall shear stress τ_w is $\tau_w lL$. The momentum equation for a steady, incompressible flow in a straight pipeline of constant bore (Fig. 9-5) is

$$(p_1 - p_2)a - \rho g a L \frac{\partial z}{\partial x} - \tau_w lL = 0 \qquad (9\text{-}19)$$

provided that $\beta_2 \approx \beta_1$ as is nearly always the case. Using $L\, \partial z / \partial x = (z_2 - z_1)$ and $\tau_w = \frac{1}{2} f \rho \bar{V}^2$ where f is a skin friction coefficient, we obtain

$$\left(\frac{p}{\rho} + gz\right)_1 - \left(\frac{p}{\rho} + gz\right)_2 = \frac{1}{2} f \frac{L}{R} \bar{V}^2 \qquad (9\text{-}20)$$

in which $R \equiv A/l$ is known as the *hydraulic radius* of the cross-section.

The mean velocity is constant in a steady incompressible flow along a constant bore pipe. We may therefore add $\frac{1}{2}\bar{V}^2$ into each of the terms in parenthesis in Eq. (9-20) to produce Bernoulli sums for the sections 1 and 2. The equation becomes

$$B_1 - B_2 = \frac{1}{2} f \frac{L}{R} \bar{V}^2 \qquad (9\text{-}21)$$

which is one form of the Darcy–Weisbach equation. A more common form is obtained by dividing each term by g so that the expressions in parentheses become total heads. The resulting equation may be written as

$$h_f = f \frac{L}{R} \frac{\bar{V}^2}{2g} \qquad (9\text{-}22)$$

in which h_f denotes the loss of total head due to friction.

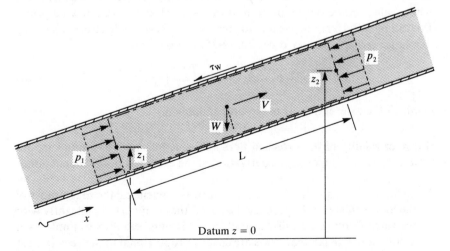

Figure 9-5 Steady flow in a simple pipe.

A comparison of Eq. (9-21) with Eq. (9-18) shows that the loss coefficient for skin friction along a pipe may be written as

$$k = f \frac{L}{R} \tag{9-23}$$

For the special case of a pipe of circular section and diameter D, the hydraulic radius is $R = A/l = D/4$ and the loss coefficient is

$$k = 4f \frac{L}{D} \tag{9-24}$$

Since circular section pipes are so common, some authors choose to define an alternative skin friction coefficient $\lambda = 4f$ and thereby eliminate the factor 4 in Eq. (9-24). This practice has merit, but it leads to a less elegant form of Eq. (9-23) and it implies the definition $\lambda = \tau_w/\frac{1}{8}\rho V^2$ which is at variance with the usual practice for defining stress coefficients (see section 5-6).

9-3-3 Extended Bernoulli Equation

The total change in the Bernoulli sum along any particular streamline is the sum of all the individual contributions due to local or distributed losses. Since each contribution can be expressed as $\frac{1}{2}k_i \bar{V}_i^2$, the total change between any two points 1 and 2 can be written as

$$B_1 - B_2 = \frac{1}{2}\Sigma k_i \bar{V}_i^2 \tag{9-25}$$

and this is known as an *extended Bernoulli equation*. Most of the coefficients must be obtained empirically, but once they are known the equation is as useful as its inviscid flow counterpart (9-5). Often, it is used in circumstances where values of the Bernoulli sums at the points 1 and 2 are known a priori. For example B/g is equal to the static piezometric head in the reservoirs in Fig. 9-3. Since $Q = a_i \bar{V}_i$ in any pipe, Eq. (9-25) can be written as

$$B_1 - B_2 = \frac{1}{2}Q^2 \Sigma \left(\frac{k_i}{a_i^2} \right) \tag{9.26}$$

in which the rate of flow Q is the only unknown.

Range of validity of the extended Bernoulli equation The following simplifications have been made in the derivation of the extended Bernoulli equation:

1. So far, no attention has been paid to factors influencing the magnitudes of the loss coefficients. In practice, however, they can vary significantly with the rate of flow, especially at low Reynolds numbers. For example, skin friction forces in laminar flows are usually proportional to the velocity, not

its square. It follows that the skin friction coefficient defined by $f \equiv \tau_w / \frac{1}{2} \rho \bar{V}^2$ is inversely proportional to the velocity.

2. The general equation (9-25) may be used for compressible or incompressible flows with an appropriate meaning implied for the Bernoulli sum. Experiments show that the numerical values of the various loss coefficients are influenced only slightly by compressibility, except at high Mach numbers. However, changes in density can influence the loss coefficients indirectly by causing changes in the Reynolds number.

   The particular forms of the skin friction expressions developed in section 9-3-2 are valid only for incompressible flows. However, they may also be used for compressible flows when the length L is infinitesimal. In this case, the change in the Bernoulli sum can be obtained by integrating along the pipe.

3. In the form presented herein the extended Bernoulli equation applies only to steady flows. When the flow is unsteady, account must also be taken of the acceleration $\partial V / \partial t$ in Eq. (9-1). For sufficiently gentle accelerations this simply leads to the inclusion of the additional term $\int (\partial V / \partial t)\, ds$ in equations such as (9-25) and (9-26). In rapidly accelerating flows, however, account must also be taken of the influence of unsteadiness on the numerical values of the loss coefficients themselves.

4. Changes in the Bernoulli sum also occur in pumps and turbines where work is done on or by the fluid. Using ΔB_p to denote changes due to this cause, a general expression for the extended Bernoulli equation is

$$B_1 - B_2 = \int_1^2 \frac{\partial V}{\partial t}\, ds + \tfrac{1}{2}\Sigma k_i \bar{V}_i^2 + \Delta B_P \qquad (9\text{-}27)$$

5. In this section, the discussion has focused on flows in pipes. Nevertheless, the concepts are equally valid for flows in channels. So are most of the equations provided that the pressures and elevations are evaluated at the centroids of the flow sections.

9-3-4 Bernoulli and Energy Equations

The various forms of the Bernoulli equations are commonly referred to as energy equations. Since each term has the dimensions of energy per unit mass, this practice is understandable. However, it should be recognized that only so-called 'mechanical' energy terms are present. No explicit allowance can be made for thermal effects.

A fuller comparison of Bernoulli and energy equations is given in section 11-3-1.

9-4 MEASUREMENT OF FLOWRATE AND VELOCITY

The measurement of pressure is discussed in section 3-6. We often also need to measure velocities or their counterpart, the flow rate. In a laboratory it is sometimes possible to measure a steady rate of flow of a liquid by allowing it to discharge into a tank and recording the increase of liquid volume in the tank in some time interval. Alternatively, the increase of mass can be measured in a weigh-tank. In both cases, it is possible to achieve high accuracy by using a sufficiently large tank and measuring over a long time interval.

Unfortunately, this simple method can rarely be used in everyday engineering practice. Even in a laboratory, it is not suitable for use in a closed system. Also it is not appropriate for the measurement of non-steady flows. We now consider a few of the most important alternative methods of flow measurement.

9-4-1 Weirs

Figure 9-6 illustrates a liquid flow through a sharp-edged triangular weir (sometimes called a vee notch) in the side of a tank. As with an orifice, a contraction of the 'jet' occurs, but in this case the flow does not become uniaxial close to the outlet.

Although significant errors are unavoidable it is informative to use the Bernoulli equation to gain an insight into the relationship between the flowrate Q and the height H_0 of the upstream liquid surface. The Bernoulli equation shows that the velocity at a depth H below the upstream surface is $V = \sqrt{2gH}$ if the local gauge pressure is zero. At the same depth the cross-sectional area of an elemental strip of width b and depth δH in the plane of the weir is $b \, \delta H = 2(H_0 - H) \tan \theta \, \delta H$. If we now assume (wrongly) that the velocity vectors at this section are horizontal, that there is no surface drawdown (wrong again), and that atmospheric pressure prevails throughout the section (also wrong) then the volumetric flux is

$$Q = \int_0^{H_0} 2 \tan \theta (2gH)^{1/2}(H_0 - H) \, dH \qquad (9\text{-}28)$$

On integration,

$$Q = 2 \tan \theta (2g)^{1/2} [\tfrac{2}{3}H_0 H^{3/2} - \tfrac{2}{5}H^{5/2}]_0^{H_0} \qquad (9\text{-}29)$$

On introducing a coefficient of discharge C_d to account for the erroneous assumptions, we finally obtain

$$Q = \tfrac{8}{15} \tan \theta (2g)^{1/2} C_d H_0^{5/2} \qquad (9\text{-}30)$$

Remarkably, experiment shows that the coefficient C_d is nearly constant at sufficiently high Reynolds numbers provided that H_0 exceeds about 20 mm (so

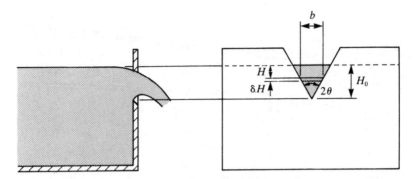

Figure 9-6 Flow over a triangular weir.

that the influence of surface tension is small). A value of 0.58 may be used for sharp-edged weirs with vertex angles between 45° and 120°.

Sharp-edged weirs of other shapes are sometimes used, but the triangular shape is probably the most useful because it can be used over a wide range of flow rates. This is because the head varies with $Q^{2/5}$ whereas the head above a rectangular weir, say, varies with $Q^{2/3}$. Notice, however, that the triangular weir is more sensitive to errors in the measurement of the height of the liquid surface.

In rivers and man-made channels, flow measurement controls commonly consist of weirs formed by building blockages in the bed of the channel. When the blockage is sufficiently high, the water flows over it and discharges in a manner resembling the flow over a sharp-crested weir.

Another flow control used in channels is the venturi flume. In the simplest case, it is the width rather than the depth of the channel that is restricted (although we sometimes constrict them both). At some rates of flow, the device acts similarly to the venturi meter described in the following section, but at most rates of flow it acts more like a large weir.

Henderson (1966) gives a good account of flow measurement in channels.

9-4-2 Venturi Meter

The rate of flow along a pipe can be determined by measuring the pressure difference caused by a restriction. The *Venturi meter* depicted in Fig. 9-7 is a convergent–divergent tube which can be installed in a pipeline and through which the whole of the flow passes. Pressure tappings are provided at various points around the circumference just upstream of the convergent section and at the throat. The pressure difference between these positions is typically measured using a differential U-tube manometer.

The incompressible Bernoulli equation (9-3) may be written for the points

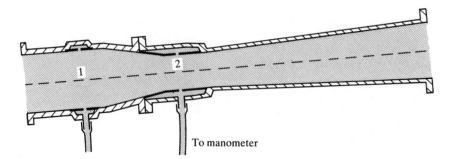

Figure 9-7 Typical Venturi meter (flow from left to right).

1 and 2 on the axial streamline as

$$h^*_1 + \frac{V_1^2}{2g} = h^*_2 + \frac{V_2^2}{2g} \tag{9-31}$$

in which $h^* = p/w + z$ is the piezometric head. Assuming one-dimensional flow, the continuity equation is $a_1 V_1 = a_2 V_2$ and so

$$h^*_1 - h^*_2 = \left\{ \left(\frac{a_1}{a_2} \right)^2 - 1 \right\} \frac{V_1^2}{2g} \tag{9-32}$$

After substituting measured values for the areas and for the piezometric head difference between the points 1 and 2, we obtain the velocity V_1. The flow rate is then given by

$$Q = C_d a_1 V_1 \tag{9-33}$$

in which the coefficient of discharge C_d (typically 0.98) is introduced because the flow is not truly inviscid or one-dimensional.

Notice that the rate of flow is dependent upon the piezometric head difference, not the static head difference (see section 3-6). This makes it possible to use the meter in any orientation without loss of accuracy. The piezometric head difference may be measured in any convenient manner. However, differential U-tube manometers are commonly used for this purpose and the reason can be inferred from a comparison of Eqs (3-35) and (9-32). By inspection, the required measurement (Δh^*) is obtained directly.

A similar device known as a *Venturi flume* can be used to measure rates of flow in open channels (see Example 9-5, p. 258). In practice, however, they are used in an ingenious manner that makes it possible to estimate the rate of flow by means of a single depth measurement alone. There is no need to measure two depths—which would be analogous to the Venturi meter.

Orifice meters An *orifice meter* is a simple alternative to the Venturi meter. As illustrated in Fig. 9-8, it consists of an orifice plate which may be located at any

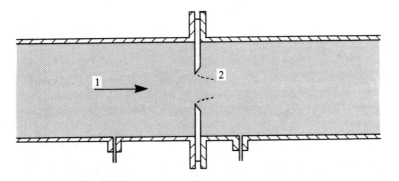

Figure 9-8 Typical orifice meter.

convenient flange in the pipe. Its principal advantages over the Venturi meter are its simplicity and lower cost. Its main disadvantages are that it offers a greater resistance to the flow and that it is more easily damaged—which leads to an unknown change in its coefficient of discharge.

For incompressible flows, Eq. (9-32) and (9-33) may again be used, with the point 2 chosen at the vena contracta downstream of the orifice. In practice, the area of the orifice is used instead of the (unknown) area of the vena contracta in Eq. (9-32) and the coefficient of discharge is adjusted accordingly.

9-4-3 Pitot-static Tube

Each of the above devices is designed to measure a rate of flow. Sometimes, however, we need to measure *velocities* at points in a flowfield. The *pitot-static tube* shown in Fig. 9-9 is a simple yet accurate instrument that is commonly used for this purpose. It consists of two tubes, one within the other, and these are aligned parallel to the flow in the region of measurement. The outer tube communicates with the flow through small holes in its circumference and so the pressure within it is equal to the static pressure p of the flow. The inner (pitot) tube causes a *stagnation point* in the flow at its mouth, that is a point in the flow at which the local velocity is zero. The point B on the leading edge of the aerofoil shown in Fig. 9-12 is also a stagnation point.

When the fluid is incompressible, the Bernoulli sums at the points 1 and 2 are equal because both are equal to the value just upstream of the instrument. Since the elevation is the same at both points (or negligibly different), the difference between the pressures in the two tubes is equal to $\frac{1}{2}\rho V^2$ where V is the required velocity. This difference is sometimes called the *dynamic pressure* and $V^2/2g$ is the *dynamic head* or *kinetic head*. At the other end of the instrument, the tubes are separated and the difference between their pressures is measured, usually with a differential manometer.

When the flow is compressible, the difference between the stagnation

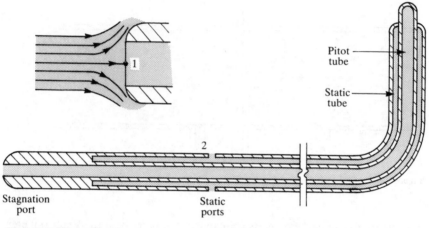

Figure 9-9 Pitot-static tube.

pressure and the static pressure is not simply $\frac{1}{2}\rho V^2$. Instead, it satisfies a more complex relationship which can be found from the homentropic form of Eq. 9-11 (i.e. with $N = \gamma$). Even this expression is inadequate when the flow is supersonic becuse a shock front forms upstream of the tube.

9-4-4 Specialized Techniques for Measuring Velocity

Flow visualization A pitot-static tube measures the velocity at one position in a steady flow-field. When more widespread information is required, use can be made of one of a wide variety of *flow visualization* techniques that are available. The frontispiece gives an indication of the power of photographic techniques. Another method utilises the fact that the path of a moving particle appears as a streak on a photograph taken with a sufficiently long exposure time. If small particles are distributed in the flow, then the lengths of their respective streaks on the photograph give a measure of their velocities. The advantage of this method is that simultaneous measurements are obtained for the velocities at many positions in the flow-field. In water, the particles may be tiny hydrogen bubbles generated by electrolysis at a very thin wire mounted in the flow.

Laser–Doppler anemometer When it is not possible to insert a mechanical device into the flow, the *laser–Doppler* technique is sometimes employed. This makes use of the tiny particles such as dust that are always present in a real fluid. Light from a laser beam is focused onto a tiny point in the flow, and is reflected by particles passing through that point. Because of the velocity of the particles, the *frequency* of the reflected light differs from that which would be obtained by reflection from a stationary particle.

Hot-wire anemometer This is a useful device for the measurement of rapidly varying velocities. It consists of a very thin wire—typically 10 μm in diameter —mounted between two rigid, electrically conducting supports. The wire is heated electrically and held in the flow on the end of a suitable probe. The flow cools the wire at a rate which increases with the velocity, and so the electrical current required to hold its temperature constant is a measure of the fluid velocity.

The main advantages of the hot wire are its ability to respond to very-high-frequency fluctuations and its suitability for a wide range of velocities (but not for very small velocities). Also, being small, it does not greatly disturb the flow. Its disadvantages are its high initial cost, its delicate nature and its need for regular cleaning to prevent deposits collecting on it and altering the cooling effect of the flow. In liquids, *hot films* are often used. These are similar to hot wires, but they are more robust.

9-5 APPLICATIONS OF BERNOULLI EQUATIONS

Example 9-1: Flow through an orifice

(a) Determine the volumetric flux through the 0.002 m² vena contracta at the outlet of the tank of plan area 6 m² shown in Fig. 9-10a when $H = 1$ m.

(b) Hence determine the time taken for the liquid surface level to fall from $H = 1$ m to $H = 0.5$ m.

SOLUTION

(a) Since the flow is nearly steady, the Bernoulli equation (9-3) may be applied along a typical streamline between the point 1 in the liquid surface and the point 2 at the centroid of the vena contracta (see comment 3 below). By choosing the centroid of the vena contracta as the arbitrary datum for the elevation z, we obtain

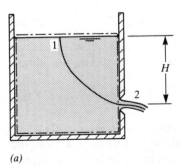

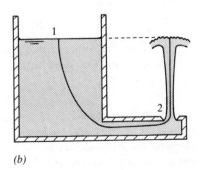

(a)   (b)

Figure 9-10 Flow through an orifice.

$$\frac{p_{AT}}{\rho g}+\frac{V_1{}^2}{2g}+H=\frac{p_{AT}}{\rho g}+\frac{V_2{}^2}{2g}+0 \qquad (9\text{-}34)$$

in which p_{AT} denotes the atmospheric pressure. The velocity V_1 may be neglected in comparison with V_2 and so $H=V_2{}^2/2g$, or

$$V_2=\sqrt{2gH} \qquad (9\text{-}35)$$

The flow through the vena contracta is nearly uniaxial. When H is much larger than the diameter of the jet, V_2 is a good approximation for the *mean* velocity through the vena contracta. Therefore the volumetric flux is given with sufficient accuracy by

$$Q=a_2V_2 \qquad (9\text{-}36)$$

When $H=1\,\mathrm{m}$, Eq. (9-35) gives $V_2=4.43\,\mathrm{m/s}$ and so $Q=8.86\,\mathrm{l/s}$.

(b) Although the flow is very nearly steady when viewed over a short interval, its long-term effect is not steady. The continuity equation (6-5) may be written as

$$\frac{\partial V_{cv}}{\partial t}=0-Q \qquad (9\text{-}37)$$

in which V_{cv} denotes the volume of liquid inside the control volume shown in Fig. 9-10a. The rate of *increase* of volume is $A\,dH/dt$ in which A denotes the plan area of the tank. Also $Q=a_2V_2=a_2\sqrt{2gH}$, and so Eq. (9-37) becomes

$$\frac{dH}{dt}=kH^{1/2} \qquad (9\text{-}38)$$

in which $k=-(a_2/A)\sqrt{2g}=-0.001\,48\,\mathrm{m^{1/2}/s}$ in this example. Equation (9-38) may be written as

$$H^{-1/2}\,dH=k\,dt \qquad (9\text{-}39)$$

and integrated to give

$$2(H_f{}^{1/2}-H_i{}^{1/2})=k(t_f-t_i) \qquad (9\text{-}40)$$

in which the suffices f and i denote the final and initial conditions respectively. On substituting $H_i=1\,\mathrm{m}$ and $H_f=0.5\,\mathrm{m}$, we obtain $(t_f-t_i)=396\,\mathrm{s}$.

Comments on Example 9-1

1. Because of viscous and surface tension effects, Eq. (9-35) slightly over-estimates the true velocity at the centroid of the vena contracta. Also, the flow is not quite uniaxial and the mean velocity in the jet is not quite equal to the value at the centroid. To allow for these effects, a *velocity coefficient*

C_v may be introduced as

$$\bar{V}_2 = C_v V_2 \tag{9-41}$$

Its value is usually very close to unity, typically being about 0.98.

2. A more important difficulty is hidden in the statement of the problem. In a practical case, it is the size of the *orifice*, not the vena contracta, that is known. The area of the vena contracta is conventionally related to the area of the orifice a_o by $a_{vc} = C_c a_o$ in which C_c is a *coefficient of contraction*. The numerical value of C_c for a sharp-edged circular orifice is found to be about 0.63 when the Reynolds number (defined for this purpose as $\rho D V_2/\mu$ in which D is the orifice diameter and μ is the dynamic viscosity of the fluid) exceeds about 10 000.

The expression for the volumetric flow rate through the orifice is usually written as

$$Q = C_d a_o \sqrt{2gH} \tag{9-42}$$

in which a_o denotes the area of the orifice and C_d is known as a *coefficient of discharge*. By inspection, $C_d = C_v C_c$. Its numerical value is usually about 0.61.

3. Strictly, Eq. (9-12) should have been used in the analysis instead of Eq. (9-3) because the flow is not quite steady. This would have necessitated the evaluation of $\int_1^2 (\partial V = \partial t)\, ds$ along the streamline and we should verify that this term is small. We begin by noting that the velocities at the vena contracta when $H = 1$ m and $H = 0.5$ m are approximately 4.43 m/s and 3.13 m/s respectively. Therefore the average magnitude of the acceleration at the vena contracta during this period is $(4.43 - 3.13)/396$ m/s^2 $= 0.003\,28$ m/s^2. Now the accelerations along most of the streamline are much smaller than this (because the area of flow is much greater). Therefore the magnitude of $\int_1^2 (\partial v/\partial s)\, ds$ along a 1 m long streamline from the surface to the vena contracta is much smaller than 0.003 28 m^2/s^2. As expected, this is negligible in comparison with $\frac{1}{2} V_2^2$.

4. The above analysis can also be applied to the tank shown in Fig. 9-10b. However, it is more difficult to locate the position of the vena contracta (and hence $z = 0$) in this case.

Example 9-2: Flow through a submerged orifice Estimate the time required for the equalization of the levels in the two tanks shown in Fig. 9-11. The cross-sectional plan areas are $A_1 = 6$ m^2 and $A_2 = 12$ m^2, the initial difference in levels is 1 m, the diameter of the orifice is 75 mm and the coefficient of discharge may be taken as 0.61.

SOLUTION Once again, the Bernoulli equation may be applied along a typical streamline between the surface point 1 and the centroid of the vena contracta 2. By assuming that the pressure at the vena contracta is equal to the local hydrostatic pressure (see comment 2 below), we obtain

$$\frac{p_{AT}}{\rho g}+0+H_1=\frac{p_{AT}+\rho g H_2}{\rho g}+\frac{V_2{}^2}{2g}+0 \tag{9-43}$$

Therefore

$$V_2{}^2=2g(H_1-H_2) \tag{9-44}$$

and the volumetric flux through the orifice is

$$Q=C_d a_0\sqrt{2g(H_1-H_2)} \tag{9-45}$$

Application of the continuity equation to the contents of the two tanks in turn in a similar manner to Example 9-1 yields

$$A_1\frac{\mathrm{d}H_1}{\mathrm{d}t}=-Q \quad \text{and} \quad A_2\frac{\mathrm{d}H_2}{\mathrm{d}t}=+Q \tag{9-46}$$

which may be combined to give

$$\frac{\mathrm{d}}{\mathrm{d}t}(H_1-H_2)=-\left(\frac{1}{A_1}+\frac{1}{A_2}\right)Q \tag{9-47}$$

On substituting for Q from (9-45) and writing $H=H_1-H_2$, we obtain

$$\frac{\mathrm{d}H}{\mathrm{d}t}=kH^{1/2} \tag{9-48}$$

in which

$$k=-\left(\frac{1}{A_1}+\frac{1}{A_2}\right)C_d a_0\sqrt{2g} \tag{9-49}$$

Equation (9-48) is identical to (9-38) and so the solution is

$$2(H_f{}^{1/2}-H_i{}^{1/2})=k(t_f-t_i) \tag{9-50}$$

In this example, $k=-0.002\,98$ m$^{1/2}$/s and so the time for the difference in levels to reduce from 1 m to zero is $(t_f-t_i)=670$ s.

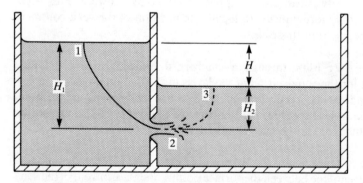

Figure 9-11 Equalization of levels.

Comments on Example 9-2

1. Even though the orifice is submerged, the Bernoulli equation may be applied with good accuracy between the points 1 and 2. However, it must not be applied between the points 2 and 3 because the liquid expands as a free jet and the influence of shear stresses is important.

2. At the vena contracta, the flow is nearly uniaxial. It therefore follows from the vortex equation developed in Chapter Ten that there is negligible variation in pressure along a horizontal line through the jet, normal to the direction of flow. Since the gauge pressure elsewhere on this line is equal to $\rho g H_2$, the gauge pressure in the vena contracta is also $\rho g H_2$.

Example 9-3: Compressible flow Verify that the velocity through the vena contracta in Example 7-3 is 262 m/s if homentropic conditions prevail upstream.

SOLUTION Since homentropic conditions exist, we may use Eq. (9-11) with $N = \gamma$. For a horizontal streamline through the centroid of the vena contracta,

$$\frac{\gamma}{\gamma-1}\frac{p_1}{\rho_1}+0+0=\frac{\gamma}{\gamma-1}\frac{p_2}{\rho_2}+\tfrac{1}{2}V_2{}^2+0 \qquad (9\text{-}51)$$

From Example 7-3 (p. 188) we know that $p_1 = 150\,\text{kPa}$, $p_2 = 100\,\text{kPa}$, $\rho_1 = 1.67\,\text{kg/m}^3$, $\rho_2 = 1.25\,\text{kg/m}^3$ and $\gamma = 1.40$. Therefore $V_2 = 262\,\text{m/s}$.

Comments on Example 9-3

1. The analysis of compressible flows is by no means always as straightforward as this. However, the use of isothermal and polytropic relationships is often possible and this considerably reduces the complexity involved in the introduction of the additional variable ρ in comparison with incompressible flow analyses.

2. It is always necessary to consider the possibility of heat transfers when dealing with gases because changes in temperature directly influence the relationship between pressure and density.

Example 9-4: Flow past an aerofoil Figure 9-12 depicts a steady flow past an aircraft wing, viewed relative to the aircraft. Far upstream, the air pressure, density and velocity are 0.5 bar, $0.75\,\text{kg/m}^3$ and 150 m/s respectively. Estimate the pressures at the points B, C and D where the velocities are zero, 275 m/s and 125 m/s respectively. Neglect viscous effects and changes in elevation. For air, use $\gamma = 1.40$.

SOLUTION Since heat transfers are negligible on the surface of an aircraft wing, we may use the homentropic relationship

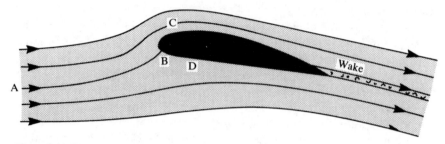

Figure 9-12 Flow past a wing aerofoil.

$$\frac{\gamma}{\gamma-1}\frac{p_A}{\rho_A}+\tfrac{1}{2}V_A{}^2=\frac{\gamma}{\gamma-1}\frac{p_1}{\rho_1}+\tfrac{1}{2}V_1{}^2 \tag{9-52}$$

between a point A far upstream and a typical point I on the aerofoil. Since p_A, ρ_A, V_A, V_1 and γ are all known, it is useful to rearrange this equation as

$$\frac{p_1}{\rho_1}=\frac{p_A}{\rho_A}+\frac{\gamma-1}{2\gamma}(V_A{}^1-V_1{}^2) \tag{9-53}$$

The pressures and densities at the points A and I are also related by the isentropic expression (2-22) as

$$\frac{p_1}{\rho_1{}^\gamma}=\frac{p_A}{\rho_A{}^\gamma} \tag{9-54}$$

Equation (9-53) and (9-54) are a pair of simultaneous equations in p_1 and ρ_1. For the particular points B, C and D they give $p_B=0.590$ bar, $p_C=0.328$ bar and $p_D=0.526$ bar.

Comments on Example 9-4
1. The pressure at C is smaller than that at D and so the wing experiences an upward *lift* force. Notice that the pressure difference (p_A-p_C) is much bigger than (p_D-p_A). It may be said that the lift is due more to low pressure on the upper surface than to high pressure on the lower surface.
2. Inviscid analyses can provide good estimates of the pressure distributions around real aerofoils, and predicted lift forces are often in close agreement with measured values. However, it is not possible to use the same method to estimate drag forces.

   Inviscid theories predict that *no* drag is experienced by any object moving at a steady speed in a fluid that extends to infinity. This is the well known *d'Alembert's paradox*. To explain drag forces, it is necessary to take account of shear stresses that modify the flow close to a solid surface.

Example 9-5: Venturi flume control Figure 9-13 depicts a free surface flow of water through a venturi flume in a 2 m wide rectangular channel.

Estimate the rate of flow along the channel if the mean velocity at the 1 m wide throat of the flume satisfies $V_c^2 = gd_c$. The depth of flow a short distance upstream of the flume is $d_1 = 2.5$ m.

SOLUTION The flow is converging both horizontally and vertically as it approaches the throat of the flume. Therefore separation effects are not expected and the Bernoulli equation may be applied between the upstream section 1 and the throat section C. For points on the bed of the channel at these sections, the pressure is $p = \rho g d$ and so the steady flow Bernoulli equation may be written as

$$d_c + \frac{V_c^2}{2g} + z_c = d_1 + \frac{V_1^2}{2g} + z_1 \qquad (9\text{-}55)$$

and rearranged using $Q = b_1 d_1 V_1$ to give

$$d_c + \frac{V_c^2}{2g} = d_1 + \frac{Q^2}{2g b_1^2 d_1^2} + (z_1 - z_c) \qquad (9\text{-}56)$$

We are told that $V_c^2 = gd_c$ at the throat and so $0.5 d_c = V_c^2 / 2g$ and $1.5 d_c = (d_c + V_c^2 / 2g)$. Therefore the rate of flow is

$$Q = b_c d_c V_c = b_c g^{1/2} d_c^{3/2} = b_c g^{1/2} \left[\frac{2}{3} \left(d_c + \frac{V_c^2}{2g} \right) \right]^{3/2} \qquad (9\text{-}57)$$

or, using (9-56)

$$Q = b_c g^{1/2} \left[\frac{2}{3} \left\{ d_1 + \frac{Q^2}{2g b_1^2 d_1^2} + (z_1 - z_c) \right\} \right]^{3/2} \qquad (9\text{-}58)$$

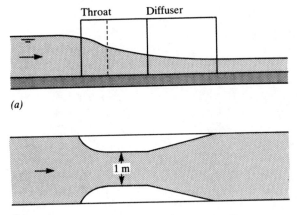

(a)

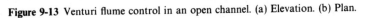

(b)

Figure 9-13 Venturi flume control in an open channel. (a) Elevation. (b) Plan.

Since $z_1 = z_c$, the only unknown parameter in Eq. (9-58) is the rate of flow Q. For the particular values in this example, the equation yields $Q = 7.17\,\mathrm{m^3/s}$.

Comments on Example 9-5

1. It is remarkable that the rate of flow can be deduced from the measurement of a single depth d_1 when the geometry of the structure is known. This makes venturi flumes and their counterparts, weirs, extremely useful methods of flow measurement.
2. At significantly lower rates of flow, the assumption that $V_c^2 = gd_c$ at the throat becomes invalid. In this case, the flume could notionally be used in a manner analogous to a venturi meter in a pipe (section 9-4-2), the rate of flow being deduced from measurements of the depths at the throat as well as upstream. In practice, however, this is relatively uncommon. We are more likely to install a weir in the bed of the flume to cause $z_c > z_1$. The above analysis is valid for this case too.
3. In the numerical example we realistically neglected skin friction and the slope of the channel. Skin friction is necessarily small in a short length of channel and venturi flumes are rarely, if ever, used in steeply sloping channels.

Example 9-6: Oscillating flow in a U-tube Figure 9-14 depicts a large U-tube containing water. The tube is used in a laboratory to provide an oscillatory flow in the horizontal region of length $L_H = 2\,\mathrm{m}$ and cross-

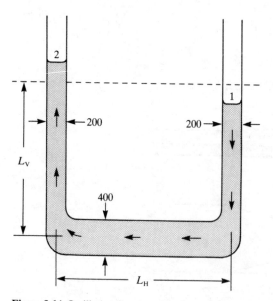

Figure 9-14 Oscillating flow in a U-tube (arrows indicate instantaneous direction of flow).

section 0.4×0.4 m. Estimate the mean length L_V of the water columns in the vertical limbs of breadth 0.2 m and width 0.4 m when the period of oscillation is 3 seconds.

SOLUTION Consider the flow at an instant when the free surface in the left hand limb is rising at a velocity V and is higher than the surface in the other limb. For incompressible flow, the unsteady Bernoulli equation (9-14) may be written for a streamline connecting the points 1 and 2 as

$$\left(\frac{p_{AT}}{\rho} + \tfrac{1}{2}V_1{}^2 + gz_1\right) - \left(\frac{p_{AT}}{\rho} + \tfrac{1}{2}V_2{}^2 + gz_2\right) = \sum_{H+V} L\frac{\partial V}{\partial t} \qquad (9\text{-}59)$$

in which the suffices H and V denote the horizontal and vertical portions of the tube. The total contribution of the two vertical limbs to the sum on the right-hand side of Eq. (9-59) is $2L_V dV/dt$ and the contribution of the horizontal portion is $\tfrac{1}{2}L_H dV/dt$ (NB: The velocity in the horizontal portion is half that in the vertical limbs because of the difference in cross-sectional area.) Therefore, since $V_1{}^2 = V_2{}^2 = V^2$, Eq. (9-59) becomes

$$z_1 - z_2 = \frac{2L_V + \tfrac{1}{2}L_H}{g}\frac{dV}{dt} \qquad (9\text{-}60)$$

Now the velocity at the point 2 is $V = dz_2/dt$. Similarly $V = -dz_1/dt$ and so, by addition,

$$\frac{d}{dt}(z_1 - z_2) = -2V \qquad (9\text{-}61)$$

By differentiating (9-60) and comparing the result with (9-61) we deduce that

$$\frac{2L_V + \tfrac{1}{2}L_H}{g}\frac{d^2V}{dt^2} = -2V \qquad (9\text{-}62)$$

This is an equation of simple harmonic motion for which the period of oscillation is

$$T = 2\pi\sqrt{\frac{2L_V + \tfrac{1}{2}L_H}{2g}} \qquad (9\text{-}63)$$

When $T = 3$ seconds, $2L_V + \tfrac{1}{2}L_H = 4.47$ m and so $L_V = 1.74$ m.

Comments on Example 9-6
1. Because the streamline is not truly composed of three straight sections, the proposed integration of $\partial V/\partial t$ is not exact. Nevertheless, the analysis gives a good approximation to the period of oscillation. In a practical case, the level of water in the vertical limbs can be adjusted by trial and error until the required result is obtained.

2. When the U-tube is in use there will be some resistance to motion, notably in the test region if the flow is obstructed. A small input of energy is therefore required to maintain a steady oscillation. Often, however, the damping of the motion is so small that its amplitude remains virtually constant for the duration of an experiment even without an external energy supply.

Example 9-7: Convergent flow in a nozzle Estimate the normal stress σ_N in the 5 mm thick walls of the 200 mm bore horizontal pipe shown in Fig. 9-15 when the mean velocity is 2 m/s. The outlet diameter of the nozzle is 50 mm and the water ($\rho = 1000 \text{ kg/m}^3$) discharges into air at atmospheric pressure. Neglect all losses and assume one-dimensional flow conditions.

SOLUTION The Bernoulli equation may be written for the points 1 and 2 on a typical streamline as

$$\frac{p_1}{\rho} + \tfrac{1}{2} V_1{}^2 = \frac{p_{AT}}{\rho} + \tfrac{1}{2} V_2{}^2 \tag{9-64}$$

in which p_{AT} denotes atmospheric pressure. By continuity, $a_1 V_1 = a_2 V_2$ and so $V_2 = 16 \text{ m/s}$ in this instance. Equation (9-64) therefore gives

$$p_1 - p_{AT} = \tfrac{1}{2}\rho(V_2{}^2 - V_1{}^2) = 510 \text{ kPa} \tag{9-65}$$

The force in the pipe walls is obtained by applying the momentum equation to the contents of the control volume shown in the figure, giving

$$(p_1 a_1 + \sigma_N a_w) - p_{AT}(a_1 + a_w) = \dot{m}(V_2 - V_1) \tag{9-66}$$

in which $a_w = \pi(200 + 5)5 \text{ mm}^2 = 0.003\,22 \text{ m}^2$ is the cross-sectional area of the wall and $a_1 = \tfrac{1}{4}\pi 200^2 \text{ mm}^2 = 0.0314 \text{ m}^2$ is the area of flow. Using $\dot{m} = \rho a_1 V_1 = 62.8 \text{ kg/s}$, we obtain

$$(\sigma_N - p_{AT}) = -4.39 \text{ MPa} \tag{9-67}$$

The negative sign indicates that the direction of the stress indicated in Fig. 9-15 is incorrect. The true stress is tensile.

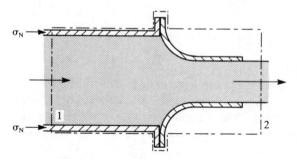

Figure 9-15 Steady flow through a nozzle.

Comments on Example 9-7

1. For a well-streamlined nozzle, the results of this analysis will be in close agreement with experiment. The only appreciable error results from non-uniformities in the assumed velocity distribution. Account can be taken of these by means of a velocity coefficient C_v similar to that described for an orifice in the comments on Example 9-1. Alternatively, if the approximate shape of the velocity distribution is known, specific allowance can be made for it in the manner used in Example 7-1 (p. 185).

2. It is the usual practice in engineering to quote all normal stresses relative to the ambient pressure as is done in Eq. (9-67). However, it is rare to do so explicitly. Usually we would simply say that the tensile stress is 4.39 MPa and expect people to assume that this is relative to the ambient conditions. The distinction is rarely important except for objects at great depths.

Example 9-8: Divergent flow in a sudden expansion Estimate the value of the energy loss coefficient k for the sudden expansion shown in Fig. 9-16a, assuming that the pressure over the whole of the surface A is equal to the pressure p_1 just upstream. Assume one-dimensional flow conditions in the upstream pipe and at the section 2 in the larger pipe.

SOLUTION The steady-flow momentum equation (7-20) may be written for the control volume shown in the figure as

$$(p_1 - p_2)a_2 = \dot{m}(V_2 - V_1) \tag{9-68}$$

in which skin friction is neglected because the pipe is short and, for simplicity, the pipe axis is assumed to be horizontal. By writing $\dot{m} = \rho a_2 V_2$, we obtain

$$p_1 - p_2 = \rho(V_2^2 - V_1 V_2) \tag{9-69}$$

and the addition of $\frac{1}{2}\rho(V_1^2 - V_2^2)$ to both sides yields

$$(p_1 + \tfrac{1}{2}\rho V_1^2) - (p_2 + \tfrac{1}{2}\rho V_2^2) = \tfrac{1}{2}\rho(V_2^2 - 2V_1 V_2 + V_1^2) \tag{9-70}$$

On division by ρ, the terms in parentheses on the left-hand side of this equation may be interpreted as Bernoulli sums and so

$$B_1 - B_2 = \tfrac{1}{2}(V_2 - V_1)^2 \tag{9-71}$$

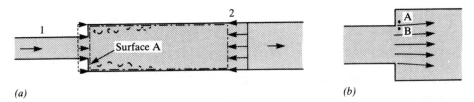

(a)　　　　　　　　　　　　　　　　　　　　*(b)*

Figure 9-16 Flow through an abrupt expansion. (a) Pressure forces. (b) Local detail.

Therefore the energy loss coefficient k based on the kinetic head $V_1^2/2g$ is

$$k \equiv \frac{B_1 - B_2}{V_1^2/2} = \left(\frac{V_2}{V_1} - 1\right)^2 = \left(\frac{a_1}{a_2} - 1\right)^2 \qquad (9\text{-}72)$$

Comments on Example 9-8

1. It is unusual to be able to estimate loss coefficients in a simple manner, largely because the flow often contracts before expanding (see Fig. 9-4). Nevertheless, this particular expression—known as the Borda–Carnot relationship—is in reasonable agreement with experiment. Other loss coefficients can be estimated from it if the extent of the initial contractions is known.

2. The assumption that the pressure over the whole of the surface A is equal to p_1 is justified in Chapter Ten where it is shown that negligible pressure variations exist normal to parallel streamlines. Thus in Fig. 9-16b, the pressures at the points A and B respectively outside and inside the bounding streamline are equal. Negligible pressure variations exist on the surface A because the induced velocities in this region are tiny.

Example 9-9: Multiple losses in a pipeline Estimate the rate of flow of water along the pipeline shown in Fig. 9-17 when the difference in the reservoir surface levels is 10 m. The pipes AC, CD and DF are 2 km, 1 km and 1 km long, their internal diameters are 0.5 m, 0.6 m and 0.4 m, and their skin friction coefficients may be assumed to be 0.007, 0.004 and 0.006 respectively. The loss coefficients for the pipe entrance, the bend, the sudden contraction and the fully open valve are 0.5, 0.4, 0.3 and 0.8 respectively, each coefficient referring to the local kinetic head in the usual manner.

SOLUTION When steady flow is established in the pipeline, the sum of all of the head losses is exactly equal to the difference between the Bernoulli heads in the reservoirs, namely 10 m. In the manner described in section 9-3-3, we first determine each loss coefficient k_i and then deduce k_i/a_i^2. Using Eq. (9-72) for the sudden expansion and noting that this implies $k \approx 1$ for the pipe outlet, we obtain the values listed in Table 9-1.

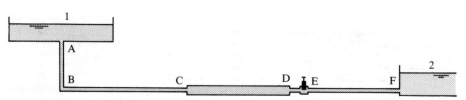

Figure 9-17 Losses in a pipeline.

Table 9.1 Pipeline losses

Origin of loss	k_i	V_i	k_i/a_i^2
Entrance, A	0.50	V_{AC}	13
Skin friction, AC	112.0	V_{AC}	2905
Bend, B	0.40	V_{AC}	10
Expansion, C	0.093	V_{AC}	2
Skin friction, CD	26.67	V_{CD}	334
Contraction, D	0.30	V_{DF}	19
Skin friction, DF	60.0	V_{DF}	3800
Valve, E	0.80	V_{DF}	51
Exit, F	1.05	V_{DF}	63

By summing the values in the final column, we obtain $\Sigma(k_i/a_i^2) = 7197$ and so the extended Bernoulli equation (9-26) gives

$$B_1 - B_2 = 7197\frac{Q^2}{2} \tag{9-73}$$

Since $B_1 - B_2 = 10$ m, the volumetric flux is $Q = 0.165$ cumec.

Comments on Example 9-9

1. The velocity in any particular pipe can be obtained from $V = Q/a$ and the value of the Bernoulli sum at any position follows from a further application of Eq. (9-26), including as many terms in the summation as necessary. For example, only the first three rows of Table 9-1 are needed to determine B_c. The local piezometric pressure is $p^* = \rho B - \frac{1}{2}\rho V^2$ and the static pressure is $p = p^* - \rho gz$.

2. If Eq. (9-73) is redeveloped, taking account only of skin friction losses, the coefficient 7197 is replaced by 7039 and the resulting error in the calculated flow rate is only about 1 per cent. For this reason, local losses are nearly always neglected when dealing with long pipelines. As indicated in section 9-3-1, this is also why local losses are sometimes called 'minor' losses. Problem 4 at the end of this chapter deals with a case where local losses could more reasonably be termed 'major'.

Example 9-10: Booster pump To provide an increased rate of flow when it is occasionally needed, a centrifugal pump is installed downstream of the point B in the pipeline shown in Fig. 9-17. Its 'head-discharge' characteristic satisfies

$$\Delta h_0 = C_1 + C_2 Q - C_3 Q^2 \tag{9-74}$$

in which $C_1 = 30$ m is the no-flow head, $C_2 = 125$ s/m^2 and $C_3 = 1000$ s^2/m^5. Estimate the rate of flow with the pump operating.

SOLUTION The installation of the pump has no influence on the loss characteristics of the remainder of the system (except for any dependence on the Reynolds number). The new rate of flow is obtained by extending Eq. (9-73) to include the pump head-discharge relationship, giving

$$(h_{01} - h_{02}) + (C_1 + C_2Q - C_3Q^2) = 7197\frac{Q^2}{2g} \tag{9-75}$$

which has two roots, namely $Q = 0.223$ cumec and $Q = -0.131$ cumec. The first of these is the required solution because the relationship (9-74) is invalid for reverse flows.

Comments on Example 9-10

1. The combination of the pump characteristic (9-74) with the system characteristic (9-73) is often carried out graphically, particularly when no simple algebraic relationship is known for the former. Figure 9-18 demonstrates the ease with which the operating point can be found. It also enables experienced engineers to spot that the specified pump will operate well away from its duty point—i.e. inefficiently.
2. The total head just upstream of the pump can be estimated when the length

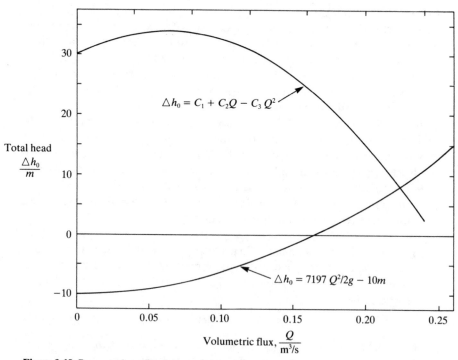

Figure 9-18 Pump and system characteristics.

AB is known. If this is 50 m, say, then the contribution of skin friction to the sum $\Sigma(k_i/a_i{}^2)$ is 73 and the overall sum upstream of the pump is $13+73+10=96$ (see Table 9-1). Hence the total head is

$$h_{0B}=h_{01}-\frac{Q^2}{2g}\Sigma\left(\frac{k_i}{a_i{}^2}\right)=9.76\ \text{m} \qquad (9\text{-}76)$$

in which the elevation $z=0$ is chosen at the surface level in the downstream reservoir. The velocity in the pipe BC is $V=Q/a=1.14\ \text{m/s}$ and so the piezometeric head at the inlet to the pump is

$$\left(\frac{p}{\rho g}+z\right)_{B}=h_{0B}-\frac{V_{BC}{}^2}{2g}=9.69\ \text{m} \qquad (9\text{-}77)$$

The corresponding static pressure head is $(9.69\ \text{m}-z_B)$ and this must not fall below about minus 7 m if the pump is to operate satisfactorily (i.e. without the occurence of severe cavitation).

PROBLEMS

1 Figure 9-19 depicts blood $(\rho=1250\ \text{kg/m}^3)$ flowing along a 7.5 mm diameter artery which has a 5 mm branch. The gauge pressure and the mean velocity upstream of the branch are 6 kPa and 0.6 m/s respectively. The velocity downstream of the branch is 0.4 m/s. Estimate the gauge pressures downstream of the branch and inside the branch (a) assuming no losses and (b) assuming a loss of $\frac{1}{4}\rho V_3{}^2$ on entry to the branch.

[6.125 kPa, 6.098 kPa; 6.125 kPa, 6.035 kPa]

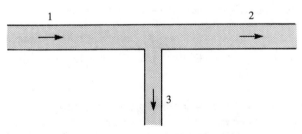

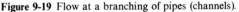

Figure 9-19 Flow at a branching of pipes (channels).

2 Suppose that Fig. 9-19 represents water $(\rho=1000\ \text{kg/m}^3)$ flowing along a 3 m wide rectangular channel which has a 2 m wide branch. The depth and velocity upstream of the branch are 1.5 m and 2.3 m/s and the downstream velocity is 1.3 m/s. Estimate the depths downstream of the branch and inside the branch (a) assuming no losses and (b) assuming a total head loss of $\bar{V}_3{}^2/2g$ on entry to the branch.

[1.68 m, 1.71 m; 1.68 m, 1.67 m]

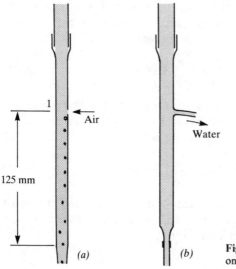

(a)   (b)   **Figure 9-20** Influence of an exit restriction on pressures further upstream.

3 Figure 9-20 depicts a 10 mm diameter flexible tube hanging vertically downwards from the end of a domestic water tap.

  (a) Estimate the gauge pressure at the section 1 when the rate of flow is 10 l/min and hence show that air would leak in through a hole in the wall of the pipe at this section. Neglect skin friction.

$$[-1.23 \, \text{kPa}]$$

  (b) Suppose that the end of the tube is gradually restricted (by your fingers, perhaps) as shown in Fig. 9-20b. If the rate of flow is maintained at 10 l/min, what will be the outlet cross-section when water begins to leak out of the hole? Neglect skin friction and surface tension (somewhat unreasonably).

$$[63.2 \, \text{mm}^2]$$

4 Repeat Example 9-9 for pipe lengths AC = 20 m, CD = 10 m and DF = 10 m.

$$[0.927 \, \text{m}^3/\text{s}]$$

5 A 2.5 m diameter tunnel ($f = 0.004$) connects a reservoir to a hydro-electric power station. The tunnel is 5 km long and there is a surge tank 0.5 km from the station as shown in Fig. 9-21. The elevation of the surface level in the reservoir is 60 m greater than that of the turbines. Estimate the free surface level in the surge tank when (a) there is no flow and (b) the flow is steady and the pressure head at inlet to the turbines is 48 m. Also estimate the power available to the turbines.

$$[60 \, \text{m}, \, 49.2 \, \text{m}, \, 6.27 \, \text{MW}]$$

6 The orifice meter shown in Fig. 9-22 is used to measure the rate of airflow along a 150 mm diameter pipe. The diameter of the orifice is

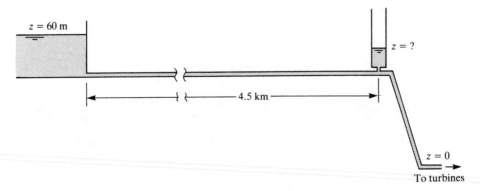

Figure 9-21 Water supply for a hydroelectric power station.

60 mm and the contraction coefficient to the vena contracta just downstream is 0.63. What pressure difference $(p_1 - p_2)$ will be measured when $p_1 = 150\,\text{kPa}$, $\rho_1 = 1.5\,\text{kg/m}^3$ and (a) $V_1 = 5\,\text{m/s}$ and (b) $V_1 = 20\,\text{m/s}$? Neglect compressibility effects.

[1.83 kPa, 29.2 kPa]

7 Repeat Problem 6 allowing for compressibility. Take $\gamma = 1.4$.

[1.85 kPa, 41.8 kPa]

8 The water depth upstream of a 0.35 m high, broad-crested weir in a rectangular section channel is 1.6 m. Estimate the rate of flow per unit width (normal to the page in Fig. 9-23a) when the depth over the weir is 1.0 m.

[2.84 m²/s]

9 The water depth upstream of a 0.55 m high, broad-crested weir in a rectangular section channel is 1.6 m. The weir is acting as a control section (Fig. 9-23b) for which the relationship $V^2 = gd$ may be used. Estimate the rate of flow per unit width.

[2.29 m²/s]

10 The venturi flume shown in Fig. 9-24 has a 0.25 m high weir at the throat where the width of flow is 0.8 m. Determine the upstream depth in the 1.2 m wide rectangular section channel when the rate of flow is $2\,\text{m}^3/\text{s}$. Assume that $V^2 = gd$ somewhere in the throat.

[1.48 m]

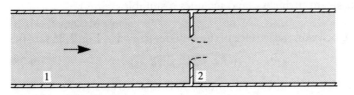

Figure 9-22 Simple orifice meter.

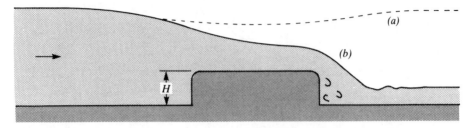

Figure 9-23 Two types of flow over a weir. (a) Restriction only. (b) Control section.

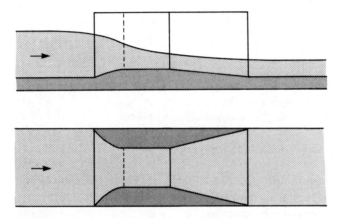

Figure 9-24 Venturi flume with a weir at the throat.

11 Steam enters the inlet guide vanes of an axial flow turbine at a pressure of 15 MPa, a temperature of 550 °C and a density of 43.7 kg/m³. The flow is wholly axial at inlet but there is a strong circumferential velocity at outlet (ready for the inlet to the rotor). The turbine passages are tapered slightly to ensure that the axial velocity component remains constant at 75 m/s even though the density decreases downstream. Assuming a polytropic flow with $N = 1.37$, estimate the pressure drop across the guide vanes if the outlet circumferential component of velocity is (a) 150 m/s and (b) 300 m/s. (Problem 8 at the end of Chapter Eight gives additional information about steam turbines—and includes a suitable figure.)

[0.49 MPa, 1.87 MPa]

12 The velocity on the surface of the cylinder shown in Fig. 9-25 satisfies

$$V_\theta = -2 V_\infty \sin \theta \tag{9-78}$$

where V_∞ denotes the undisturbed free-stream velocity. By evaluating the velocity at $\theta = 0$, $\frac{1}{2}\pi$, π and $\frac{3}{2}\pi$, determine the corresponding

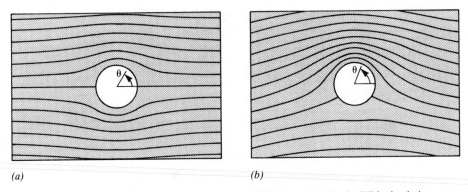

(a)   (b)

Figure 9-25 Irrotational flows past a cylinder. (a) Without circulation. (b) With circulation.

pressures relative to the free-stream pressure. Also determine the locations of the positions on the cylinder surface at which the pressure is equal to the free stream pressure. Use $V_\infty = 20 \text{ m/s}$ and $\rho = 1.2 \text{ kg/m}^3$.

[240 Pa, -720 Pa, 240 Pa, -720 Pa; 30°, 150°, 210°, 330°]

13   Repeat Problem 13 for the cylinder in Fig. 9-25b. The velocity on its surface satisfies

$$V_\theta = -V_\infty(1 + 2\sin\theta) \qquad (9\text{-}79)$$

[0, -1920 Pa, 0, 0; 0°, 180°, 270°]

14   The cross-sectional area of the diffuser shown in Fig. 9-26 is designed to ensure that the velocity varies linearly between the sections 1 and 2. In a computer controlled acceleration, the pressure p_2 is continually adjusted to keep it equal to p_1. If $a_2 = 2a_1$, show that the induced acceleration satisfies

$$\frac{dV_1}{dt} = \frac{V_1^2}{2L} \qquad (9\text{-}80)$$

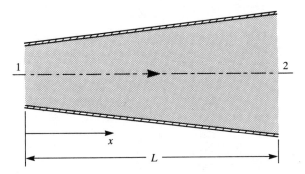

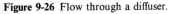

Figure 9-26 Flow through a diffuser.

Hence show that the time required to increase the velocity from V_0 to $2V_0$ at the section 1 is L/V_0. Regard the flow as incompressible and inviscid.

FURTHER READING

Ackers, P., White, W.R., Perkins, J.A. and Harrison, A.J.M. (1978) *Weirs and Flumes for Flow Measurement*, Wiley.
Cheremisinoff, N.P. (1979) *Applied Fluid Flow Measurement: Fundamentals and Technology*, Marcel Dekker.
Durst, F., Melling, A. and Whitelaw, J.H. (1981) *Principles and Practice of Laser–Doppler Anemometry*, 2nd edn, Academic Press.
Haywood, A.T.J. (1979) *Flowmeters*, Macmillan.
Henderson, F.M. (1966) *Open Channel Flow*, Macmillan.
Jain, A.K. (1976) Accurate explicit equation for friction factor, *J. Hyd. Div.*, ASCE, **102**, 674–677.
Miller, D.S. (1978) *Internal Flow Systems*, BHRA, The Fluid Engineering Centre.
Ower, E. and Pankhurst, R.C. (1977) *The Measurement of Air Flow*, 5th edn, Pergamon.

TEN

VORTEX EQUATIONS

10-1 ROTATIONAL AND IRROTATIONAL FLOWS

In most natural flows, individual streamlines are curved. There are a few exceptions where streamlines are straight, or nearly so, but these tend to be in man-made structures such as pipes and channels. Usually, therefore, individual fluid particles follow curved paths as they make their way downstream, and often the paths are quite complex.

We now enquire whether the individual particles rotate as they follow these tortuous paths. That is, do they like to face in the direction of travel as implied by Fig. 10-1a or do they continue to look in some predetermined direction irrespective of their direction of travel as implied by Fig. 10-1b? Alternatively, perhaps they rotate (Fig. 10-1c) or maybe they do none of these things, preferring instead to behave as shown in Fig. 10-1d.

In fact, all of these types of behaviour are possible, but type (a) is unlikely to occur in practice except when the whole flow behaves like a solid body (in which case it could be viewed as stationary relative to suitably chosen axes). Type (b) behaviour is termed *irrotational* because the individual particles do not rotate. It is an important type of flow partly because it happens to be especially amenable to analysis, but also because it is a good approximation to many external flows (section 1-2), except close to solid surfaces. Anything other than type (b) behaviour is termed *rotational*.

10-1-1 Vorticity

The *vorticity* at any point in a fluid flow is defined to be twice the local rate of rotation. Its components about the x-, y- and z-axes were defined mathemati-

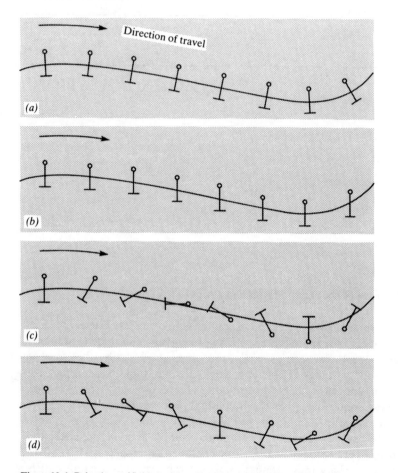

Figure 10-1 Behaviour of fluid particles traversing a pathline (diagrams show successive positions of the same particle). (a) Facing direction of travel. (b) Irrotational. (c) Clockwise rotation. (d) Oscillating.

cally in section 4-6-1 as

$$\xi \equiv \frac{\partial w}{\partial y} - \frac{\partial v}{\partial z} \qquad (10\text{-}1)$$

$$\eta \equiv \frac{\partial u}{\partial z} - \frac{\partial w}{\partial x} \qquad (10\text{-}2)$$

$$\zeta \equiv \frac{\partial v}{\partial x} - \frac{\partial u}{\partial y} \qquad (10\text{-}3)$$

By inspection, the vorticity about any particular axis involves gradients of the two components of velocity in a plane normal to the axis. In Figs 4-18 and

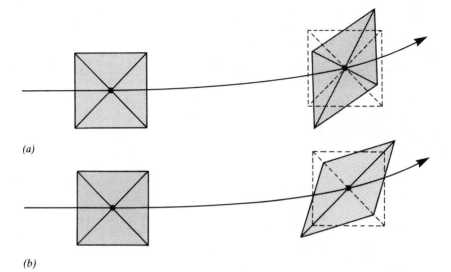

(a)

(b)

Figure 10-2 Rotation and distortion of a fluid particle. (a) Rotational flow. (b) Irrotational flow.

10-2*a*, the element distorts and rotates as it moves downstream. In Fig. 10-2*b*, it distorts, but there is no net rotation and so the flow is irrotational.

Fluid elements distort as they flow, whether or not the flow is rotational. They are capable of distorting indefinitely. However, distortion involves shear strain and this implies the existence of shear stress. For most of this chapter we shall assume that the shear stresses are sufficiently small to have negligible influence on the flow. It is shown in Example 10-9 (p. 291) that this implies that the vorticity of individual particles can never change. In particular, an irrotational flow-field will remain irrotational for all time in the absence of fluid shear.

It is useful to develop an expression relating the vorticity to the degree of curvature of streamlines. For this purpose consider the streamlines shown in Fig. 10-3 in which the origin of the natural coordinates is chosen at the typical point P where the velocity (along the streamline) is V. At the adjacent point Q, the magnitude of the velocity is $V_Q = V + \partial V/\partial s \, \delta s$ and this has a component $\delta V_n = -(V + \partial V/\partial s \, \delta s) \sin \delta\theta$ in the original n-direction. By inspection, $\sin \delta\theta$ will tend to $\delta s/R$ as δs approaches zero, R being the local radius of streamline curvature in the s-n plane. Since δV_n may alternatively be written as $\partial V_n/\partial s \, \delta s$, we obtain

$$\frac{\partial V_n}{\partial s} \delta s = -\left(V + \frac{\partial V}{\partial s} \delta s\right) \frac{\delta s}{R} \tag{10-4}$$

Therefore in the limit when the second-order term in δs vanishes,

$$\frac{\partial V_n}{\partial s} = -\frac{V}{R} \tag{10-5}$$

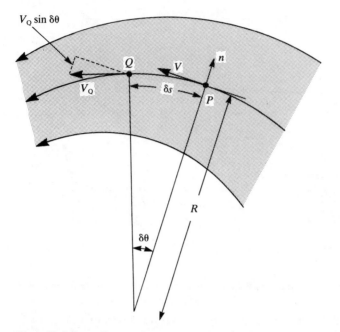

Figure 10-3 Streamline curvature.

and the vorticity component in the s-n plane is

$$\zeta = \frac{\partial V_n}{\partial s} - \frac{\partial V_s}{\partial n} = -\left(\frac{V}{R} + \frac{\partial V}{\partial n}\right)$$

(10-6)

The use of this expression is illustrated in Example 10-1 (p. 279).

10-1-2 Circulation

The terms rotational and irrotational are used to describe the motion of individual fluid particles. However, it is also useful to view motion on a macroscopic scale and the concept of *circulation* provides a method of assessing the collective effect of the rotation of many particles.

Consider the arbitrarily chosen closed loop shown in Fig. 10-4. At the typical point P the local velocity is V and its component tangential to the curve is V_1. The circulation Γ is defined as the line integral of the velocity around the curve, namely

$$\Gamma \equiv \oint V_1 \, dl$$

(10-7)

in which the symbol $\oint$ indicates that the integration must be carried out for the complete loop.

The circulation is a measure of the total amount of rotation within the chosen loop. For example, if the loop is chosen at a radius r from the axis of a

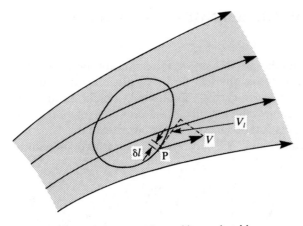

Figure 10-4 Circulation around an arbitrary closed loop.

rigid body rotating with an angular velocity Ω, the relevant velocity is $V_1 = r\Omega$ and the length of the curve is $2\pi r$ so the circulation is $\Gamma = 2\pi r^2 \Omega$. By inspection, this is proportional to the angular velocity Ω and to the area of the surface within the loop (πr^2).

In a truly irrotational flow field, no particle rotates and so $\Gamma = 0$ for all possible loops. However, it is sometimes useful to hypothesize the existence of one or more discontinuities at which circulation exists. In Fig. 10-5 such discontinuities are imagined at the points A and B and their magnitudes are denoted by Γ_A and Γ_B. If there is no rotation anywhere else in the flow field then the circulation around the typical curves 1, 2, 3 and 4 will be

$$\Gamma_1 = \Gamma_A; \qquad \Gamma_2 = 0; \qquad \Gamma_3 = \Gamma_B; \qquad \Gamma_4 = \Gamma_A + \Gamma_B \qquad (10\text{-}8)$$

The circulation is the same for all curves enclosing the same discontinuities.

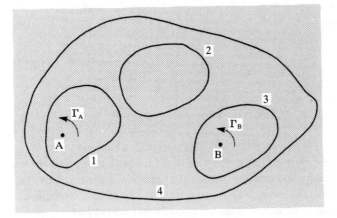

Figure 10-5 Circulation around line vortices (vorticity discontinuities).

10-2 VARIATION OF THE BERNOULLI SUM ACROSS STREAMLINES

In Chapter Nine the Bernoulli sum was shown to remain constant *along* any particular streamline in a steady, inviscid flow. No attempt was made to determine whether the same or different values would obtain on different streamlines. In fact, the same value applies if the flow is irrotational and different values apply if it is rotational. We shall now prove this.

The first two Euler equations (7-88) may be written in terms of the natural coordinates as

$$-g\frac{\partial z}{\partial s} - \frac{1}{\rho}\frac{\partial p}{\partial s} = \frac{\partial V_s}{\partial t} + V_s\frac{\partial V_s}{\partial s} \tag{10-9}$$

$$-g\frac{\partial z}{\partial n} - \frac{1}{\rho}\frac{\partial p}{\partial n} = \frac{\partial V_n}{\partial t} + V_s\frac{\partial V_n}{\partial s} \tag{10-10}$$

in which the only body force is due to gravity and z denotes the elevation. The remaining terms in (7-88) vanish because their coefficients are zero in the natural coordinate system. After adding $V_s\,\partial V_s/\partial n$ to both sides of the second equation and rearranging the terms, we can write the Euler equations as

$$\frac{1}{\rho}\frac{\partial p}{\partial s} + \frac{\partial}{\partial s}(\tfrac{1}{2}V_s^2) + g\frac{\partial z}{\partial s} = -\frac{\partial V_s}{\partial t} \tag{10-11}$$

$$\frac{1}{\rho}\frac{\partial p}{\partial n} + \frac{\partial}{\partial n}(\tfrac{1}{2}V_s^2) + g\frac{\partial z}{\partial n} = -\frac{\partial V_n}{\partial t} - V_s\left(\frac{\partial V_n}{\partial s} - \frac{\partial V_s}{\partial n}\right) \tag{10-12}$$

Equation (10-11) is identical to Eq. (9-1), which was integrated to derive the Bernoulli equation for the special cases of incompressible, isothermal and polytropic flows in which an analytical relationship exists between the pressure and the density. For such flows, the equations tangential and normal to a streamline become

$$\frac{\partial B}{\partial s} = -\frac{\partial V_s}{\partial t} \tag{10-13}$$

$$\frac{\partial B}{\partial n} = -\frac{\partial V_n}{\partial t} - V_s\zeta \tag{10-14}$$

which are differential forms of Bernoulli and vortex equations.

By inspection, the Bernoulli sum remains constant both along and across streamlines in a steady, irrotational flow. In a steady, rotational flow, the Bernoulli sum is constant along any streamline, but not across streamlines. That is, different values of B are obtained on different streamlines.

10-2-1 Steady Incompressible Flows

In the special case of steady, incompressible flows, Eq. (10-10) can be written as

$$\frac{\partial}{\partial n}(p+\rho gz)= -\rho V_s \frac{\partial V_n}{\partial s} \tag{10-15}$$

in which $p+\rho gz$ is the piezometric pressure p^*. Therefore, using (10-5), we obtain

$$\frac{\partial p^*}{\partial n} = \frac{\rho V_s^2}{R} \tag{10-16}$$

which is an expression for the rate of change of piezometric pressure normal to a streamline. It is valid for rotational as well as irrotational flows and its use is illustrated in Examples 10-1 and 10-2 below.

10-2-2 Parallel Flow

When adjacent streamlines are parallel to one another, $V_n = 0$ and Eq. (10-10) shows that the lateral pressure distribution satisfies

$$\frac{\partial p}{\partial n} = -\rho g \frac{\partial z}{\partial n} - \frac{\partial V_n}{\partial t} \tag{10-17}$$

Thus only hydrostatic variations in pressure exist normal to streamlines whenever $\partial V_n/\partial t = 0$—as is nearly always the case for carefully chosen axes. This is why the piezometric pressure is always assumed to be uniform within uniaxial flow sections.

10-3 APPLICATIONS OF VORTEX EQUATIONS

Example 10-1: Tornado As a first approximation, a tornado may be regarded as a steady irrotational cylindrical vortex with a vertical axis. Develop an expression for the pressure distribution in a horizontal plane through a tornado, treating the air as an incompressible fluid.

SOLUTION The streamline pattern is depicted in Fig. 10-6a. Natural coordinate axes s and n are chosen at a typical point P on a circular streamline and the origin of n is conveniently chosen at the centre of the vortex. Equation (10-6) may be written (for $\zeta = 0$) as

$$\frac{dV}{dn} + \frac{V}{n} = 0 \tag{10-18}$$

in which the total derivative d/dn is used instead of the partial derivative $\partial/\partial n$ because the velocity is a function of n alone. This equation may be

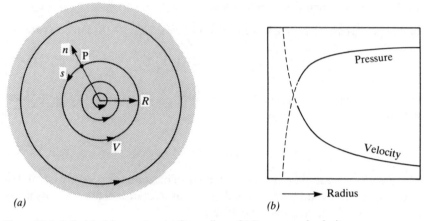

Figure 10-6 Cylindrical free vortex. (a) Streamlines. (b) Pressure and velocity.

rearranged as $dV/V + dn/n = 0$ and integrated to give $\ln(V) + \ln(n) = K$ in which K is a constant of integration. Since $\ln(V) + \ln(n)$ is equal to $\ln(Vn)$, we obtain the hyperbolic relationship

$$Vn = C \tag{10-19}$$

in which C is a constant.

The pressure distribution is obtained by noting that the Bernoulli sum is constant throughout the irrotational flow field. If atmospheric conditions prevail at $n = \infty$, the pressure at any other point satisfies

$$\frac{p}{\rho} + \tfrac{1}{2}V^2 + gz = \frac{p_{AT}}{\rho} + \tfrac{1}{2}V_\infty^2 + gz_\infty \tag{10-20}$$

Equation (10-19) shows that the velocity is zero at $n = \infty$. Therefore, the required pressure distribution in a horizontal plane is

$$p - p_{AT} = -\tfrac{1}{2}\rho V^2 = -\frac{\rho C^2}{2n^2} \tag{10-21}$$

This expression could alternatively have been derived by integrating Eq. (10-16) after substituting $V_s = C/n$ and $R = n$.

Comments on Example 10-1

1. The predicted velocity and pressure distributions are depicted in Fig. 10-6b. They are reasonably accurate at radii exceeding about $C/100$ metres. For smaller values of n, the velocity exceeds 100 m/s and homentropic relationships describe the pressure distribution more closely (see Problem 2 at the end of this chapter).

2. At very small radii the large velocity gradients predicted by this analysis

($V_s \to \infty$ and $\partial V_s/\partial n \to -\infty$ as $n \to 0$) would imply high shear stresses in a real fluid. In practice, however, it is much easier for the fluid to rotate like a solid body and so the central cores of real vortices are rotational.

3. The *circulation* around a typical circular streamline of radius r is the product of its circumference $2\pi r$ and the tangential velocity V, i.e. $\Gamma = 2\pi V r = 2\pi C$. The circulation around *any* closed loop containing the axis is $2\pi C$ and that $\Gamma = 0$ around all other closed loops (see problem 4).

4. An irrotational vortex is sometimes called a *free vortex*.

Example 10-2: Spiral vortex The velocity components at a radius r on a typical streamline of the spiral vortex shown in Fig. 10-7 are

$$V_r = -\frac{C_1}{2\pi r}; \qquad V_\theta = \frac{C_2}{2\pi r} \qquad (10\text{-}22)$$

in which C_1 and C_2 are constants.

Show that the angle between a tangent to a streamline and a radial line through the axis is the same everywhere and determine an expression for the piezometric pressure distribution.

SOLUTION The direction of the velocity V_s at any point satisfies

$$\tan\phi = \frac{V_\theta}{V_r} = -\frac{C_2}{C_1} \qquad (10\text{-}23)$$

in which ϕ is a constant.

The piezometric pressure distribution is obtained from Eq. (10-16) which gives

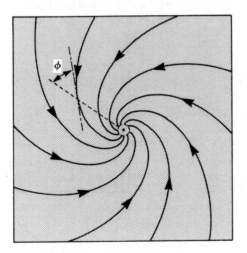

Figure 10-7 Spiral flow towards a drain.

$$\frac{\partial p^*}{\partial r} = \frac{\rho(V_r^2 + V_\theta^2)}{r} = \frac{\rho(C_1^2 + C_2^2)}{4\pi^2 r^3} \tag{10-24}$$

On integrating and choosing $p^* = p^*_{AT}$ at $r = \infty$, we obtain

$$p^* - p^*_{AT} = -\frac{\rho(C_1^2 + C_2^2)}{8\pi^2 r^2} \tag{10-25}$$

Comments on Example 10-2

1. The spiral vortex is a reasonable approximation to the flow towards a drain in a sink or bathtub. The shapes of the velocity and pressure distribution are the same as those shown in Fig. 10-6 for a free vortex. The circulation around any loop enclosing $r = 0$ is equal to C_2 and the rate of flow towards $r = 0$ is C_1 (per unit width normal to the page).
2. The direction of circulation in a sink vortex is influenced by disturbances before the plug is removed. In the absence of all other disturbances, it is governed by the earth's rotation. Contrary to popular belief, the earth's rotation is not usually the dominant factor. It has no effect at the equator and it is only one revolution per day at the poles.

Example 10-3: Rotational line vortex The tank of liquid depicted in Fig. 10-8a spins about its vertical axis at a constant rate Ω. Derive an equation describing the surface profile after sufficient time has elapsed for steady-state conditions to prevail. Above the liquid surface the air is at atmospheric pressure.

SOLUTION The *forced vortex* resembles a rigid body rotation. An observer rotating with the tank would see no movement of the fluid particles. An observer who does not rotate with the tank will find that the tangential velocity at any radius r from the vertical axis of rotation

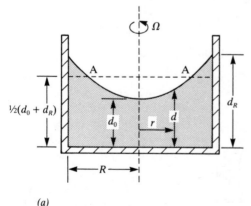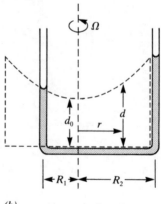

(a)   (b)

Figure 10-8 Cylindrical forced vortex.

satisfies $V = r\Omega$. Equation (10-16) becomes $dp^*/dr = \rho\Omega^2 r$, which may be integrated to give

$$p^* - p^*_0 = \tfrac{1}{2}\rho\Omega^2 r^2 = \tfrac{1}{2}\rho V^2 \qquad (10\text{-}26)$$

in which p^*_0 denotes the piezometric pressure at $r = 0$.

Equation (10-26) is valid in any *horizontal* plane. Additionally, we know that the piezometric pressure does not vary in the vertical direction when all the streamlines are horizontal (see section 10-2-2), so we may evaluate it at any convenient elevation. On the base of the tank, the pressure at a typical radius is $p_{AT} + \rho gd$ and so $p - p_0 = \rho g(d - d_0)$. Therefore Eq. (10-26) shows that the equation of the surface profile is

$$d - d_0 = \frac{r^2\Omega^2}{2g} = \frac{V^2}{2g} \qquad (10\text{-}27)$$

Comments on Example 10-3

1. Equation (10-26) may be applied to any incompressible fluid undergoing a rigid-body rotation irrespective of the shape of the container or the orientation of the axis of rotation. Equation (10-27) is less general, but it is valid for liquids with a free surface rotating about a vertical axis. For example, it correctly describes the levels in the U-tube shown rotating about a non-central axis in Fig. 10-8b. An observer rotating with the fluid would regard it as being at rest and would attribute the radial variations of the piezometric pressure and the surface level to an inertial body force associated with the rotation of the reference axes.

2. In the special case of cylindrical tanks, the principal depths are easily found because the volume of a paraboloid is equal to the area of its base multiplied by half its maximum height. Therefore in Fig. 10-8a the original depth in the non-rotating tank was $\tfrac{1}{2}(d_0 + d_R)$. As a numerical example, suppose that the original depth in a 0.3 m diameter tank is 0.4 m. When the tank and its contents rotate steadily at 16 rad/s, Eq. (10-27) shows that $d_R - d_0 = 0.15^2 \times 16^2/19.62$ m $= 0.294$ m. Since $\tfrac{1}{2}(d_0 + d_R) = 0.4$ m, we obtain $d_0 = 0.253$ m and $d_R = 0.547$ m.

Example 10-4: Centrifugal pump impeller Estimate the rise in piezometric pressure between the inlet and outlet sections of the pump impeller shown in Fig. 10-9. The outer and inner diameters are $D_O = 0.75$ m and $D_1 = 0.35$ m and the corresponding widths are $b_O = 85$ mm and $b_1 = 175$ mm. The absolute velocity at outlet is 26.5 m/s at 35° to the tangent and there is no swirl at inlet. The fluid is water ($\rho = 1000$ kg/m³) and the impeller rotates at 850 rev/min.

SOLUTION Consider first the pressure difference that would exist if the impeller was spinning at the specified rate but there was no flow through

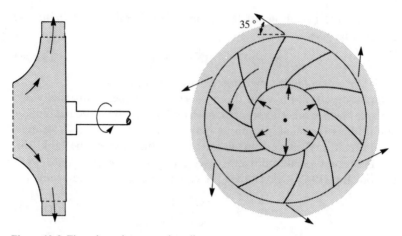

Figure 10-9 Flow through a pump impeller.

it. This would be equivalent to a solid-body rotation, and an observer rotating with the impeller would regard the fluid as being at rest. The pressure difference between the inner and outer radii would be found by integrating Eq. (10-16) in the same manner as in Example 10-3, giving

$$(p^*_o - p^*_1)_{\text{rotation}} = \tfrac{1}{2}\rho\Omega^2(r_o^2 - r_1^2) = 436 \,\text{kPa} \qquad (10\text{-}28)$$

In fact, our observer will see a steady flow through the impeller. At outlet, the area is $a_o = \pi D_o b_o = 0.200 \,\text{m}^2$ and the radial component of flow is $V_{ro} = 26.5 \sin 35° \,\text{m/s} = 15.2 \,\text{m/s}$. Therefore the volumetric flux is $Q = a_o V_{ro} = 3.04 \,\text{m}^3/\text{s}$. At inlet, the area is $a_1 = \pi D_1 b_1 = 0.192 \,\text{m}^2$ and so the radial velocity of flow is $V_{r1} = Q/a_1 = 15.8 \,\text{m/s}$.

Stationary and rotating observers see the same radial components of flow, but they see different tangential components. At the outlet, a stationary observer sees a tangential velocity component of $V_{\theta o} = 26.5 \cos 35° \,\text{m/s}$, but our rotating observer sees

$$V'_{\theta o} = V_{\theta o} - r_o\Omega = -11.7 \,\text{m/s} \qquad (10\text{-}29)$$

in which the prime denotes conditions relative to the rotating axes. At inlet, a stationary observer sees no tangential velocity component, but the rotating observer sees

$$V'_{\theta 1} = 0 - r_1\Omega = -7.79 \,\text{m/s} \qquad (10\text{-}30)$$

The total velocities at inlet and outlet relative to our rotating observer are therefore

$$V'_1 = \sqrt{V_{r1}^2 + V'^2_{\theta 1}} = 17.6 \,\text{m/s} \qquad (10\text{-}31)$$

and

$$V'_o = \sqrt{V_{ro}^2 + V'^2_{\theta o}} = 19.2 \,\text{m/s} \qquad (10\text{-}32)$$

The change in the piezometric pressure due to a change in velocity along a streamline can be determined from the steady-flow Bernoulli equation. For the steady flow through the impeller we obtain

$$(p^*_o - p^*_1)_{flow} = \tfrac{1}{2}\rho(V_1'^2 - V_o'^2) = -29.4\,\text{kPa} \tag{10-33}$$

The net increase in the piezometric pressure across the impeller is the algebraic sum of (10-28) and (10-33), namely 407 kPa.

Comments on Example 10-4

1. The advantage of choosing axes relative to the impeller is that steady flow conditions are observed. To carry out the analysis relative to stationary axes would be very difficult because the observed flow would be highly unsteady. It is often the case in fluid (or solid) mechanics that an analysis can be greatly simplified by a careful choice of the reference axes.
2. Much of the work done on the flow by the impeller appears as kinetic energy at outlet. In this particular example, the outlet velocity is 26.5 m/s and this is far greater than the velocity that will exist in the pipe downstream of the pump. The designer therefore faces the challenge of designing the volute and diffuser to 'convert' as much as possible of the available kinetic head to pressure head. In this example the available kinetic head, namely $V_o^2/2g = 35.8$ m, is almost as large as the pressure head increase in the impeller, so it is clearly very important to have a well-designed volute.

Example 10-5: Flow around a river bend Figure 10-10 depicts a flow of water around a bend in a 5 m wide rectangular channel. In the approach to the bend, the uniform depth of flow is 1.5 m and the mean velocity is 2.5 m/s. Estimate the depths on the inside and outside walls of the channel at the middle of the bend where the inside and outside radii are 20 m and 25 m respectively. Assume that the flow at this section behaves as a free vortex and neglect skin friction effects.

SOLUTION A typical streamline is depicted in the figure. The Bernoulli equation for points in the inlet section and the mid-bend section may be written as

$$d_1 + \frac{V_1^2}{2g} = d + \frac{V^2}{2g} \equiv \frac{B}{g} \tag{10-34}$$

in which the elevation datum $z = 0$ is chosen on the bed where $p = \rho g d$. The rate of flow through the elemental strip of area $d\,\delta r$ shown in Fig. 10-10 is

$$\delta Q = d\,\delta r V = (B - \tfrac{1}{2}V^2)\frac{V}{g}\,\delta r \tag{10-35}$$

Writing $V = C/r$ for the free vortex behaviour in the bend, we obtain

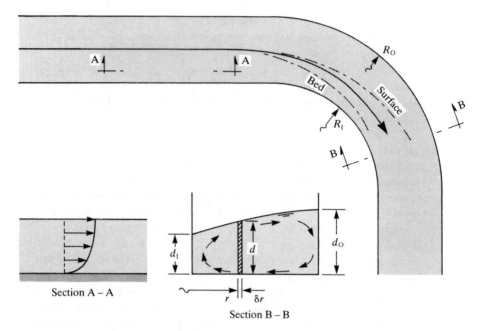

Section A – A

Section B – B

Figure 10-10 Secondary flow in a river bend.

$$Q = \int_{R_1}^{R_0} \left(B - \frac{C^2}{2r^2} \right) \frac{C}{gr} \, dr = \frac{B}{g} \ln \left(\frac{R_0}{R_1} \right) + \frac{C^3}{4g} \left(\frac{1}{R_0^2} - \frac{1}{R_1^2} \right) \quad (10\text{-}36)$$

Now the rate of flow at the inlet section is $18.75 \, \text{m}^3/\text{s}$. Therefore Eq. (10-36) may be solved to give $C = 56.28 \, \text{m}^2/\text{s}$. It follows from $V = C/r$ that the velocities at the inside and outside of the bend are

$$V_1 = 2.81 \, \text{m/s} \quad \text{and} \quad V_0 = 2.25 \, \text{m/s}$$

and Eq. (10-34) therefore shows that the corresponding depths are

$$d_1 = 1.41 \, \text{m} \quad \text{and} \quad d_0 = 1.56 \, \text{m}$$

Comments on Example 10-5
1. It is well known that rivers scour the outside of bends and deposit silt on the inside. The reason for this can be traced back to the velocity distribution in the approaching flow (section A–A). By inspection, particles near the surface travel faster than the mean velocity and particles near the bed travel more slowly. Now all particles in the shaded element of section B–B experience the same value of $\partial p/\partial r$, which causes the radial acceleration V^2/R. For surface particles, the large value of V implies a large value of R; for bed particles, the small V implies a small R. Thus surface particles follow a more gentle curve than average and bed particles follow a tighter curve.

That is, surface particles have an outward velocity component and bed particles have an inward component.

2. In practice, the circulation depicted in section B–B persists for a considerable distance downstream of the bend. This is why irrigation channels must not be sited with their intakes close to the 'inside' bank of a river. The circulation is an example of a so-called *secondary flow*. Taken together with the axial component of flow, it leads to corkscrew-like pathlines as particles move downstream.

Example 10-6: Hydrocyclone A conical container is mounted with its axis vertical as shown in Fig. 10-11. Water enters the top of the container through a pipe mounted tangentially and induces a spiral vortex about the vertical axis. Discharge takes place axially at the top (95 per cent) and at the bottom (5 per cent). Describe the probable behaviour of particles with densities of (a) $\rho = \rho_w$, (b) $\rho > \rho_w$ and (c) $\rho < \rho_w$ as they pass through the container.

SOLUTION

(a) A typical water particle ($\rho = \rho_w$) will enter the container at a radius R_1 with a velocity V_1. As it spirals towards the axis, its tangential velocity component will increase to satisfy $V_0 r = V_1 R_1$. Its radial velocity component will also increase as it approaches one of the outlets. We cannot deduce which outlet it will use, but the statement of the problem implies that the upper outlet is more likely.

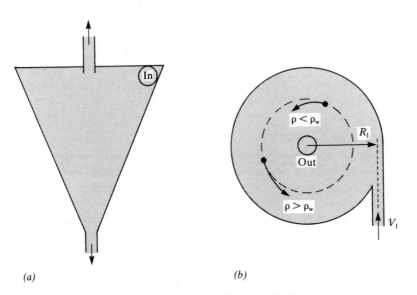

(a)　　　　　　　　　*(b)*

Figure 10-11 Principle of a hydrocyclone. (a) Elevation. (b) Plan.

(b) Particles heavier than water ($\rho > \rho_w$) are likely to behave differently. They have a natural preference for the lower outlet because of their negative buoyancy, but this effect will be important only if they have time to descend significantly before reaching the axis. So do they?

Any particle of whatever density will experience the radial piezometric pressure gradient given by Eq. (10-16). At any radius r,

$$\frac{\partial p^*}{\partial r} = \rho_w \frac{V_w^2}{r} \qquad (10\text{-}37)$$

where the suffix w denotes that it is the water that determines the pressure field. The effect of this pressure gradient on a solid particle of density ρ_s at the same radius will be to cause a pathline curvature R_s, satisfying

$$\frac{\partial p^*}{\partial r} = \rho_s \frac{V_s^2}{R_s} \qquad (10\text{-}38)$$

By comparing (10-37) with (10-38), we see that the radius of curvature of the solid particle will be

$$R_s = \frac{\rho_s}{\rho_w} \left(\frac{V_s}{V_w} \right)^2 r \qquad (10\text{-}39)$$

By inspection, if the velocity of the particle is the same as the local water velocity, the radius of curvature of its pathline will be $R_s = (\rho_s/\rho_w)r$. If $\rho_s > \rho_w$, the radius of curvature will exceed the streamline radius and so the particle will migrate to the outside of the container. This effect will persist as the particle descends and so it is likely to exit through the lower outlet.

(c) The preceding analysis is equally valid for particles that are less dense than the water. In this case it leads to the conclusion that the particles will tend to rise and approach the axis more quickly than the water. Thus they will tend to leave through the upper outlet.

Comments on Example 10-6

1. The hydrocyclone—sometimes called a vortex separator—is a simple and fairly efficient method of extracting particles from fluids. Notice particularly that the device has no moving parts. However, there are some disadvantages. Firstly, some of the fluid is discharged with the waste. Secondly, it is difficult to 'recover' the high kinetic energy of the fluid at outlet, so there will be an energy loss.

2. The preceding discussion is based on inviscid principles. In reality, boundary layers develop on the surfaces of the vessel and secondary flow mechanisms develop. These allow some particles to short-circuit the route to the axis and to contaminate the upper outlet flow.

Example 10-7: Rotation with radial flow A 2 m long horizontal pipe rotates about a vertical axis as shown in Fig. (10-12) and accelerates from rest at a rate of 3 rad/s². The pipe contains water and is connected to reservoir 1 at atmospheric pressure. Estimate the pressure head at the end 2 of the pipe when the speed of rotation reaches 30 rad/s assuming (a) that the ends of the pipe are closed and (b) that valves are slowly opened at the ends of the pipe to create a velocity of flow along it satisfying $V_f = At$ where $A = 0.6$ m/s².

SOLUTION The radial distribution of the Bernoulli sum satisfies Eq. (10-14) which, using (10-6) for the vorticity ζ, may be written as

$$\frac{\partial B}{\partial r} = -\frac{\partial V_r}{\partial t} + \frac{V_s^2}{R} + V_s \frac{\partial V_s}{\partial r} \tag{10-40}$$

(a) When the ends of the pipe are closed, the radial velocity V_r is permanently zero and $\partial V_r/\partial t = 0$. In this case, $V_s = r\Omega$ and the right-hand side of Eq. (10-40) is simply $2r\Omega^2$. The equation can be integrated to give

$$[B_2 - B_1] = [r^2 \Omega^2]_1^2 = r_2{}^2 \Omega^2 \tag{10-41}$$

For the specified conditions, this leads to

$$\frac{p_2}{\rho g} - \frac{p_1}{\rho g} = \frac{V_2{}^2}{2g} = 45.9 \text{ m} \tag{10-42}$$

(b) When there is a velocity of flow V_f along the pipe, the streamlines are not circular and $V_s \neq r\Omega$. Indeed, the notional velocity at very small radii is almost wholly radial. Nevertheless, for sufficiently small velocities of flow, we may use the approximations $V_r \approx V_f$ and $V_s \approx r\Omega$ in most of the pipe. In this case, Eq. (10-40) becomes

$$\frac{\partial B}{\partial r} = -A + 2r\Omega^2 \tag{10-43}$$

which may be integrated to give

$$[B_2 - B_1] = -Ar_2 + r_2{}^2 \Omega^2 \tag{10-44}$$

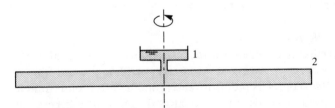

Figure 10-12 Rotating, horizontal pipe.

This leads to

$$\frac{p_2}{\rho g} - \frac{p_1}{\rho g} = -\frac{Ar_2}{g} + \frac{V_2{}^2}{2g} = 45.8 \text{ m} \qquad (10\text{-}45)$$

Comment on Example 10-7
In an accurate analysis, the axes s and n apply along and normal to streamlines. In this particular example, the velocity of flow along the pipe exceeds the tangential component of flow wherever $r < 0.2r_2$. At larger radii, the tangential component becomes increasingly dominant, but the radius of curvature of a typical streamline will nevertheless exceed the local radius of rotation of the pipe. Hence the use of the latter radius in the analysis will tend to increase the assumed value of the term $V_s{}^2/R$ in (10-40). In contrast, the approximation $V_s \approx V_\theta$ will tend to reduce the assumed value of this term. Similar arguments apply to all other terms in the equation.

Example 10-8: Geostrophic winds Because the sun's radiation falls more intensely on the equator than the poles, there is a tendency for air to rise at the equator and to sink at the poles. Convective circulations of terrestrial proportions are induced in the atmosphere to compensate for this effect, as is shown in a highly idealized form in Fig. 10-13a. Show that a necessary consequence is that winds close to the surface of the earth in the polar regions will tend to have an easterly component.

SOLUTION Consider a particle of mass m in the atmosphere at a radial distance r_A from the earth's axis of rotation. In the absence of the convective circulation, its tangential component of velocity would be $V_{\theta A} = r_A \Omega$ where $\Omega = 1$ rev/day is the rate of rotation and its angular momentum about the axis would be $J = mr_A{}^2 \Omega$. As the particle moves towards the equator, its radial distance from the axis increases to r_B, say, and the same angular momentum will imply a tangential velocity of

$$V_{\theta B} = \frac{J}{mr_B} = \frac{r_A{}^2 \Omega}{r_B} \qquad (10\text{-}46)$$

Now this is smaller than the local velocity of the earth, namely $r_B \Omega$, so an observer on the surface will detect a wind. By inspection of Fig. 10-13b, its direction is from east to west (and from pole to equator).

Comments on Example 10-8
1. On average, this effect is stronger near the poles than near the equator. By inspection, an air particle moving at a constant meridional speed from a pole to the equator changes its radial distance from the axis rapidly at first and then more gradually. Close to the equator, the effect should be quite small.

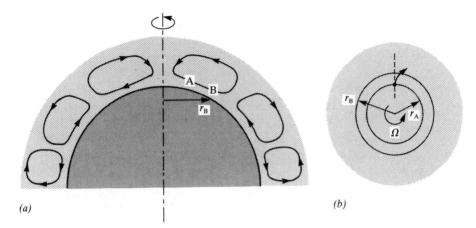

Figure 10-13 Idealized circulation pattern in the atmosphere. (a) The northern hemisphere. (b) The North Pole.

2. The initial condition assumed in the solution corresponds to rigid-body rotation. However, the same qualitative result is obtained whatever initial tangential velocity is assumed. That is, movement towards the equator is associated with an increase in the east-to-west component of flow.

3. At moderate latitudes, the terrestrial circulation implies surface velocities from the equator towards the poles. In this case the same line of reasoning predicts winds with a west-to-east component.

4. Smaller-scale motions caused by mountain ranges etc. or caused by local convection phenomena introduce great complexity into the detailed structure of atmospheric flows. The effect described above is merely a backcloth against which other effects must be considered.

Example 10-9: Helmholtz equation Show that the vorticity of a fluid particle can never change in an inviscid, incompressible, two-dimensional flow.

SOLUTION When the only body force is due to gravity, the first two Euler equations (7-88) may be written as

$$-g\frac{\partial z}{\partial x}-\frac{1}{\rho}\frac{\partial p}{\partial x}=\frac{\partial u}{\partial t}+u\frac{\partial u}{\partial x}+v\frac{\partial u}{\partial y} \qquad (10\text{-}47)$$

and

$$-g\frac{\partial z}{\partial y}-\frac{1}{\rho}\frac{\partial p}{\partial y}=\frac{\partial v}{\partial t}+u\frac{\partial v}{\partial x}+v\frac{\partial v}{\partial y} \qquad (10\text{-}48)$$

in which z denotes the elevation. By inspection, the left-hand sides of these equations can be made equal by differentiating the first with respect to y

and the second with respect to x. After doing this and subtracting the results, we obtain

$$
\frac{\partial^2 u}{\partial y\,\partial t}+u\frac{\partial^2 u}{\partial y\,\partial x}+\left\{\frac{\partial u}{\partial y}\frac{\partial u}{\partial x}+\frac{\partial u}{\partial y}\frac{\partial v}{\partial y}\right\}+v\frac{\partial^2 u}{\partial y^2}=
$$

$$
\frac{\partial^2 v}{\partial x\,\partial t}+u\frac{\partial^2 v}{\partial x^2}+\left\{\frac{\partial v}{\partial x}\frac{\partial u}{\partial x}+\frac{\partial v}{\partial x}\frac{\partial v}{\partial y}\right\}+v\frac{\partial^2 v}{\partial x\,\partial y} \qquad (10\text{-}49)
$$

The continuity equation (6-54) is $\partial u/\partial x+\partial v/\partial y=0$ and so the terms collected in brackets in Eq. (10-49) are equal to zero. The remaining terms can be rearranged as

$$
\frac{\partial}{\partial t}\left(\frac{\partial v}{\partial x}-\frac{\partial u}{\partial y}\right)+u\frac{\partial}{\partial x}\left(\frac{\partial v}{\partial x}-\frac{\partial u}{\partial y}\right)+v\frac{\partial}{\partial y}\left(\frac{\partial v}{\partial x}-\frac{\partial u}{\partial y}\right)=0 \qquad (10\text{-}50)
$$

Each term in parentheses is simply the vorticity ζ and so

$$
\left(\frac{\partial}{\partial t}+u\frac{\partial}{\partial x}+v\frac{\partial}{\partial y}\right)\zeta=0 \qquad (10\text{-}51)
$$

which is the two-dimensional form of the *Helmholtz equation*. Since $\partial/\partial t+u\,\partial/\partial x+v\,\partial/\partial y$ denotes differentiation along a particle path, the equation shows that the vorticity of an individual particle can never change.

Comments on Example 10-9

1. This result is open to the following physical interpretation. Recall that it is not possible to create shear stresses in an inviscid fluid and so tangential forces cannot be applied to inviscid fluid particles. If such a particle in a two-dimensional flow is spinning at any instant, it will continue to spin for all time. If it is not spinning, it will never spin. If no particles in a particular flow are spinning, then the whole flow is irrotational and it will remain so.
2. The result cannot be generalized to three-dimensional flows because rotations do not add vectorially. In a three-dimensional flow, distortions of fluid particles along an axis lead to changes in vorticity about that axis.

PROBLEMS

1 Show that the radius of the air core in a whirlpool satisfies

$$
\frac{R}{R_1}=\left(\frac{h_1}{h}\right)^{1/2} \qquad (10\text{-}52)
$$

where h denotes the surface depression as shown in Fig. 10-14 and the point 1 is somewhere on the free surface.

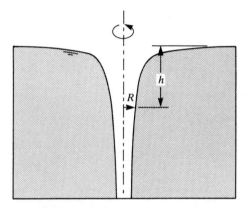

Figure 10-14 Air core in a whirlpool.

2 The velocity at a radius r in a tornado is $V = C/r$. Determine the radius at which the pressure is theoretically equal to zero, assuming (a) incompressible flow and (b) homentropic flow ($p/\rho^{1.4} = \text{constant}$).

$$[r_0{}^2 = \tfrac{1}{2}C^2\rho/p_\infty; \tfrac{1}{7}C^2\rho_\infty/p_\infty]$$

3 The velocity distribution in the flow past the cylinder of radius R shown in Fig. 9-25b satisfies

$$V_r = V_\infty \cos\theta\left(1 - \frac{R^2}{r^2}\right); \qquad V_\theta = -V_\infty \sin\theta\left(1 + \frac{R^2}{r^2}\right) - \frac{RV_\infty}{r}$$

$$(10\text{-}53)$$

Show that the difference in pressure between two points at the same radius in the directions $\theta = \tfrac{1}{2}\pi$ and $\theta = \tfrac{3}{2}\pi$ is

$$p_{\frac{3}{2}\pi} - p_{\frac{1}{2}\pi} = 2\rho V_\infty \frac{R}{r}\left(1 + \frac{R^2}{r^2}\right) \qquad (10\text{-}54)$$

Also show that the magnitudes of the radial pressure gradients on the surface of the cylinder at $\theta = \tfrac{1}{2}\pi$ and $\theta = \tfrac{3}{2}\pi$ differ by a factor of 9.

4 Verify that the circulation around any circle centred on the origin of the cylinder in Fig. 9-25b is equal to $-2\pi RV_\infty$. Also verify that the circulation is zero around any segment such as that shown in Fig. 10-15. (*NB*: None of the surfaces of the segment is parallel to the direction of flow.)

5 Flow leaves a reservoir over a spillway crest and accelerates downhill (Fig. 10-16). At a point A on the surface of the spillway, the velocity is 16 m/s. Estimate the radius of curvature of the spillway if the pressure on its surface is equal to atmospheric pressure. For simplicity, assume that all streamlines have the same radius of curvature as the spillway surface. (*NB*: The pressure on the surface of a flat spillway of slope α would be $\rho g d \cos^2 \alpha$).

$$[34.8 \text{ m}]$$

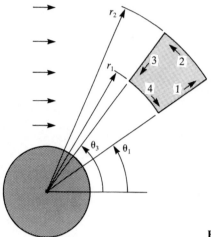

Figure 10-15 Irrotational flow with circulation.

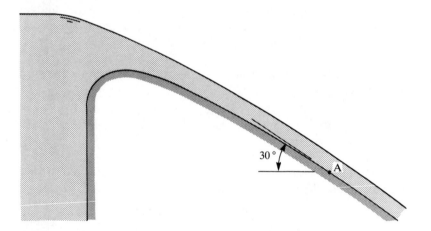

Figure 10-16 Flow over a spillway.

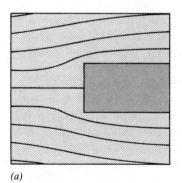

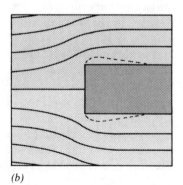

(a)   *(b)*

Figure 10-17 Flow past a bluff body. (a) Ideal flow. (b) Real flow.

6 In an irrotational flow analysis, the flow past the leading edge of a bluff body is as shown in Fig. 10-17a. Explain why the real flow will resemble Fig. 10-17b.

7 Explain why the pressure on the bed of a channel just downstream of a sluice gate is greater than would be predicted by assuming a hydrostatic pressure distribution at all sections (see Fig. 8-17).

8 A 0.5 m tall can of 0.8 m diameter rotates with its axis vertical as shown in Fig. 10-18. The can contains water and the outer surface level is at the top of the sides. What is the speed of rotation of the can and its contents when the depth at the axis is 0.2 m?

[6.07 rad/s]

9 When the can of water is rotating as described in Problem 8, 0.1 kg water is released gently into the centre of the can, causing an equal mass to be displaced over the sides. By how much does the energy of the displaced water exceed the energy of the supply?

[0.589 J]

10 Suppose that the same can has radial fins to ensure that the water rotates at the same rate as the can at all radii. What power must be supplied to maintain the speed of rotation if the rate of inflow at the central axis is 20 kg/s? Compare this configuration with a centrifugal pump impeller.

[118 W]

11 The L-shaped tube shown in Fig. 10-19 is closed at end A and open to the atmosphere ($p = 100$ kPa) at end B. The height of the vertical limb is 0.2 m and the length of the horizontal limb is 0.35 m. At what speed of rotation about the vertical axis will the pressure at A reduce to the cavitation pressure, 2 kPa, thus causing the level in the vertical limb to rise on further acceleration?

[40.4 rad/s]

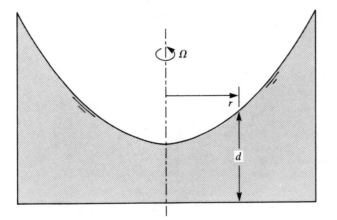

Figure 10-18 Forced rotation in a tank.

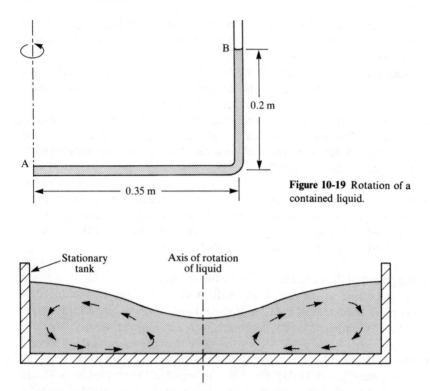

Figure 10-19 Rotation of a contained liquid.

Figure 10-20 Secondary flow in a liquid rotating in a stationary tank.

12 A cylindrical tank of water is initially at rest with its axis vertical. The water is then swirled around the tank by a paddle so that the conditions resemble a free vortex except very close to the axis and very close to the solid surfaces. Show that a secondary flow must develop as shown in Fig. 10-20 and hence explain why particulate matter usually settles at the centre of a teacup or a washing-up bowl.

FURTHER READING

Baker, C.J. (1980) Theoretical approach to prediction of local scour around bridge piers, *J. Hydr. Res.*, **18** (1), 1–12.

Bear, J. (1988) *Dynamics of Fluids in Porous Media*, Dover.

Hawthorne, W.R. (1951) Secondary circulation in fluid flow, *Proc. R. Soc.*, A206, 374–387.

Vallentine, H.R. (1969) *Applied Hydrodynamics*, Butterworths.

Van Dyke, M. (1982) *An Album of Fluid Motion*, Parabolic Press.

ELEVEN

ENERGY

11-1 FIRST LAW OF THERMODYNAMICS

All real bodies have energy in one form or another. In particular, all matter contains **internal** energy which can be used, for example, to raise the temperature of something else that is initially cooler. Moving bodies are said to possess *kinetic* energy and objects in gravitational and electromagnetic fields can be regarded as having *potential* energy.

It is found by experiment that the total energy of a system remains permanently constant unless the system takes part in either heat or work interactions with its surroundings. When no such interactions exist, the system is thermodynamically *isolated* and any change in one form of its energy must be counterbalanced by a change in one or more of its other forms. When heat or work interactions do take place, the increase ΔE in the total energy of the system satisfies

$$Q - W = \Delta E \qquad (11\text{-}1)$$

in which Q denotes the heat *received* by the system and W is the work *done* by it.

11-1-1 Heat

When a 'hot' object is placed in a bowl of 'cold' water, energy is transferred until the temperatures equalize. During the process, the internal energy of the object decreases and the internal energy of the bowl of water increases by an equal amount. The mechanism by which the energy transfer takes place is called a *heat interaction* and its magnitude is the amount of energy transferred. Notice that we cannot say that the object 'loses' heat or that the bowl of water 'gains' heat. It is internal energy, not heat, that is lost and gained.

Definition These ideas can be generalized to allow us to define heat as the energy that is transferred between two systems *as a consequence of their temperature difference* when they are brought into communication.

Adiabatic processes Processes which take place without loss or gain of heat are termed *adiabatic*. They can occur only when the system under consideration is in thermal equilibrium with its surroundings at every point on its boundary. In practice, temperature differences nearly always exist in processes of engineering interest, but the *rate* of heat transfer can often be reduced to a very low level by the use of thermal insulation. Real processes can therefore be made to approximate to adiabatic behaviour.

11-1-2 Work

Figure 11-1 shows a cylinder in which a high-pressure gas A is separated from a low-pressure gas B by a close-fitting piston that is held in position by an external restraint (not shown). If the restraint is suddenly removed the piston will move back and forth until it occupies a stable position in which the pressure is the same in both gases. During the process the internal energy of gas A will decrease and the internal energy of gas B will increase by an equal amount (neglecting for present purposes any heat interactions with the piston or the cylinder). The mechanism by which the energy transfer takes place is called a *work interaction*, and its magnitude is the amount of energy transferred.

If F_A is used to denote the average pressure force on the face of the piston in contact with the gas A while it moves through an elemental distance δx, then the magnitude of the work done at this face is $F_A \, \delta x$. In the same interval, the magnitude of the work done at the other face of the piston can be denoted by

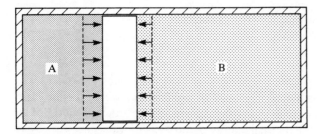

Figure 11-1 Example of a work interaction (pressure forces do work when the piston moves).

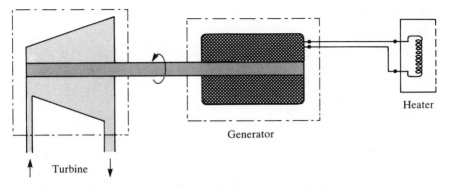

Figure 11-2 'Conversion' of fluid energy into work and then into heat.

$F_B \, \delta x$. Any difference between F_A and F_B will result in a change in the energy of the piston.

Definition It is nearly always possible to regard work as the product of a force and the distance moved in its line of action. However, the following thermodynamic definition is more general:

Work is done by a system during a given operation when the sole effect external to the system could have been the raising of a weight.

The advantage of the more general definition is illustrated in Fig. 11-2, which is a schematic representation of a fluid turbine, an electrical generator on the same shaft and an electrical heater unit. The turbine receives heat and does work as a result of shear stresses in the rotating shaft. Work is also done by pressure forces where the fluid enters and leaves the control volume. The generator has work done on it by the shaft and it supplies the energy required by the heater unit.

We now query whether the energy transfer from the generator is heat or work. Most people correctly recognize intuitively that it is work, but the notion of 'force × distance' is not very helpful in this context. The above definition is much more satisfactory because we can imagine the electrical output to be connected to an electrical motor that could be used to raise a weight. Assuming a perfectly efficient motor, the sole effect of this arrangement would be the raising of a weight. Therefore the generator does work even in the actual process where its output is used to provide heat.

Choice of reference axes The amount of work done by a force moving through a distance is dependent upon the choice of reference axes. Two observers moving relative to each other will not ascribe the same velocity to the motion of a system boundary. Nevertheless, both observers can apply Eq. (11-1) successfully in their own frame of reference.

11-1-3 Energy

Internal energy A system that is at rest and experiences no field or surface tension effects has only internal energy. This property is described in section 2-9 and the manner in which heat and work can affect it is illustrated above. For pure substances, it is convenient to use the *specific internal energy u*, namely the internal energy per unit mass. The total internal energy of a system is $\int u \, dm$.

Kinetic energy Newton's second law of motion shows that the force needed to accelerate a particle of mass m at a rate of $\delta V / \delta t$ is $F = m \, \delta V / \delta t$. During the time interval δt, the force moves through a distance $\delta x = V \, \delta t$ and so the work done on the particle is $\delta W = F \, \delta x = m V \, \delta V = m \, \delta(\frac{1}{2} V^2)$. The total work done in accelerating the particle from rest to a velocity V_0 is therefore $\frac{1}{2} m V_0^2$. At the end of the process, the particle is said to have a kinetic energy of $\frac{1}{2} m V_0^2$ because it could do this amount of work on another particle while being brought to rest. The total kinetic energy of a system composed of elemental masses δm moving with different velocities is $\int \frac{1}{2} V^2 \, dm$ in which no account needs to be taken of differences in the directions of motion of the various particles.

Unlike internal energy, the kinetic energy of a system is not an independent property of the matter contained within it. Its value depends upon the reference axes chosen by the observer. For example, relative to a passenger on a train or an airliner, the vehicle has no kinetic energy.

Potential energy When a particle of mass δm is raised through a height z in a gravitational field, its gravitational potential energy increases by an amount $gz \, \delta m$. This is equal to the work done when a force $g \, \delta m$ moves through a distance z. It is not necessary that the movement should take place vertically, but simply that the final elevation should exceed the initial elevation by the amount z. Notice that the change in potential energy is independent of the medium (air, water, etc.) in which the particle exists. The potential energy is a function of the gravitational field alone. In the study of magnetohydro-dynamics, potential energies due to magnetic and electric fields are also considered.

The concept of potential energy is introduced for convenience only. Natural phenomena can be described equally accurately without mention of potential energies provided that the work done by the fields is considered explicitly. Nevertheless, it is *conventional* to adopt the former approach.

Free-surface energy When a free surface exists between two fluids (liquid–liquid or liquid–gas), account should be taken of the surface tension in the interface. Consider the circular area of an interface depicted in Fig. 11-3 and

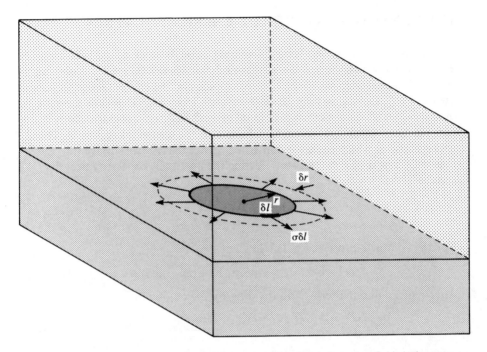

Figure 11-3 Surface tension on a circular element in an interface between two fluids (at least one must be a liquid).

suppose that the radius of the element increases by an amount δr during some process. The surface tension forces exerted on the element by its surroundings are depicted schematically in the figure. On a typical elemental length δl of the circumference, an elemental force $\delta F = \sigma \, \delta l$ moves through a distance δr and so the work done is $\sigma \, \delta l \, \delta r$. The total work done on the element of circumference $2\pi r$ is $\delta W = 2\pi r \sigma \, \delta r$. Since the increase in the surface area of the element is $\delta a = 2\pi r \, \delta r$, we find that the work done per unit area is $\delta W / \delta a = \sigma$. Therefore we may regard the surface tension coefficient σ as a free surface energy per unit area.

Total energy The total energy of a system is the sum of all the above contributions. When the only potential energy is due to gravity,

$$E = \int (u + \tfrac{1}{2} V^2 + gz) \, dm + \int \sigma \, da \qquad (11\text{-}2)$$

For a pure substance, the *specific* total energy in the absence of surface tension and fields other than gravity is

$$e = u + \tfrac{1}{2} V^2 + gz \qquad (11\text{-}3)$$

11-2 GENERAL FORM OF THE ENERGY EQUATION

Figure 11-4 depicts a general control volume through which a fluid is flowing. The *system* to which the first law of thermodynamics will be applied is chosen as the combined contents of the control volume and the elemental region I at the instant t_0. During an interval δt, the system receives an elemental amount of heat δQ and does an elemental amount of work δW as it moves to occupy its new position, namely the control volume and the elemental region O. The average rate at which energy is convected across the control surface into the region O, say, is $\delta E_O / \delta t$ and so we define the energy flux $\dot{E}$ at a typical flow section as

$$\dot{E} \equiv \frac{dE}{dt} \tag{11-4}$$

For the complete process, Eq. (11-1) gives

$$\delta Q - \delta W = (E_{cv} + \delta E_O)_{t_0 + \delta t} - (E_{cv} + \delta E_I)_{t_0} \tag{11-5}$$

After rearranging the terms and dividing by δt, we obtain

$$\frac{\delta Q}{\delta t} - \frac{\delta W}{\delta t} = \frac{(E_{t_0 + \delta t} - E_{t_0})_{cv}}{\delta t} + \frac{\delta E_{O, t_0 + \delta t}}{\delta t} - \frac{\delta E_{I, t_0}}{\delta t} \tag{11-6}$$

and in the limit, as δt approaches zero, this becomes the energy equation:

$$\dot{Q} - \dot{W} = \frac{\partial E_{cv}}{\partial t} + \dot{E}_O - \dot{E}_I \tag{11-7}$$

in which $\dot{Q} \equiv \delta Q / \delta t$ and $\dot{W} \equiv \delta W / \delta t$ are the *rates* of heat received and work done respectively.

Energy flux By regarding the mass flux $\dot{m}$ across the control surface at a typical flow section as the sum of elemental fluxes $\delta \dot{m}$, each with its own value of specific total energy, we may write the energy flux for the section as

$$\dot{E} = \int e \, d\dot{m} \tag{11-8}$$

Flow work Work *can* be done by normal and shear forces over the whole of the system boundary. However, work *is* done only where the boundary moves in the direction of the force. No work is done at a stationary boundary (e.g. a rigid pipe wall), even though both normal and shear forces may be present.

It is useful to regard the total work done as the sum of (1) the work done at sections where fluid crosses the control surface, W_f, and (2) the work done over the rest of the surface, W_s. This procedure is convenient because the former—known as *flow work*—always exists in a flow process whereas the latter—often

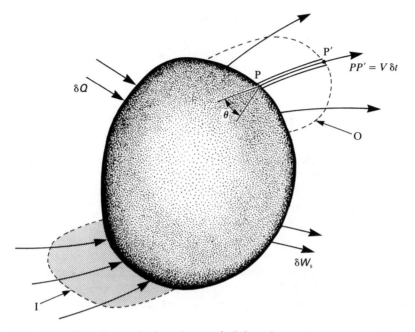

Figure 11-4 General control volume (system shaded at t_0).

called *shaft work*—is sometimes zero. Consider the flow work done by the system in Fig. 11-4 as it displaces its surroundings from the region O. The pressure force exerted on the downstream fluid by an elemental area δa of the system boundary is $p\,\delta a$ and the component of this force in the direction of flow is $p\,\delta a\cos\theta$. During the interval δt, the boundary element moves a distance $V\,\delta t$ and so the work done is $pV\cos\theta\,\delta a\,\delta t$ and the *rate* of work is $pV\cos\theta\,\delta a$. Now $\rho V\cos\theta\,\delta a$ is the elemental mass flux through the original element δa of the control surface (see Eq. (6-6)). Therefore the contribution of this element to the rate of flow work may be written as

$$\delta\dot{W}_f = \frac{p}{\rho}\,\delta\dot{m} \tag{11-9}$$

By substituting Eqs (11-8) and (11-9) into Eq. (11-7), we obtain a useful general form of the energy equation:

$$\dot{Q} - \dot{W}_s = \frac{\partial E_{cv}}{\partial t} + \int_O \left(e + \frac{p}{\rho}\right)\mathrm{d}\dot{m} - \int_I \left(e + \frac{p}{\rho}\right)\mathrm{d}\dot{m} \tag{11-10}$$

in which $\dot{W}_s$ denotes the rate of work done at all points on the boundary other than the flow sections.

The most common form of the energy equation is obtained by using Eq. (11-3) and noting that the sum $u + p/\rho$ is equal to the specific enthalpy h

(section 2-9). For any flow section, this yields

$$\int \left(e + \frac{p}{\rho}\right) d\dot{m} = \int (h + \tfrac{1}{2}V^2 + gz)\, d\dot{m} \qquad (11\text{-}11)$$

and the energy equation becomes

$$\dot{Q} - \dot{W}_s = \frac{\partial E_{cv}}{\partial t} + \int_O (h + \tfrac{1}{2}V^2 + gz)\, d\dot{m} - \int_I (h + \tfrac{1}{2}V^2 + gz)\, d\dot{m} \qquad (11\text{-}12)$$

Range of validity of the energy equation Equation (11-7) is valid for any control volume. The net rate at which heat is received $\dot{Q}$ is the algebraic sum of the heat flux over the whole of the control surface *including* the flow sections (where conduction can be significant in some types of compressible flow). The rate of work done $\dot{W}_s$ is summed over the whole of the control surface *except* the flow sections. In practice, $\dot{W}_s$ is usually composed wholly of work due to shear forces because work due to normal forces occurs only where mass crosses the control surface. A small error is introduced by the use of the pressure p in Eq. (11-9). Strictly, the work done at a flow section should be determined by considering the streamwise components of the local normal and shear stresses.

Equation (11-11) is further restricted by the use of Eq. (11-3), which is valid only for pure substances and only in the absence of surface tension and fields other than gravity. Nevertheless, these limitations are of little practical importance because the required conditions are usually satisfied.

11-3 UNIAXIAL AND ONE-DIMENSIONAL FLOWS

When the flow at an inlet or outlet section is uniaxial and the control surface is chosen normal to the direction of flow, we may replace $d\dot{m}$ by $\rho V\, da$ and write (11-11) as

$$\int \left(e + \frac{p}{\rho}\right) d\dot{m} = \int (h + \tfrac{1}{2}V^2 + gz)\rho V\, da \qquad (11\text{-}13)$$

Following the procedure used in the derivation of the uniaxial momentum equation, we define a *kinetic energy flux coefficient* α by

$$\alpha \equiv \frac{1}{\bar{\rho} a \bar{V}^3} \int \rho V^3\, da \qquad (11\text{-}14)$$

and we define *mean* values of p, u, h and z by

$$\bar{p} \equiv \frac{1}{a\bar{V}} \int pV\, da \qquad \bar{u} \equiv \frac{1}{\bar{\rho} a \bar{V}} \int u\rho V\, da$$

$$\bar{h} \equiv \frac{1}{\bar{\rho} a \bar{V}} \int h\rho V\, da \qquad \bar{z} \equiv \frac{1}{\bar{\rho} a \bar{V}} \int z\rho V\, da \qquad (11\text{-}15)$$

As a consequence of these definitions, (11-13) becomes

$$\int \left(e+\frac{p}{\rho}\right)\mathrm{d}\dot{m}=\dot{m}(\bar{\bar{h}}+\tfrac{1}{2}\alpha\bar{V}^2+g\bar{\bar{z}}) \tag{11-16}$$

in which $\dot{m}=\bar{\rho}a\bar{V}$.

In Eq. (11-15) the use of the double bar signifies that the velocity V appears as a weighting factor in the definition. This distinguishes the parameters from other mean values such as $\bar{p}$ and $\bar{z}$ defined by Eq. (7-11). In common with $\bar{p}$, $\bar{\rho}$ and $\bar{z}$, however, no use is made of these definitions in this book. They are presented only for completeness so that unnecessary errors need not be introduced if readers wish to analyse flows in which significant variations in p, ρ, h, etc. exist at flow sections. In the vast majority of cases, it will be acceptable to approximate these parameters by their values at the centroid of the flow section.

This disclaimer does not apply to the mean velocity $\bar{V}$ or to the momentum and kinetic energy flux coefficients β and α. The first of these is commonly used and the latter two are occasionally useful. Example 11-1 (p. 307) gives additional information, including an approximate relationship between β and α.

Range of validity The only additional limitation of Eq. (11-16) in comparison with Eq. (11-11) is that the flow must be uniaxial. Appropriate mean values should be defined separately for each inflow and outflow section.

One-dimensional flow At a one-dimensional flow section, $\alpha=1$ and Eq. (11-16) simplifies to

$$\int \left(e+\frac{p}{\rho}\right)\mathrm{d}\dot{m}=\dot{m}(h+\tfrac{1}{2}V^2+gz) \tag{11-17}$$

in which it is nearly always sufficiently accurate to use the values of h and z applicable at the centroid.

Steady flow When the flow is steady and there is only one inlet and one outlet, the uniaxial energy equation is

$$\dot{Q}-\dot{W}_{\mathrm{s}}=\dot{m}\{(h+\tfrac{1}{2}\alpha\bar{V}^2+gz)_{\mathrm{o}}-(h+\tfrac{1}{2}\alpha\bar{V}^2+gz)_{\mathrm{i}}\} \tag{11-18}$$

When the flow at these sections is also one-dimensional, this reduces to

$$\dot{Q}-\dot{W}_{\mathrm{s}}=\dot{m}\{(h+\tfrac{1}{2}V^2+gz)_{\mathrm{o}}-(h+\tfrac{1}{2}V^2+gz)_{\mathrm{i}}\} \tag{11-19}$$

11-3-1 Comparison with the Bernoulli equation

It is instructive to develop the energy equation in a form that can be compared easily with the Bernoulli equation considered in Chapter Nine because there are certain similarities between the two. For an incompressible flow, the sum $e + p/\rho$ can be written as

$$e + \frac{p}{\rho} = u + \frac{p}{\rho} + \tfrac{1}{2}V^2 + gz = u + B \qquad (11\text{-}20)$$

where

$$B \equiv \frac{p}{\rho} + \tfrac{1}{2}V^2 + gz \qquad (11\text{-}21)$$

denotes a Bernoulli sum. At a uniaxial flow section, the appropriate mean value of the Bernoulli sum is

$$\bar{B} \equiv \frac{p}{\rho} + \tfrac{1}{2}\alpha \bar{V}^2 + gz \qquad (11\text{-}22)$$

in which mean values such as (7-11) are implied. Therefore the energy equation (11-10) may be expressed for uniaxial flows as

$$\dot{Q} - \dot{W}_s = \frac{\partial E_{cv}}{\partial t} + \dot{m}_o(\bar{B} + u)_o - \dot{m}_l(\bar{B} + u)_l \qquad (11\text{-}23)$$

For the special case of a steady flow with only one inflow and one outflow section, at each of which the flow is one-dimensional,

$$\dot{Q} - \dot{W}_s = \dot{m}\{(B + u)_o - (B + u)_l\} \qquad (11\text{-}24)$$

and so the change in the Bernoulli sum between inlet and outlet satisfies

$$(B_O - B_I) = \frac{\dot{Q} - \dot{W}_s}{\dot{m}} - (u_O - u_I) \qquad (11\text{-}25)$$

All of these expressions are valid for an incompressible flow in the presence or absence of shear stresses (whether or not these contribute towards $\dot{W}_s$). Some of them are restricted to flows where the inlet and outlet sections are one-dimensional, but no restriction is placed on the events taking place *within* the control volume. In principle, there could be substantial heat and work interactions and there could even be phase changes.

In contrast, the Bernoulli equation (9-5) (i.e. $B_O = B_I$) is valid only for a steady, inviscid, incompressible flow along a streamline on which no discontinuous events occur. This does not preclude the possibility of heat, but it does preclude the existence of shaft work or phase changes. In these circumstances, both equations are valid and their combination shows that the heat flux satisfies

$$\dot{Q} = \dot{m}(u_O - u_I) \qquad (11\text{-}26)$$

That is, heat transfers in an incompressible flow have the effect of raising the specific internal energy of the fluid and thereby its temperature (since $\Delta u = c_v \Delta T$).

In the case of *compressible* flows, the Bernoulli sum does not satisfy

Eq. (11-21) and so the subsequent equations do not apply. In particular, Eq. (11-26) is not applicable; heat transfers influence the pressure and density of the fluid as well as its temperature.

11-4 APPLICATIONS OF THE ENERGY EQUATION

Example 11-1: Kinetic energy flux coefficient Evaluate the kinetic energy flux coefficient α for a steady, incompressible, laminar flow in a circular pipe of radius R. The velocity V at a radius r satisfies

$$V = K(R^2 - r^2) \tag{11-27}$$

in which K is a constant.

SOLUTION For an incompressible fluid Eq. (11-14) gives

$$\alpha a \bar{V}^3 = \int V^3 \, da \tag{11-28}$$

Following closely the procedure used in Example 7-1 to evaluate the momentum flux correction coefficient β, we substitute πR^2 for a, $2\pi r \, \delta r$ for δa, (11-27) for V and $\frac{1}{2}KR^2$ for $\bar{V}$ to obtain

$$\tfrac{1}{8}\alpha\pi K^3 R^8 = 2\pi K^3 \int_0^R (R^6 r - 3R^4 r^3 + 3R^2 r^5 - r^7) \, dr \tag{11-29}$$

Therefore $\qquad \tfrac{1}{16}\alpha R^8 = \dfrac{R^8}{2} - \dfrac{3R^8}{4} + \dfrac{R^8}{2} - \dfrac{R^8}{8} = \tfrac{1}{8}R^8 \tag{11-30}$

and so for this particular flow $\alpha = 2$.

Comments on Example 11-1

1. The comments made about the momentum coefficient β following Example 7-1 apply even more forcibly to the coefficient α. In a steady *turbulent* flow along a circular section pipe, α varies from about 1.08 at a Reynolds number of 4000 to about 1.03 at a Reynolds number of 10 million.
2. It is sometimes useful to make use of an algebraic relationship between β and α. In the author's experience the expression

$$\alpha \simeq 3\beta - 2 \tag{11-31}$$

is approximately valid for steady laminar and turbulent flows in pipes and channels. It is exactly true for the Poiseuille flow considered above.
3. The velocity distribution in a turbulent flow is complex but, for the present purpose, the power law relationship

$$\frac{V}{V_0} = \left(1 - \frac{r}{R}\right)^N \tag{11-32}$$

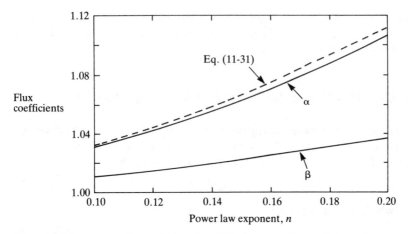

Figure 11-5 Momentum flux coefficient β and kinetic energy flux coefficient α in a power law velocity distribution.

gives a sufficiently accurate representation of a steady flow in a circular section pipe. It agrees quite well with experiment, but the exponent N varies with the Reynolds number, being about $\frac{1}{5}$ when $Re = 4000$ and approximately $\frac{1}{10}$ when $Re = 10^7$. Figure 11-5 shows the variation of β and α with n for this expression. The broken line indicates values of α predicted by Eq. (11-31) based on the correct values of β (see Problem 12 at the end of this chapter).

Example 11-2: Work done by a steam turbine An adiabatic turbine receives steam at a pressure of 15 MPa and a temperature of 550 °C. It exhausts the steam to a condenser at a pressure of 4 kPa and a specific enthalpy of 2.19 MJ/kg K. Neglecting changes in kinetic and potential energies, determine the shaft work done per kg of steam.

SOLUTION The state of any pure substance is fully specified by any two independent properties (see section 2-9). Since the inlet pressure and temperature of the steam are known, its specific enthalpy may be obtained from suitable thermodynamic charts. For instance, Haywood (1972) gives $h_1 = 3.45$ MJ/kg. By neglecting changes in kinetic and potential energies in comparison with the change in specific enthalpy, Eq. (11-19) may be written for the control volume in Fig. 11-6 as

$$\dot{Q} - \dot{W}_s = \dot{m}(h_O - h_1) \tag{11-33}$$

Since the turbine is adiabatic, $\dot{Q} = 0$ and so the shaft work done per unit mass of steam is

$$\dot{W}_s/\dot{m} = h_1 - h_O = 1.26 \text{ MJ/kg} \tag{11-34}$$

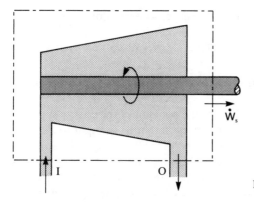

Figure 11-6 Steam turbine.

Comments on Example 11-2

1. It is common practice to neglect changes in kinetic and potential energies when analysing flows involving substantial amounts of heat or work. Suppose for example that the velocity increases from 20 m/s to 200 m/s and that the elevations of the inlet and outlet sections differ by 10 m. In this case, $\Delta(\frac{1}{2}V^2)=0.02$ MJ/kg and $\Delta(gz)=0.0001$ MJ/kg.

2. The turbine outlet pressure of 4 kPa is very much smaller than atmospheric pressure (≈ 100 kPa). This is typical of the operation of a steam turbine. The work done by the steam in expanding from atmospheric pressure to the condenser pressure greatly exceeds the work that must be done to repressurize the condensed water when it is subsequently fed to a boiler before recycling through the turbine. The principal disadvantages are that the condenser must be very large (to cope with the great volume of the expanded steam) and that there is a risk of infiltration of air in the sub-atmospheric region.

Example 11-3: Gas compressor A gas compressor receives air ($R=287$ J/kg K, $c_p=1004$ J/kg K) from the atmosphere at a pressure of 100 kPa and a temperature of 20 °C and delivers it at a pressure of 600 kPa. The process is polytropic with $N=1.5$ and the required shaft power is 245 kW per kilogram of air. Determine the rate of heat loss from the compressor per kilogram of air.

SOLUTION At inlet, the density is $\rho_1=p_1/RT_1=1.19$ kg/m³. Since the process is polytropic with $N=1.5$, the density at outlet is

$$\rho_O=\rho_1\left(\frac{p_O}{p_1}\right)^{1/n}=3.92 \text{ kg/m}^3 \qquad (11\text{-}35)$$

Therefore the outlet temperature is $T_O=p_O/R\rho_O=533$ K $=260$ °C. Using the steady-flow energy equation (11-19) and neglecting the influence of velocity and elevation, the heat received per kilogram of air is

$$\frac{\dot{Q}}{\dot{m}} = \frac{\dot{W}_s}{\dot{m}} + c_p(T_0 - T_1) = -4.26 \, \text{kW/kg} \tag{11-36}$$

Since this is negative, heat is emitted, not received, by the compressor.

Comment on Example 11-3 Since less than 2 per cent of the energy supplied to the compressor is wasted as heat the process might be thought to be more than 98 per cent efficient. However, it is theoretically possible to design a machine in which there is negligible heat loss and in which the process is isentropic (i.e. polytropic with $N = \gamma = 1.4$, say). In this case, the density and temperature of the air at outlet would be $\rho_0 = 4.27 \, \text{kg/m}^3$ and $T_0 = 489 \, \text{K}$ and the power required by the compressor would be $\dot{W}_s = 197 \, \text{kW/kg}$. By comparing powers supplied to the actual machine and the real machine, we obtain an isentropic efficiency of 80.4 per cent.

Example 11-4: Aero-engine reheat To generate additional thrust during take-off or during acceleration from subsonic to supersonic speeds, the exhaust gases in some aero-engines are heated by burning fuel in the jet pipe (the duct downstream of the turbine). As a first approximation, the process depicted in Fig. 11-7 may be regarded as a one-dimensional flow with heat addition (see comment 2). Determine the additional thrust and the rate of heat supplied when the velocity V_2 is increased to 700 m/s by reheat if the outlet velocity and temperature are 500 m/s and 600 °C respectively in the absence of reheat. Treat the fluid as a perfect gas for which $c_p = 1.15 \, \text{kJ/kg K}$ and $R = 287 \, \text{J/kg K}$ and assume that the outlet pressure is 100 kPa. The mass flux is the same for both cases, but the nozzle outlet area is increased from its usual value of $0.5 \, \text{m}^2$ to $0.7 \, \text{m}^2$ during reheat.

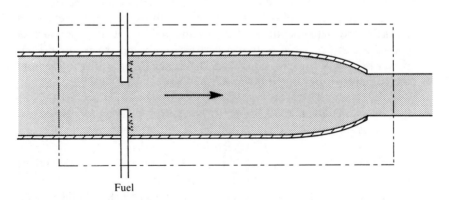

Fuel

Figure 11-7 Aero engine reheat.

SOLUTION

(a) *Without reheat* Using the equation of state (1-11) and noting that $T_{2A}=600\,°C\approx 873$ K, we deduce that the outlet density is

$$\rho_{2A}=\frac{p_{2A}}{RT_{2A}}=0.399\ \text{kg/m}^3 \qquad (11\text{-}37)$$

and so the mass flux through the duct is

$$\dot{m}=\rho aV=99.8\ \text{kg/s} \qquad (11\text{-}38)$$

(b) *With reheat* The additional thrust ΔF due to reheat is the increase in the outlet momentum flux, namely

$$\Delta F=\dot{m}(V_{2B}-V_{2A})=20.0\ \text{kN} \qquad (11\text{-}39)$$

The outlet density and temperature are determined from the mass flux and the equation of state respectively as

$$\rho_{2B}=\frac{\dot{m}}{a_{2B}V_{2B}}=0.204\ \text{kg/m}^3 \qquad (11\text{-}40)$$

and

$$T_{2B}=\frac{p_{2B}}{R\rho_{2B}}=1711\ \text{K} \qquad (11\text{-}41)$$

The heat flux is deduced from the steady-flow energy equation which, for the two conditions, gives

$$0=\dot{m}\{(h+\tfrac{1}{2}V^2)_{2A}-(h+\tfrac{1}{2}V^2)_1\} \qquad (11\text{-}42)$$

and

$$\dot{Q}=\dot{m}\{(h+\tfrac{1}{2}V^2)_{2B}-(h+\tfrac{1}{2}V^2)_1\} \qquad (11\text{-}43)$$

in which changes in elevation are neglected and the suffix 1 denotes the common upstream conditions. By subtraction, and using (2-16) to replace Δh by $c_p\Delta T$, we obtain

$$\dot{Q}=\dot{m}\{c_p(T_{2B}-T_{2A})+\tfrac{1}{2}(V_{2B}{}^2-V_{2A}{}^2)\}=108\ \text{MW} \qquad (11\text{-}44)$$

Comments on Example 11-4

1. A considerable increase in thrust is obtained by this means, but the process is very inefficient because the high temperature of the exhaust gases ($\approx 1440°C$) implies that a great deal of energy is 'wasted'. However, it can be economical to choose this design when the additional thrust is required only temporarily. The alternative is to provide a larger engine that would not run at full power in cruise conditions.

2. It is not strictly correct to regard the burning of fuel in an airstream as a form of heat. Indeed, for the control volume shown in Fig. 11-7, the flow is adiabatic except for conduction through the walls. The 'heat' is really a release of chemical energy during the combustion process.

3. The specific heat capacity of the mixture will differ from that of the original exhaust gases and so the above analysis is only approximate.

Example 11-5: Gas mixtures Humid air at a temperature of 26 °C and containing 0.0085 kg water vapour per kilogram of dry air flows steadily along a duct into a region where a fine spray of liquid water at a temperature of 30 °C is added. After leaving the region, the temperature of the air is 19 °C. There is no external heat or work and the process takes place at a constant pressure of 100 kPa. Estimate the mass of liquid water evaporated per kilogram of dry air and evaluate the partial pressures of the dry air and water vapour at outlet.

The specific enthalpies of saturated water and steam at 19 °C are 79.8 kJ/kg and 2536 kJ/kg, and the specific heat capacities of air, water and steam at constant pressure may be taken as 1.01 kJ/kg, 4.19 kJ/kg and 1.82 kJ/kg respectively. The gas constants for steam and dry air are 463 J/kgK and 287 J/kgK respectively. Treat the flow as one-dimensional.

SOLUTION The steady-flow energy equation (11-19) may be written for one-dimensional flow through the control volume shown in Fig. 11-8 as

$$\dot{Q} - \dot{W}_s = \dot{m}_a(h_2 - h_1)_a + \dot{m}_{s1}(h_2 - h_1)_s + \Delta\dot{m}_w(h_{s2} - h_{w3}) \quad (11\text{-}45)$$

in which the suffixes a, w and s denote air, liquid water and steam respectively. Since $\dot{Q} = W = 0$, the equation may be rearranged as

$$\frac{\Delta\dot{m}_w}{\dot{m}_a}(h_{s2} - h_{w3}) = (h_1 - h_2)_a + \frac{\dot{m}_{s1}}{\dot{m}_a}(h_1 - h_2)_s \quad (11\text{-}46)$$

Using $\Delta h = c_p \Delta T$, we obtain $h_{a1} - h_{a2} = 1.01(26 - 19)\,\text{kJ/kg} = 7.07\,\text{kJ/kg}$ and $h_{s1} - h_{s2} = 12.7\,\text{kJ/kg}$. Also, $h_{w3} - h_{w2} = c_p(T_3 - T_2) = 46.1\,\text{kJ/kg}$ and so the specific enthalpy of the liquid water spray is $h_{w3} = (79.8 + 46.1)\,\text{kJ/kg} = 125.9\,\text{kJ/kg}$. By substituting these values into (11-46) and noting that $h_{s2} = 2536\,\text{kJ/kg}$ and $\dot{m}_{s1}/\dot{m}_a = 0.0085$, we find that the added water mass is $\dot{m}_w/\dot{m}_a = 0.002\,98$ kg water/kg air.

At exit, the *partial pressure* of the dry air is related to the density and temperature by the equation of state $p_{a2} = \rho_{a2}R_aT_2$. A similar expression

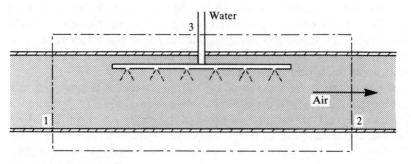

Figure 11-8 Air humidifier.

may be used for the steam and the ratio of the two pressures is

$$\frac{p_{s2}}{p_{a2}} = \frac{(\rho RT)_{s2}}{(\rho RT)_{a2}} \tag{11-47}$$

Since both the steam and the dry air occupy the whole volume, the ratio of their densities is equal to the ratio of their masses. That is p_{s2}/p_{a2} $= (\dot{m}_{s1} + \Delta\dot{m}_w)/\dot{m}_a = 0.0085 + 0.00298 = 0.001148$ at the section 2. The ratio of the gas constants is $R_s/R_a = 463/287$ and so (11-47) shows that the ratio of the partial pressures of the steam and dry air is $p_{s2}/p_{a2} = 0.0185$. Since the sum of the partial pressures is equal to the total pressure, 100 kPa, we obtain $p_{s2} = 1.82$ kPa and $p_{a2} = 98.18$ kPa.

Comments on Example 11-5
1. The ratio of the masses of steam and dry air in a sample of humid air is known as the *specific humidity*. In everyday conditions, its numerical value is typically about 1 per cent.
2. A more commonly used method of describing the water content of air is by use of the *relative humidity* which is the ratio of (a) the partial pressure of the steam and (b) the pressure of *saturated* steam at the same temperature. For example, the pressure of saturated steam at 19 °C is found from tables of the properties of steam to be 2.20 kPa. Therefore the relative humidity of the air at outlet in the present example is 1.82/2.20, that is about 83 per cent. The relative humidity at inlet is about 40 per cent.
3. Strictly, steam does not behave as a perfect gas. However, for the purposes of humidity calculations, errors resulting from this assumption are very small when the temperature is less than about 50 °C. Even at higher temperatures, it is rarely necessary to allow for the small reductions in the specific heat capacity.

Example 11-6: Pressurized surge tank Figure 11-9 is a schematic representation of a pressurized cylindrical surge tank with a vertical axis. When the liquid surface rises above its mean level, the gas above it is compressed and so resists the rise more strongly than in the corresponding case with a free atmospheric surface. The increase in pressure is accompanied by a rise in temperature of the gas and therefore by heat fluxes $\dot{Q}_1$ and $\dot{Q}_2$ through the tank walls and into the liquid respectively. Develop a relationship between the pressure of the gas and the height H shown in Fig. 11-9. Assume that $\dot{Q}_2 = 0$ and that $\dot{Q}_1 = \lambda A(T_{AT} - T)$ in which λ is a heat transfer coefficient, A is the surface area of the tank, and T_{AT} and T are the external and internal temperatures respectively.

SOLUTION Consider a time interval δt in which the liquid surface rises by an amount δz and the gas pressure and temperature rise by δp and δT respectively. The work done *on* the gas at the liquid surface is the product

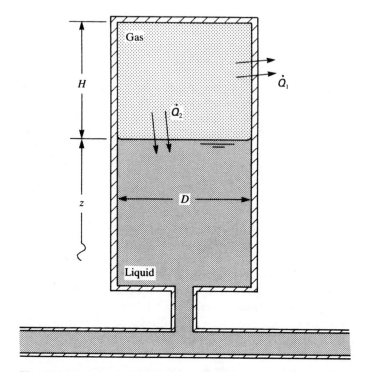

Figure 11-9 Pressurized surge tank.

of the force $\frac{1}{4}\pi D^2 p$ and the distance δz. Since $\delta H = -\delta z$, the work done *by* the gas is $\delta W = \frac{1}{4}\pi D^2 p\ \delta H$. The heat received is $\delta Q = \lambda A (T_{AT} - T)\ \delta t$ in which the surface area $A = \frac{1}{4}\pi D^2 + \pi DH$. The increase in energy of the gas is $\delta E = m\ \delta u = mc_v\ \delta T$ where m denotes the mass of gas and changes in energy other than the internal energy are assumed to be negligible. Equation (11-1) gives, for the gas as the chosen system,

$$\lambda \pi D(\tfrac{1}{4}D + H)(T_{AT} - T)\ \delta t - \tfrac{1}{4}\pi D^2 p\ \delta H = mc_v\ \delta T \qquad (11\text{-}48)$$

Since the density of gas is $\rho = m/V = m/\frac{1}{4}\pi D^2 H$, the equation of state may be written as $\frac{1}{4}\pi D^2 Hp = mRT$, which may be differentiated to give

$$\tfrac{1}{4}\pi D^2 (H\ \delta p + p\ \delta H) = mR\ \delta T \qquad (11\text{-}49)$$

By combining (11-48) and (11-49) to eliminate $m\ \delta T$, we obtain

$$\tfrac{1}{4}\pi D^2 \frac{c_v}{R}(H\ \delta p + p\ \delta H) = \lambda \pi D(\tfrac{1}{4}D + H)(T_{AT} - T)\ \delta t - \tfrac{1}{4}\pi D^2 p\ \delta H$$

$$(11\text{-}50)$$

Since $R/c_v = \gamma - 1$ (section 2.9), multiplication by $(\gamma - 1)/\frac{1}{4}\pi D^2$ yields

$$H\ \delta p + p\ \delta H = \lambda(\gamma - 1)\left(1 + \frac{4H}{D}\right)(T_{AT} - T)\ \delta t - (\gamma - 1)p\ \delta H \quad (11\text{-}51)$$

By combining the terms in $p\,\delta H$ and dividing throughout by pH we obtain

$$\frac{\delta p}{p}+\gamma\frac{\delta H}{H}=\lambda(\gamma-1)\left(1+\frac{4H}{D}\right)\frac{(T_{AT}-T)}{pH}\delta t \qquad (11\text{-}52)$$

The right-hand side of this equation cannot be integrated explicitly. However, in numerical solutions, the coefficient of δt can be evaluated at any instant $t=t_1$, say, at which the values of p, H and T are known. The equation may then be integrated approximately to give

$$\ln\left(\frac{p_2}{p_1}\right)+\gamma\ln\left(\frac{H_2}{H_1}\right)\simeq\lambda(\gamma-1)\left(1+\frac{4H_1}{D}\right)\frac{(T_{AT}-T_1)}{p_1 H_1}(t_2-t_1)$$

$$(11\text{-}53)$$

This may be expressed alternatively as

$$p_2 H_2{}^{\gamma}\simeq p_1 H_1{}^{\gamma}e^{\eta} \qquad (11\text{-}54)$$

in which e is the base of natural logarithms and η denotes the expression on the right-hand side of Eq. (11-53).

Comments on Example 11-6

1. In a practical application Eq. (11-54) is solved simultaneously with a second relationship between p_2 and H_2 determined from the dynamics of the liquid flows in the pipeline. Its use is straightforward because the numerical value of the right-hand side is known. In the special case of adiabatic conditions (unlikely to be a realistic assumption), $\eta=0$ and $p_2 H_2{}^{\gamma}=p_1 H_1{}^{\gamma}$.
2. The processes in a pressurized surge tank are often approximated by a polytropic relationship $p/\rho^{N}=\text{constant}$. This procedure is unsatisfactory because the index N is both unknown and variable. Its value at any instant depends upon the direction of the heat flux and this depends upon the liquid level at which thermal equilibrium exists ($T=T_{AT}$). Furthermore, for any particular heat flux, the value of the index N depends upon whether the liquid level is rising or falling and the rate at which is is so doing.
3. Simplifications in the above analysis include neglecting heat transfers with the liquid and neglecting evaporation of the liquid. Neither of these effects can be quantified easily, partly because in practice they always occur together. Another simplification is hidden by assuming that a value can be obtained for the heat transfer coefficient λ. Fortunately, the errors resulting from these simplifications are sufficiently small for realistic simulations to be achieved in most applications.

Example 11-7: Efficiency of a centrifugal pump
A centrifugal pump delivers $0.15\,\text{m}^3/\text{s}$ water from a low-level reservoir to a high-level reservoir in which the free-surface levels are respectively 5 m and 30 m above the pump (Fig. 11-10). The 0.3 m bore suction and delivery pipes are 250 m

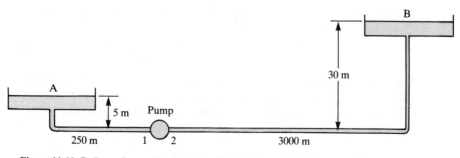

Figure 11-10 Delivery from a low head to a high head.

and 3000 m long and their skin friction coefficient is $f=0.006$. Estimate the hydraulic efficiency of the pump and the water temperature rise induced by it. The density of the water is $1000\,\text{kg/m}^3$, its specific heat capacity is $c_v = 4190\,\text{J/kg K}$ and the gravitational acceleration is $9.81\,\text{m/s}^2$.

SOLUTION The cross-sectional area of the pipelines is $a = \frac{1}{4}\pi D^2 = 0.0707\,\text{m}^2$ and the mean velocity along them is $\bar{V} = Q/a = 2.122\,\text{m/s}$. Using the Darcy–Weisbach equation (9-22), the reductions in the total heads along the 250 m and 3000 m pipelines are 4.59 m and 55.1 m water respectively. Since the reservoir levels are 5 m and 30 m above the pump, the total heads (B/g) at the pump inlet and outlet must be

$$\frac{B_1}{g} = 5\,\text{m} - 4.59\,\text{m} = 0.41\,\text{m}; \qquad \frac{B_2}{g} = 30\,\text{m} + 55.1\,\text{m} = 85.1\,\text{m}$$

$$(11.55)$$

respectively where the pump itself is chosen as the arbitrary datum $z=0$.

The shaft power is 165 kW and the mechanical efficiency is 96 per cent. Therefore the power supplied to the water in the pump impeller is $|W_s| = 0.96 \times 165\,\text{kW} = 158.4\,\text{kW}$. This has led to an increase of $(B_2 - B_1)/g = 84.7\,\text{m}$ in the total head which implies a useful power of $\dot{m}(B_2 - B_1) = 124.6\,\text{kW}$. The hydraulic efficiency of the unit is therefore $124.6/158.4 = 79$ per cent.

The increase in the internal energy of the water can be deduced from Eq. (11-25). Assuming adiabatic conditions $(\dot{Q}=0)$, the increase is

$$u_2 - u_1 = \frac{|\dot{W}_s|}{\dot{m}} - (B_2 - B_1) = 225\,\text{J/kg K} \qquad (11.56)$$

Since the specific heat capacity of the water is $c_v = 4190\,\text{J/kg K}$, the temperature rise of the water as it flows through the pump is

$$\Delta T = \frac{\Delta u}{c_v} = 0.054\,\text{K} \qquad (11.57)$$

Comments on Example 11-7

1. When a pump operates close to its optimum design flow rate, losses in its impeller are small, being mainly due to skin friction. Most of the inefficiency results from the flows in the volute and in the diffuser ring if one is present. The purpose of these items is to receive fluid at high speed from the impeller and to deliver it at lower speed but higher pressure at the pump outlet. In practice, all such devices are relatively inefficient; it is quite an achievement to design a pump with a diffuser efficiency in excess of 50 per cent.

2. The total head at inlet to the pump is 0.41 m. Since the velocity is 2.12 m/s and the elevation is $z = 0$, the pressure head is $(0.41 - 2.12/19.62)$ m $= 0.02$ m. In practice, sub-atmospheric pressures will occur within the unit and the possibility of cavitation, though unlikely, cannot be discounted. It would be better to install the pump closer to the upstream reservoir if this would ensure a higher static pressure at inlet.

Example 11-8: Surface tension Figure 11-11 depicts a cylinder containing water in which there is a spherical air bubble. Neglecting the compressibility of the water, show that the work done by the piston when it slowly compresses the contents of the cylinder is smaller than the resulting work done on the air bubble. For what radius of bubble is the latter twice as large as the former when the pressure in the water is (*a*) atmospheric pressure, 100 kPa and (*b*) 1 kPa? The surface tension coefficient of the water–air interface is $\sigma = 0.075$ N/m (i.e. 0.075 J/m^2).

SOLUTION If the pressure in the liquid is denoted by p_L, the work done by the piston in reducing the volume of the contents by an amount δV is $p_L \, \delta V$. Since the liquid is to be regarded as incompressible, the whole of the volume change is accounted for by the air. Its pressure at any instant is $p_L + 2\sigma/r$ where r is the bubble radius and σ is the surface tension coefficient (section 2.8) and so the work done on the air is $\{p_L + 2(\sigma/r) \, \delta V\}$.

The work done on the air will be double that done by the piston when

$$\left(p_L + \frac{2\sigma}{r} \right) \delta V = 2 p_L \, \delta V \tag{11-58}$$

that is when $r = 2\sigma/p_L$. For liquid pressures of 100 kPa and 1 kPa, this gives bubble radii of 0.0015 mm and 0.15 mm respectively.

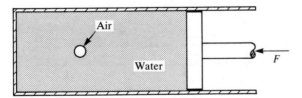

Figure 11-11 Compression of an air bubble in water.

Comments on Example 11-8

1. The same result can be obtained without using the knowledge that the internal pressure exceeds the external pressure by $2\sigma/r$. A reduction r in the radius of a sphere causes a reduction $\delta A = \delta(4\pi r^2) = 8\pi r\ \delta r$ in its surface area and a reduction $\delta V = \delta(\frac{4}{3}\pi r^3) = 4\pi r^2\ \delta r$ in its volume. By inspection, therefore, $\delta A = 2\delta V/r$. Now σ is the free-surface energy per unit area. Therefore, the free-surface energy of the bubble air–water interface reduces by $\sigma\ \delta A = 2\sigma\ \delta V/r$ when the bubble volume reduces by δV. The energy of the air must therefore increase by this amount in addition to the external work $p_L\ \delta V$.

2. Neglecting surface tension effects would introduce an error of 1 per cent with bubbles fifty times as large as those in this example. The phenomenon is significant in cavitating flows where low pressures are experienced. Examples include steady flows over spillways and unsteady flows involving water-hammer.

3. In practice the air bubble will contain water vapour. This could influence the analysis by condensation during the compression.

Example 11-9: Flow with large heat transfer Figure 11-12 depicts a steady flow of air ($R = 287\ \text{J/kg K}$, $c_p = 1004\ \text{J/kg K}$) along a horizontal pipe of constant area. The pipe surroundings are at very high temperature and the resulting heat flow to the air causes it to accelerate rapidly to sonic speed at outlet.

The velocities at the sections A, B, C and D are 200 m/s, 300 m/s, 385 m/s and 450 m/s respectively. The pressure and temperature at the point A are 250 kPa and 125 °C. Estimate the temperatures at B, C and D and the heat influx in the sections AB, BC and CD. Neglect skin friction.

SOLUTION The density at A is $\rho = p/RT = 2.188\ \text{kg/m}^3$ and the densities at B, C and D follow from continuity which requires that ρV is constant in a steady flow along a pipe of constant area. The pressure at each location can be obtained from the momentum equation

$$(p - p_A)a = \dot{m}(V - V_A) \tag{11-59}$$

and the temperatures follow from the equation of state $T = p/R\rho$. Finally, the rate of heat transfer can be deduced from the steady flow energy equation, namely

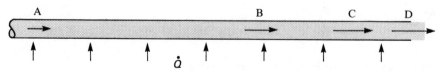

Figure 11-12 Acceleration due to heat influx.

$$\dot{Q} = \dot{m}\{(h + \tfrac{1}{2}V^2) - (h_1 + \tfrac{1}{2}V_1^2)\} \qquad (11\text{-}60)$$

The numerical values obtained in this manner are:

V, m/s:	200	300	385	450
ρ, kg/m³:	2.188	1.459	1.137	0.972
p, kPa:	250	206.2	169.1	140.6
T, °C:	125	219.5	245.1	230.7

$$\dot{Q}_{AB} = 134.9 \, \text{kJ/kg}, \qquad \dot{Q}_{BC} = 54.8 \, \text{kJ/kg}, \qquad \dot{Q}_{CD} = 12.7 \, \text{kJ/kg}.$$

Comments on Example 11-9

1. The most interesting feature of these results is in the last interval. By inspection, heat addition to the flow has resulted in a temperature decrease. This intuitively outrageous result is accounted for by a large increase in kinetic energy.
2. Skin friction has been neglected in the analysis for simplicity only. There is no difficulty in principle in including it in Eq. (11-59).

11-5 DIFFERENTIAL FORM OF THE UNIAXIAL ENERGY EQUATION

The uniaxial energy equation is now developed for control volumes of elemental length such as those depicted in Fig. 11-13. Using h, ρ, $\bar{V}$, etc. to denote values at the mid-section, we express Eq. (11-16) as

$$\int \left(e + \frac{p}{\rho}\right) d\dot{m} = \dot{m}(h + \tfrac{1}{2}\alpha \bar{V}^2 + gz) \pm \tfrac{1}{2}\frac{\partial}{\partial x}\{\dot{m}(h + \tfrac{1}{2}\alpha \bar{V}^2 + gz)\} \, \delta x$$

$$(11.61)$$

in which h and z should strictly be mean values defined in (11-15) with the velocity as a weighting factor. The plus and minus signs apply at the outlet and inlet sections respectively.

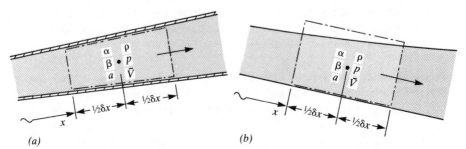

Figure 11-13 Elemental control volumes for uniaxial flows. (a) Pipe flow. (b) Channel flow.

The rate of change of energy within the control volume is

$$\frac{\partial E_{cv}}{\partial t} = \frac{\partial}{\partial t}\{(u + \tfrac{1}{2}\beta\bar{V}^2 + gz)\rho a\,\delta x\} \tag{11-62}$$

in which the implied average values are obtained by averaging over a cross-section *without* introducing the velocity as a weighting factor. No concept of flux is associated with averaging *within* the control volume. As usual, however, we shall assume that all parameters except the velocity may be assumed equal to their values at the centroid.

The energy equation is obtained by substituting Eqs (11-61) and (11-62) into Eq. (11-10). Using $\dot{q}$ and $\dot{w}_s$ to denote the rates of heat and work *per unit length*, we obtain

$$\boxed{\dot{q} - \dot{w}_s = \frac{\partial}{\partial t}\{\rho a(u + \tfrac{1}{2}\beta\bar{V}^2 + gz)\} + \frac{\partial}{\partial x}\{\dot{m}(h + \tfrac{1}{2}\alpha\bar{V}^2 + gz)\}} \tag{11-63}$$

which may alternatively be expressed in *conservation form* as

$$\frac{\partial}{\partial t}\{\rho a(u + \tfrac{1}{2}\beta\bar{V}^2 + gz)\} = \frac{\partial}{\partial x}\{\dot{q} - \dot{w}_s - \dot{m}(h + \tfrac{1}{2}\alpha\bar{V}^2 + gz)\} \tag{11-64}$$

These are useful forms of the energy equation for numerical computations, but alternative expressions are often preferred for analytical purposes. By replacing u by $h - p/\rho$ and using the continuity equation in a similar manner to that followed for the momentum equation in section 7-5, the energy equation becomes

$$\dot{q} - \dot{w}_s = -\frac{\partial}{\partial t}(pa) + \rho a\frac{\partial}{\partial t}(h + \tfrac{1}{2}\beta\bar{V}^2 + gz)$$

$$+ \dot{m}\frac{\partial}{\partial x}(h + \tfrac{1}{2}\alpha\bar{V}^2 + gz) + \tfrac{1}{2}(\alpha - \beta)\bar{V}^2\frac{\partial\dot{m}}{\partial x} \tag{11-65}$$

which is less cumbersome than it appears at first sight because the last term may usually be neglected. This is exactly correct for all steady flows and for one-dimensional unsteady flows because either $\partial\dot{m}/\partial x$ or its coefficient is zero. It is a good approximation in slowly changing flows because $\partial\dot{m}/\partial x$ is small (see Eq. 6-45).

Range of validity Equations (11-64) and (11-65) are less general than the corresponding integral formulations because they must be applied to flows that are uniaxial at all sections. Strictly, additional terms should be included in (11-65) to allow for differences between $\bar{h}$ and $\bar{\bar{h}}$ and between $\bar{z}$ and $\bar{\bar{z}}$, but it is most unlikely that these terms would ever be of practical significance.

One-dimensional flow When the flow is one-dimensional, $\alpha = \beta = 1$ and

$$\dot{q} - \dot{w}_s = -\frac{\partial}{\partial t}(pa) + \rho a \left(\frac{\partial}{\partial t} + V\frac{\partial}{\partial x}\right)(h + \tfrac{1}{2}V^2 + gz) \qquad (11\text{-}66)$$

For the special case of a particle path, $dx/dt = V$ and the sum $\partial/\partial t + V\partial/\partial x$ is often replaced by D/Dt to denote differentiation following the fluid. In this case, the one-dimensional energy equation becomes

$$\dot{q} - \dot{w}_s = -\frac{\partial}{\partial t}(pa) + \rho a \frac{D}{Dt}(h + \tfrac{1}{2}V^2 + gz) \qquad (11\text{-}67)$$

Steady flow When the flow is steady, $\partial/\partial t = 0$ and the continuity equation (6-45) requires that $\partial\dot{m}/\partial x = 0$. The energy equation (11-65) therefore reduces to

$$\dot{q} - \dot{w}_s = \dot{m}\frac{\partial}{\partial x}(h + \tfrac{1}{2}\alpha\bar{V}^2 + gz) \qquad (11\text{-}68)$$

PROBLEMS

1 Two streams of water $(c_v = 4190 \text{ J/kg K})$ meet at a pipe junction as shown in Fig. 11-14. The conditions in the two pipes are as shown below and the velocity of the well-mixed stream in the outlet pipe is 1.5 m/s. Estimate the temperature of the water in the outlet pipe.

Pipe	Diameter	Velocity	Temperature
1	50 mm	2 m/s	10 °C
2	25 mm	1 m/s	60 °C

[15.6 °C]

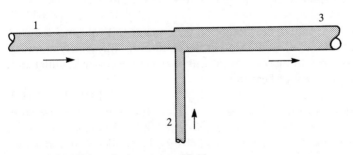

Figure 11-14 Mixing of two streams of fluid.

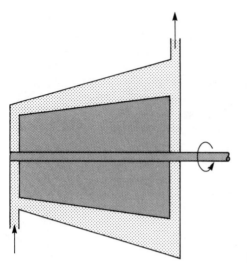

Figure 11-15 Idealized gas turbine.

2 Repeat Problem 1 for two streams of air ($c_p = 1004\,\text{J/kg K}$) as shown below. The velocity in the outlet pipe is $210\,\text{m/s}$.

Pipe	Diameter	Velocity	Temperature	Density
1	50 m	200 m/s	10 °C	3.7 kg/m³
2	25 m	100 m/s	60 °C	3.14 kg/m³

[11.3 °C]

3 Natural gas ($R = 520\,\text{J/kg K}$, $c_p = 2230\,\text{J/kg K}$) flows at a rate of $100\,\text{kg/s}$ along a 1.2 m diameter horizontal pipeline. The gas temperature of 10 °C is constant along the pipe because of heat flow from the surrounding ground. The upstream and downstream pressures are 900 kPa and 100 kPa respectively. Estimate the total rate of heat inflow along the pipe.

[836 kW]

4 Suppose that the rate of heat inflow to the pipeline in Problem 3 reduces and is insufficient to maintain isothermal conditions, the density at outlet being $0.865\,\text{kg/m}^3$. In this case, what will be the temperature at outlet and what will be the rate of heat inflow?

[319 kW, 7.7 °C]

5 At inlet to the gas turbine shown in Fig. 11-15 the temperature, pressure and velocity of the gas ($c_p = 1150\,\text{J/kg K}$) are 1000 °C, 300 kPa and 30 m/s. At outlet they are 730 °C, 100 kPa and 80 m/s. The outlet elevation is 2 m above the inlet. Assuming adiabatic conditions when the machine reaches steady state, estimate the power supplied to the shaft per unit mass of gas.

Also show that the error incurred by neglecting the change in kinetic energy and potential energy would be less than 1 per cent.

[313 kW/kg]

6 Figure 11-16 depicts the velocity distribution alongside a train in a railway tunnel. Some of the air displaced by the train is pushed 'forwards' along the tunnel and some is forced 'backwards' towards the rear of the vehicle. (a) For the control volume shown in the figure, state whether the heat and work are positive or negative at (i) the tunnel surface and (ii) the train surface. (b) Repeat the problem for axes moving with the train.

[(a) tunnel: $Q-$ve, $W=0$; train: $Q+$ve, $W-$ve
(b) tunnel: $Q-$ve, $W-$ve; train: $Q+$ve, $W=0$]

7 A centrifugal pump is designed to deliver $0.1\,\text{m}^3/\text{s}$ water ($\rho = 1000\,\text{kg/m}^3$, $c_v = 4190\,\text{J/kg K}$) against a head of 40 m when rotating at 1320 rev/min. (a) If the hydraulic efficiency of the unit at its design point is 79 per cent, what power must be supplied by the shaft?

[49.7 kW]

(b) When a valve far downstream of the pump is nearly closed, the actual rate of flow reduces to $0.01\,\text{m}^3/\text{s}$ and the power supplied by the shaft is 27.5 kW. If the head difference across the pump is now 45 m, estimate the increase in temperature of the water between inlet and outlet.

[0.55°C]

8 When the valve described in Problem 7 is closed fully while the pump continues to operate, the required shaft power is found to be 25 kW. Estimate the rate of increase of temperature of the pump assuming adiabatic conditions. The masses of the pump and the contained water are 50 kg and 40 kg respectively and their specific heat capacities are $c_v = 450\,\text{J/kg K}$ and $4190\,\text{J/kg K}$.

[0.132 K/s]

9 The nozzle shown in Fig. 11-17 is connected to a large tank containing air ($R = 287\,\text{J/kg K}$, $\gamma = 1.4$) at a pressure of 100 kPa and a temperature of

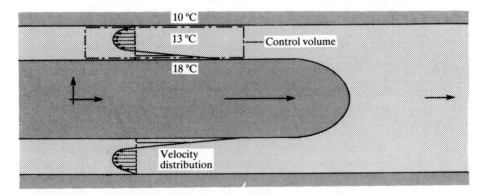

Figure 11-16 Velocity distribution alongside a railway train in a tunnel.

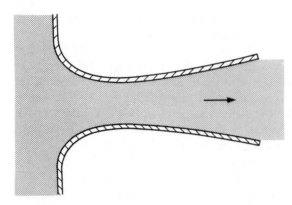

Figure 11-17 Convergent–divergent nozzle.

300 K. The external pressure is gradually lowered and air begins to exhaust through the nozzle. Determine the values of T, ρ and V at the outlet when the pressure is (a) 90 kPa, (b) 75 kPa and (c) 60 kPa. Assume that the flow is homentropic.

$$[(a)\ 291\ \text{K},\ 1.077\ \text{kg/m}^3,\ 134\ \text{m/s}$$
$$(b)\ 276\ \text{K},\ 0.946\ \text{kg/m}^3,\ 218\ \text{m/s}$$
$$(c)\ 259\ \text{K},\ 0.806\ \text{kg/m}^3,\ 286\ \text{m/s}]$$

10 Repeat Problem 9 for outlet pressures of (a) 52.83 kPa, (b) 45 kPa and (c) 35 kPa. Hence verify that the maximum rate of air flow through any nozzle with a specified outlet diameter is obtained with an external pressure of 52.83 kPa. Also verify that the outlet momentum flux continues to increase even though the mass flux is smaller in cases (b) and (c).

(*NB*: This is an example of supersonic flow. The specified conditions at the exit are attainable only if the nozzle has an appropriately sized throat. Moreover, the implied size is dependent upon the outlet pressure.)

$$[(a)\ 250\ \text{K},\ 0.736\ \text{kg/m}^3,\ 317\ \text{m/s}$$
$$(b)\ 239\ \text{K},\ 0.657\ \text{kg/m}^3,\ 351\ \text{m/s}$$
$$(c)\ 222\ \text{K},\ 0.549\ \text{kg/m}^3,\ 395\ \text{m/s}]$$

11 Repeat Problem 9(c) assuming a non-uniform outlet velocity distribution for which $\alpha = 1.07$.

$$[259\ \text{K},\ 0.806\ \text{kg/m}^3,\ 277\ \text{m/s}]$$

12 The velocity distribution in a steady turbulent flow along a rectangular-section channel may be approximated by the power law distribution

$$\frac{V}{\hat{V}} = \left(\frac{y}{d}\right)^N \tag{11-69}$$

where $\hat{V}$ denotes the maximum velocity (Fig. 11-18). Show that the mean velocity is $\bar{V} = \hat{V}/(1+N)$ and hence show that the momentum and kinetic energy flux coefficients are given by

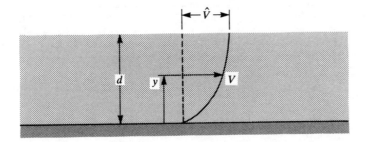

Figure 11-18 Velocity distribution in a free surface flow.

$$\beta = \frac{(1+N)^2}{1+2N} \quad \text{and} \quad \alpha = \frac{(1+N)^3}{1+3N} \tag{11-70}$$

Hence estimate the magnitudes of the coefficients when $N = 0.1$ and $N = 0.2$.

[1.008, 1.024; 1.029, 1.080]

13 When the Reynolds number is sufficiently large, the power law exponent N in Eq. (11-69) is small. Show that when $N \ll 1$ the flux coefficients satisfy $\alpha \approx 3\beta - 2$.

FURTHER READING

Cravalho, E.G. and Smith, J.L. (1981) *Engineering Thermodynamics*, Pitman.

Haywood, R.W. (1972) *Thermodynamic Tables in SI (metric) Units*, 2nd edn, Cambridge University Press.

Haywood, R.W. (1980) *Analysis of Engineering Cycles*, Pergamon.

Kay, J.M. and Nedderman, R.M. (1985) *Fluid Mechanics and Transfer Processes*, Cambridge University Press.

Shapiro, A.H. (1953) *The Dynamics and Thermodynamics of Compressible Fluid Flow*, 2 vols, Ronald Press.

Woods, L.C. (1985) *Thermodynamics of Fluid Systems*, Oxford University Press.

TWELVE

ENTROPY

12-1 SECOND LAW OF THERMODYNAMICS

The continuity, momentum and energy principles provide relationships that must be satisfied in any fluid flow. However, they do *not* tell us in which direction a flow will occur. For example, the momentum equation requires that the pressure force on a fluid element flowing steadily in a horizontal direction must counterbalance the shear force, but it does not stipulate which force is directed upstream and which is directed downstream. Nevertheless, we know from experience that the shear force will oppose the flow.

The natural law governing the direction of processes in space and time is known as the second law of thermodynamics (section 1-3-4). It applies to any process undergone by any system, but it is usually presented in forms that *appear* to be less general, for example

> No system can deliver net work in a cyclic process involving heat interactions at only one temperature $\qquad$ (12-1)

Perpetual motion machines To explain the reason for this form of presentation of the law, it is useful to consider the notion of perpetual motion machines. The *first* law of thermodynamics (11-1) shows that when any system delivers work in an adiabatic process ($Q = 0$) its energy decreases. Eventually, therefore, the work must cease (or become infinitesimal). Thus the first law requires that *perpetual motion machines of the first kind* (PMM1s) cannot exist.

In practice we would be nearly as well pleased if we could build a machine which would deliver work indefinitely while receiving heat from a single source. Such a machine could be used to provide power for our domestic needs and to propel aircraft and ships while having no other effect than, say, cooling the atmosphere or the oceans. It is called a *perpetual motion machine of the second kind* (PMM2). Unfortunately, experience shows that we cannot construct a PMM2 either. Whenever we attempt to do so, our machine takes in more energy as heat than it delivers as work. We have to provide it with a heat sink into which the surplus energy may be deposited, and the sink must be at a lower temperature than the source. In practice, therefore, PMM2s are impossible—as the second law (12-1) states.

The following section contains further information about the results that are obtained when machines are used to 'convert' heat into work or vice versa. It is important to realize that these results cannot be *proved*. They are merely experimental facts, typical of the kind that led to the 'discovery' of the second law. In the succeeding section, the law is regarded as a fact and various consequences are discussed. In particular, the absolute temperature scale is defined and *entropy* is shown to be a property of a system.

12-2 REVERSIBILITY

Experience shows that the maximum possible amount of work that can be done by any machine receiving heat depends upon the temperatures of the heat source and the heat sink. Experience also shows that the *actual* work done is always less than the maximum possible value. Figure 12-1a depicts a system receiving heat Q_{1E} from a constant temperature source and delivering Q_{2E} to a

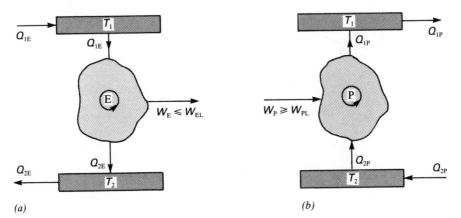

Figure 12-1 Systems operating cyclically while receiving and delivering heat. (a) System delivering work. (b) System receiving work.

constant temperature sink. The symbol Q signifies that the process is *cyclic*, that is the initial conditions are reproduced every once in a while. In each cycle the system delivers work $W_E = Q_{1E} - Q_{2E}$ and this is less than W_{EL}, the upper limit of all possible values. The suffix E denotes an engine, namely a system delivering work (e.g. a turbine).

Figure 12-1*b* illustrates the complementary case of a system that has work done on it (e.g. a pump). Once again, experience shows that there are limits to the possible values of the ratios W_p/Q_{1P} and W_P/Q_{2P}. The *minimum* possible value of W_p needed to deliver the heat Q_{1P} is denoted by W_{PL}.

Most real machines are designed to operate in a manner that involves the supply or rejection of heat at a range of temperatures during each cycle. The limiting values of W/Q for these machines are functions of the design as well as the extreme temperatures T_1 and T_2. However, the values of W_{EL} and W_{PL} used in Fig. 12-1 are the limiting values of the best possible designs. In practice these involve no heat at intermediate temperatures, and the limiting ratios are functions of the temperatures T_1 and T_2 only.

Anyone with an inquisitive disposition will naturally wish to question whether the limiting values for the two types of system are the same. To investigate this possibility, it is useful to arrange that the systems operate between the *same* temperature reservoirs with the work from the engine being used to drive the pump—via a rotating shaft, say. Figure 12-2 illustrates this configuration and indicates the boundary of a combined system that can be used to describe the result.

Whenever such an arrangement is devised in practice and operated with $\Delta W = \Delta Q = 0$, the machines eventually 'run down'. The heat Q_{1P} is found to be less than Q_{1E} and so the temperature T_1 falls. Similarly $Q_{2P} < Q_{2E}$, so T_2 rises

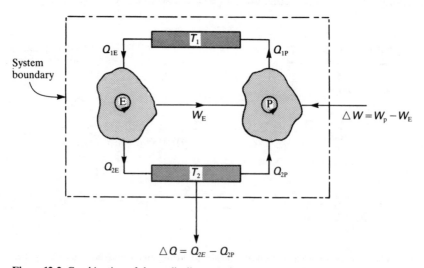

Figure 12-2 Combination of the cyclically operating systems shown in Fig. 12-1.

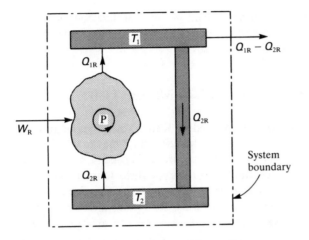

Figure 12-3 Reversible heat pump in an irreversible system.

until it becomes equal to T_1. Thereafter, nothing happens. To prevent the machines running down, we must do work ΔW on the pump and extract an equal amount of heat from the low temperature reservoir. However, by means of sufficiently ingenious designs, it is found that ΔW and ΔQ can be made indefinitely small. Therefore we deduce that the limiting ratios for the two machines are equal.

If the configuration in Fig. 12-2 is *imagined* to be operating continuously in its limiting state with $\Delta W = \Delta Q = 0$, there are no interactions with the surroundings. The two machines run on and on, but they have no external effect whatsoever. In this case, the net effect of P in each cycle is simply the reverse of the net effect of E, and we say that both machines are operating *reversibly*.

The term *reversible* is not used to imply that the output from a system is used reversibly, but simply that the processes *within* the system are reversible. For example, suppose that the engine and pump shown in Fig. 12-2 have been made to operate reversibly, but that the engine is then removed and replaced by a metal bar as shown in Fig. 12-3. In this configuration, the bar is conducting heat Q_{2R} from the high-temperature reservoir to the low-temperature one, and heat $(Q_{1R} - Q_{2R})$ is delivered elsewhere. If the pump itself behaves in exactly the same manner as before, with $W_P = W_{PL}$, then it is entirely satisfactory to say that it is working reversibly—hence the suffix R—even though the overall system depicted by the broken line does not behave reversibly.

Irreversibility All real processes are *irreversible* even though some approximate very closely indeed to reversible behaviour. In fluids, the main source of irreversibility is viscosity. This causes shear forces which always act in such a way as to oppose the development of the lateral velocity gradients that characterize a flow. Other sources of irreversibility include all diffusion processes and most other mixing processes.

12-3 DEFINITION OF ABSOLUTE TEMPERATURE

Hypothetical reversible processes were used by Kelvin to define the absolute temperature scale that bears his name. Imagine any machine operating reversibly between two reservoirs of temperature T_1 and T_2. The ratios W_R/Q_{1R} and W_R/Q_{2R} have their limiting values for these temperatures, and hence the ratio Q_{1R}/Q_{2R} also has its limiting value. Kelvin recognized that this fact could be utilized to *define* a temperature scale and, after considering several alternatives, he chose to define the ratio of any arbitrary pair of temperatures T_1 and T_2 by

$$\frac{T_1}{T_2} \equiv \frac{Q_{1R}}{Q_{2R}} \tag{12-2}$$

Since this definition gives only the *ratio* of any two temperatures, a supplementary definition is needed to give absolute numerical values. For this purpose, the internationally agreed convention is to define the temperature of the triple point of water to be exactly 273.16 K. This seemingly arbitrary number is chosen for historical reasons because it ensures that there are almost exactly 100 units (kelvins) between the freezing and boiling points of water at atmospheric pressure.

The major advantage of the absolute temperature scale defined in this manner is that it is independent of thermometric substances. Previous temperature scales depended upon such parameters as the thermal coefficient of expansion of mercury in a glass thermometer. The extrapolation of their results outside the range of their so-called 'fixed' points (e.g. the freezing and boiling points of water at atmospheric pressure) was somewhat arbitrary. For many years, the chosen supplementary definition for the absolute temperature scale was that there should be exactly 100 units between the specified fixed points, and so the temperature at the triple point of water was only approximately equal to 273.16 K. The present definition is more satisfactory because the conditions pertaining at the triple point can be reproduced more accurately than the freezing and boiling points at atmospheric pressure.

12-4 DEFINITION OF ENTROPY

Suppose that an ideal system operates reversibly during a cycle in which it exchanges elemental net heats δQ_{1R} and δQ_{2R} with reservoirs of temperature T_1 and T_2 respectively. For this cycle, Eq. (12-2) gives

$$\frac{\delta Q_{1R}}{T_1} = -\frac{\delta Q_{2R}}{T_2} \tag{12-3}$$

in which the minus sign is introduced to satisfy the convention that heat inflows are positive and heat outflows are negative. Let successive states of the system during the cycle be denoted by A, B, C and D so that the cyclic process may be represented by $A \rightarrow B \rightarrow C \rightarrow D \rightarrow A \rightarrow B$ etc. Suppose that all of the heat transfers with reservoir 1 take place during the process $A \rightarrow B \rightarrow C$ and that all of the heat transfers with reservoir 2 take place during the process $C \rightarrow D \rightarrow A$. In this case Eq. (12-3) becomes

$$\left(\frac{\delta Q_{1R}}{T_1}\right)_{ABC} + \left(\frac{\delta Q_{2R}}{T_2}\right)_{CDA} = 0 \tag{12-4}$$

Now suppose that the same system subsequently operates reversibly during a cycle in which it exchanges elemental heats δQ_{1R} and δQ_{3R} with the original reservoir 1 and with a third reservoir at a temperature T_3. Imagine that the process $A \rightarrow B \rightarrow C$ is identical to before, and that all of the heat transfers with reservoir 3 take place during the process $C \rightarrow E \rightarrow A$ that completes the new cycle. For this cycle,

$$\left(\frac{\delta Q_{1R}}{T_1}\right)_{ABC} + \left(\frac{\delta Q_{3R}}{T_3}\right)_{CEA} = 0 \tag{12-5}$$

By comparison of (12-4) with (12-5), it is seen that the ratio $(\delta Q_R)/T$ has the same value for both of the processes that change the state of the system from C to A. Moreover, the temperatures T_2 and T_3 are arbitrary and no restriction has been placed on the type of process except that it is reversible.

Definition An elemental change in the entropy S of a system between two states is defined as

$$\delta S \equiv \frac{\delta Q_R}{T} \tag{12-6}$$

Since the value of $(\delta Q_R)/T$ is independent of the temperature and of the (reversible) process, this definition is unambiguous and so the entropy is a property of the system.

A finite change in state of a system may be regarded as an infinite sum of elemental changes and so a finite change in entropy satisfies

$$\Delta S = \int \frac{dQ_R}{T} \tag{12-7}$$

Notice that the entropy is *defined* in terms of reversible processes. To evaluate the change in entropy of a system between two states, we usually consider a hypothetical reversible process, *not* the actual process that the system undergoes.

12-4-1 Real Processes

For the reversible cycles discussed above, the first law equation (11-1) gives, typically, $\delta Q_{1R} + \delta Q_{2R} = \delta W_R$ because the energy of the system is the same at the end of each cycle. Similarly, for a *real* cycle involving the same net work, $\delta Q_1 + \delta Q_2 = \delta W_R$. By subtraction, therefore,

$$(\delta Q_1 - \delta Q_{1R}) + (\delta Q_2 - \delta Q_{2R}) = 0 \tag{12-8}$$

Suppose that the system *receives* net heat from reservoir 1 and that it delivers positive work. In this case, $T_1 > T_2$ and the heat flows at the reservoirs 1 and 2 are positive and negative respectively. For reasons discussed in section 12-2, we know that when the system delivers the given amount of work $\delta W = \delta W_R$ in a real cycle, it requires more heat input than it would in a reversible cycle, that is $\delta Q_1 > \delta Q_{1R}$. Therefore the first term in parenthesis in (12-8) is positive and the second term is negative. By dividing the first term by T_1 and the second term by T_2 (which is less than T_1 for an engine), we obtain the inequality

$$\frac{(\delta Q_1 - \delta Q_{1R})}{T_1} + \frac{(\delta Q_2 - \delta Q_{2R})}{T_2} < 0 \tag{12-9}$$

A similar argument may be used to yield the same expression for a system that has work done on it, and so the relationship is valid for any elemental cyclic process.

In the cycle $A \rightarrow B \rightarrow C \rightarrow D \rightarrow A$ discussed above, all the heat transfers with reservoir 1 take place during the process $A \rightarrow B \rightarrow C$. If we imagine that the real behaviour differs negligibly from reversible behaviour during this process, then $\delta Q_1 = \delta Q_{1R}$ and Eq. (12-10) shows that

$$\frac{\delta Q_2}{T_2} - \frac{\delta Q_{2R}}{T_2} < 0 \tag{12-10}$$

for the real process $C \rightarrow D \rightarrow A$. Since the choice of the process 2 is arbitrary and since $\delta S = \delta Q_R / T$, it follows that the ratio $\delta Q / T$ in *any* real process between two states of a system satisfies

$$\frac{\delta Q}{T} < \delta S \tag{12-11}$$

For processes involving a finite change of state,

$$\int \frac{dQ}{T} < \Delta S \tag{12-12}$$

Adiabatic processes For the special case of *adiabatic* real processes, this relationship gives $\Delta S > 0$, which is the famous principle of increasing entropy of an adiabatic system.

Isentropic processes When the entropy of a system remains constant during any process, $\Delta S = 0$ and the process is termed *isentropic*. For real processes, (12-11) shows that isentropic conditions can exist only when δQ is negative, that is when heat is exhausted by the system. In this case a second system (the surroundings) must receive heat, and so its entropy will increase.

For analytical purposes we commonly imagine that the behaviour of a real system may be regarded as reversible. In this case an isentropic process is also adiabatic. Notice, however, that work interactions are not excluded in such processes; the system need not be *isolated* ($Q = W = 0$).

12-4-2 Entropy of a pure fluid substance

When the system is a pure fluid substance, a simple relationship exists between the entropy and other fluid properties. Consider a pure fluid substance of mass m, pressure p and temperature T undergoing a *reversible* process in which no change in energy occurs except in the internal energy (i.e. no change in kinetic or potential energies). For an elemental part of the process (11-1) gives

$$\delta Q_R - \delta W_R = m\, \delta u \qquad (12\text{-}13)$$

Since the system is a fluid, the existence of shear forces would imply the existence of irreversibilities and so the reversible work must be due solely to pressure forces on the system boundary. The force on an elemental area δa is $p\, \delta a$. If this force moves through a distance δx normal to the surface then the work done is $p\, \delta a\, \delta x$. Since the product $\delta a\, \delta x$ represents the contribution of this element to the total change in volume δV of the system (see Fig. 12-4), the total work done over the whole of the surface is $p\, \delta V$. For a system of mass $m = \rho V$ we may write $\delta V = \delta(m/\rho) = m\, \delta(1/\rho)$ and so the work done is $pm\, \delta(1/\rho)$. Equation (12-13) becomes

$$\delta Q_R = m\left\{\delta u + p\, \delta\!\left(\frac{1}{\rho}\right)\right\} \qquad (12\text{-}14)$$

The change in entropy during the process is $\delta S = \delta Q_R/T$ and so the change in *specific entropy* s (entropy per unit mass) satisfies

$$T\, \delta s = \delta u + p\, \delta\!\left(\frac{1}{\rho}\right) \qquad (12\text{-}15)$$

which is the relationship (2-21) presented in section 2-9. Since $h = u + p/\rho$, the expression may also be written as

$$T\, \delta s = \delta h - \frac{1}{\rho}\, \delta p \qquad (12\text{-}16)$$

In the special case of an incompressible substance, (12-15) simplifies to

$$T\, \delta s = \delta u \qquad (12\text{-}17)$$

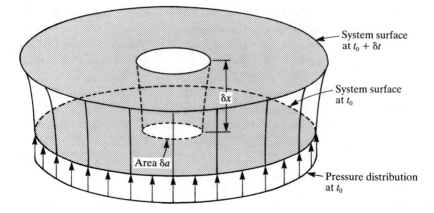

Figure 12-4 Increase in volume of a system.

Although the above derivation deals with a system undergoing a *reversible* process in which there is no change in energy other than internal energy, the result is valid for all pure fluid substances undergoing any process. The change in entropy of a system between two states is independent of the sequence of events (process) that cause the change, and so we may calculate it in any convenient manner. Strictly, however, these relationships should be used only for fluid samples that are sufficiently small for spatial pressure variations due to, say, gravity to be negligible.

12-5 GENERAL FORM OF THE ENTROPY EQUATION

Figure 12-5 depicts a general control volume through which fluid is flowing. Equation (12-11) may be applied to the system composed of the contents of the control volume and the elemental region I at the arbitrary instant t_0. During an interval δt, the system moves to its new position, namely the control volume and the elemental region O. While doing so, it receives elemental amounts of heat δQ_1, δQ_2, δQ_3, etc. at temperatures of T_1, T_2, T_3, etc. respectively. The average rate at which entropy is convected across the control surface from the region I, say, in the interval δt is $\delta S_I / \delta t$, and we define the entropy flux for a typical flow section as

$$\dot{S} \approx \frac{dS}{dt} \tag{12-18}$$

For the process undergone by the system, Eq. (12-11) gives

$$\sum \frac{\delta Q}{T} < (S_{cv} + \delta S_O)_{t_0 + \delta t} - (S_{cv} + \delta S_I)_{t_0} \tag{12-19}$$

After rearranging the terms and dividing by δt, we obtain

$$\sum \frac{1}{T} \frac{\delta Q}{\delta t} < \frac{(S_{t_0+\delta t} - S_{t_0})_{\mathrm{cv}}}{\delta t} + \frac{S_{\mathrm{O},t_0+\delta t}}{\delta t} - \frac{S_{\mathrm{I},t_0}}{\delta t} \qquad (12\text{-}20)$$

In the limit, as δt approaches zero, the entropy equation is

$$\boxed{\int \frac{\mathrm{d}\dot{Q}}{T} < \frac{\partial S_{\mathrm{cv}}}{\partial t} + \dot{S}_{\mathrm{O}} - \dot{S}_{\mathrm{I}}} \qquad (12\text{-}21)$$

in which the integration of $\mathrm{d}\dot{Q}/T$ must be carried out over the whole of the control surface.

Entropy flux By regarding the mass flux $\dot{m}$ across the control surface at a typical section as the sum of elemental fluxes $\delta \dot{m}$, each with its own value of specific entropy s, we may write the entropy flux for the section as

$$\dot{S} = \int s\,\mathrm{d}\dot{m} \qquad (12\text{-}22)$$

Range of validity of the entropy equation Equation (12-21) is applicable to any control volume, large or elemental. When it is applied to a hypothetical reversible process, the inequality becomes an equality. The integral on the left-

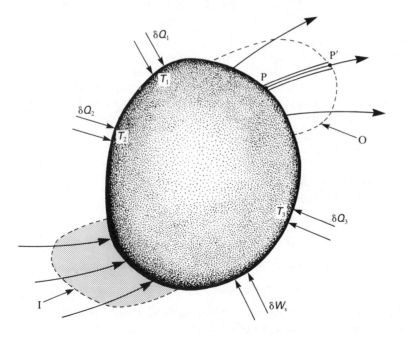

Figure 12-5 General control volume (system shaded at t_0).

hand side includes heat due to conduction and radiation at the inlet and outlet sections as well as heat around the rest of the control surface. However, in most low-speed flows account need be taken of the latter only. Notice that *work* does not appear explicitly in entropy equations; it may exist, but it has no direct influence on entropy changes.

12-6 UNIAXIAL AND ONE-DIMENSIONAL FLOWS

At a uniaxial flow section where the control surface is normal to the direction of flow, we may replace $\delta \dot{m}$ by $\rho V \, \delta a$. By defining a *mean* specific entropy for the section as

$$\bar{\bar{s}} \equiv \frac{1}{\bar{\rho} a \bar{V}} \int s\rho V \, da \tag{12-23}$$

we enable the entropy flux to be expressed as

$$\dot{S} = \int s\rho V \, da = \bar{\bar{s}} \bar{\rho} a \bar{V} \tag{12-24}$$

and so the entropy equation (12-21) becomes

$$\boxed{\int \frac{d\dot{Q}}{T} < \frac{\partial S_{cv}}{\partial t} + \dot{m}_o \bar{\bar{s}}_o - \dot{m}_i \bar{\bar{s}}_i} \tag{12-25}$$

Range of validity The only additional limitation of this expression in comparison with Eq. (12-21) is that the inflows and outflows must be uniaxial. When there is more than one inlet and/or outlet section, appropriate mean values of the specific entropy should be defined for each. In practice we make no direct use of the mean value in this book. If greater rigour is ever required it will almost certainly be sufficient to approximate $\bar{\bar{s}}$ by the value of s at the centroid of the flow section.

Adiabatic flows In the very important case of adiabatic flows, the uniaxial entropy equation simplifies to

$$\frac{\partial S_{cv}}{\partial t} + \dot{m}_o s_o - \dot{m}_i s_i > 0 \tag{12-26}$$

When it is also steady and there is only one inflow and outflow section,

$$s_o > s_i \tag{12-27}$$

and so the entropy increases in the direction of flow. These equations also apply to any other flow in which $\int d\dot{Q}/T$ is zero.

Isothermal flows It is sometimes possible to regard a flow as isothermal. In this case, T is constant and (12-25) shows that

$$\frac{\dot{Q}}{T} < \frac{\partial S_{cv}}{\partial t} + \dot{m}_O s_O - \dot{m}_I s_I \tag{12-28}$$

Steady flows For the case of a steady flow where there is only one inlet and one outlet, Eq. (12-25) becomes

$$\int \frac{\mathrm{d}\dot{Q}}{T} < \dot{m}(s_O - s_I) \tag{12-29}$$

12-7 APPLICATIONS OF THE ENTROPY EQUATION

Example 12-1: Hydraulic jump Figure 12-6 depicts a stationary hydraulic jump in a steady flow of water ($\rho = 1000\,\mathrm{kg/m^3}$) along a rectangular horizontal channel of width $b = 4$ m normal to the page. Just upstream and downstream of the jump the depths are $d_1 = 0.5$ m and $d_2 = 2.0$ m respectively. By assuming one-dimensional flow at these sections, (a) estimate the rate of flow along the channel and (b) prove that the direction of flow is from left to right.

SOLUTION

(a) The forces acting on the control volume in a horizontal direction are (i) pressure forces on the vertical faces 1 and 2, (ii) shear forces on the lateral faces and (iii) surface tension forces where the liquid surface meets the vertical faces 1 and 2. The latter two are very small in comparison with the net pressure force and so they are neglected.

   Since the flows at the sections 1 and 2 are horizontal and one-dimensional, the vertical pressure distributions satisfy the hydrostatic formula (3-11). The gauge pressure force acting on a vertical control surface in the liquid is the product of the area bd and the gauge

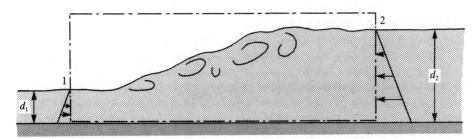

Figure 12-6 Hydraulic jump.

pressure at its centroid $\frac{1}{2}\rho gd$. Therefore the net pressure force in the direction of flow is $\frac{1}{2}\rho gbd_1{}^2 - \frac{1}{2}\rho gbd_2{}^2$. Using the steady-flow momentum Eq. (7-20) and neglecting any contribution of the air to the momentum flux, we obtain

$$\tfrac{1}{2}\rho gb(d_1{}^2 - d_2{}^2) = \dot{m}(V_2 - V_1) \tag{12-30}$$

Since the flow is steady, we may replace V by $\dot{m}/\rho bd$ to give

$$\tfrac{1}{2}\rho gb(d_1{}^2 - d_2{}^2) = \dot{m}\left(\frac{\dot{m}}{\rho bd_2} - \frac{\dot{m}}{\rho bd_1}\right) \tag{12-31}$$

This is a quadratic equation in $\dot{m}$ for which the solution is $\dot{m} = 14\,000$ kg/s. The volumetric flux $Q = \dot{m}/\rho$ is 14.0 cumec and the velocities are $V_1 = 7.00$ m/s and $V_2 = 1.75$ m/s.

(b) The flow is nearly adiabatic and there is no external work. Therefore the steady-flow energy equation (11-19) gives

$$(h + \tfrac{1}{2}V^2 + gz)_1 = (h + \tfrac{1}{2}V^2 + gz)_2 \tag{12-32}$$

irrespective of the direction of flow. Using values at the centroids of the flow sections where the elevation and pressure are $\frac{1}{2}d$ and $\frac{1}{2}\rho gd$ respectively, we may write

$$h = u + \frac{p}{\rho} = u + \tfrac{1}{2}gd \tag{12-33}$$

Equation (12-32) simplifies to

$$(u + \tfrac{1}{2}V^2 + gd)_1 = (u + \tfrac{1}{2}V^2 + gd)_2 \tag{12-34}$$

Therefore the change in the specific internal energy across the jump is

$$u_2 - u_1 = \tfrac{1}{2}(V_1{}^2 - V_2{}^2) + g(d_1 - d_2) \tag{12-35}$$

For this numerical example, $u_2 - u_1 = 8.28$ J/kg. Since this is positive, Eq. (12-17) shows that the specific entropy change $s_2 - s_1$ is also positive. Therefore the flow must be from left to right.

Comments on Example 12-1

1. Hydraulic jumps are induced at the bottom of spillways to prevent excessive scour in the downstream channel. The reduction in kinetic energy (velocity) is accompanied by an increase in potential energy (depth) and in internal energy (temperature). The temperature rise is very small. Since $\Delta u = c_v \Delta T$ and c_v is approximately 4190 J/kg K for water, the temperature increase in this example is about 0.002 °C. In nearly all liquid flows involving little external heat and work, it is reasonable to neglect the influence of temperature on the values of properties such as the specific heat capacities and viscosity.

2. Although the flow rate in this example has been estimated from measurements of the geometry alone, hydraulic jumps are not used for this purpose in practice. Weirs and Venturi flumes are preferred because the depths can be measured more accurately and because the influence of the momentum and kinetic energy coefficients is much smaller. These were both assumed equal to unity in this example, but they can differ greatly from unity in practice, especially at the downstream section.

Example 12-2: Perfect gas relationships Develop the relationships (2-22), (2-23) and (2-24) for isentropic processes of a perfect gas.

SOLUTION

(a) For any pure fluid substance undergoing any process, Eq. (12-15) gives

$$\delta s = \frac{\delta u}{T} + \frac{p}{T}\,\delta\left(\frac{1}{\rho}\right) \qquad (12\text{-}36)$$

For perfect gases, $\delta u = c_v\,\delta T$ and the equation of state is $p = \rho RT$. Therefore

$$\delta s = c_v\frac{\delta T}{T} + R\frac{\delta(1/\rho)}{(1/\rho)} \qquad (12\text{-}37)$$

which may be integrated between any two states 1 and 2 to give

$$s_2 - s_1 = c_v\ln\left(\frac{T_2}{T_1}\right) - R\ln\left(\frac{\rho_2}{\rho_1}\right) \qquad (12\text{-}38)$$

or

$$\ln\left\{\frac{T_2/T_1}{(\rho_2/\rho_1)^{R/c_v}}\right\} = \frac{s_2 - s_1}{c_v} \qquad (12\text{-}39)$$

For an isentropic process, $s_1 = s_2$. Equation (2-19) gives $R/c_v = \gamma - 1$, and so

$$\frac{T_2}{T_1} = \left(\frac{\rho_2}{\rho_1}\right)^{\gamma-1} \qquad (12\text{-}40)$$

which is equivalent to Eq. (2-23).

(b) Equation (2-24) can be developed similarly. Equation (12-16) may be written as

$$\delta s = \frac{\delta h}{T} - \frac{\delta p}{\rho T} \qquad (12\text{-}41)$$

For a perfect gas, $\delta h = c_p\delta T$ and $p = \rho RT$ and so

$$\delta s = c_p\frac{\delta T}{T} - R\frac{\delta p}{p} \qquad (12\text{-}42)$$

which may be integrated between the states 1 and 2 to give

$$s_2 - s_1 = c_p \ln\left(\frac{T_2}{T_1}\right) - R\ln\left(\frac{p_2}{p_1}\right) \tag{12-43}$$

or

$$\ln\left\{\frac{T_2/T_1}{(p_2/p_1)^{R/c_p}}\right\} = \frac{s_2 - s_1}{c_p} \tag{12-44}$$

For isentropic processes, $s_1 = s_2$. Since $R/c_p = 1 - 1/\gamma$, we obtain

$$\frac{T_2}{T_1} = \left(\frac{p_2}{p_1}\right)^{1-1/\gamma} \tag{12-45}$$

which is equivalent to Eq. (2-24).

(c) The third relationship is obtained by eliminating T_2/T_1 from (12-40) and (12-45). Since $1 - 1/\gamma$ is equal to $(\gamma - 1)/\gamma$,

$$\frac{p_2}{p_1} = \left(\frac{\rho_2}{\rho_1}\right)^{\gamma} \tag{12-46}$$

Comment on Example 12-2 Isentropic relationships are commonly used to approximate the behaviour of real processes. When the assumption of isentropic conditions is not acceptable, the relationships (12-38) and (12-43) can be very useful. These are valid for *any* process of a perfect gas.

Example 12-3: Gas turbine Estimate the maximum possible specific work for an adiabatic gas turbine that receives gas at a pressure of 275 kPa and a temperature of 850 °C and exhausts it to atmosphere at a pressure of 100 kPa. For the gas, use $\gamma = 1.33$ and $c_p = 1.15 \, \text{kJ/kg K}$.

SOLUTION The maximum possible work will be obtained when the turbine operates reversibly. Since the specified machine also operates adiabatically, it follows that the optimum conditions are isentropic. Using (12-45) with $p_1 = 275 \, \text{kPa}$, $p_2 = 100 \, \text{kPa}$ and $T_1 = 850\,°\text{C} = 1123 \, \text{K}$, we obtain $T_2 = 874 \, \text{K} = 601\,°\text{C}$.

The steady-flow energy equation (11-19) may be applied to the control volume shown in Fig. 12-7. Neglecting changes in elevation (reasonably) and velocity (a little less reasonably) we obtain

$$-\dot{W}_s = \dot{m}(h_2 - h_1) \tag{12-47}$$

Since $\Delta h = c_p \Delta T$, this gives

$$\frac{\dot{W}_s}{\dot{m}} = c_p(T_1 - T_2) = 286 \, \text{kJ/kg} \tag{12-48}$$

Comments on Example 12-3
1. If it was possible for the turbine to exhaust at zero pressure, the ideal outlet temperature would also be zero and the nominal maximum possible

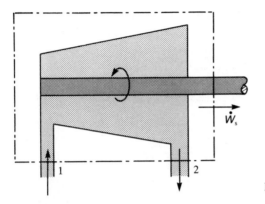

Figure 12-7 Gas turbine.

specific work would be $1150 \times 1123 \, \text{J/kg} = 1290 \, \text{kJ/kg}$. The existence of atmospheric pressure in our environment greatly limits the extent to which the energy of the gas is available to do work. With a *steam* turbine, it is possible to reduce this difficulty by discharging into a condenser at sub-atmospheric pressure (see Example 11-2, p. 308).

2. Because of irreversibility, the actual work output is less than the maximum possible. Typically, the above turbine might deliver $240 \, \text{kJ/kg}$ and we would say that its isentropic efficiency is $240/286 = 84$ per cent. For non-adiabatic machines, it would be a nonsense to compare the real operation with an isentropic process. In these cases the efficiencies are defined in terms of enthalpy changes.

3. Occasionally, inventors 'design' machines that they expect to deliver significantly more work than commercially available turbines. Equations such as (12-38) and (12-43) often allow us to check the possibility of the designs quite easily. These expressions are also useful for estimating the likely behaviour of many other processes.

Example 12-4: Room heater An inventor proposes a design for a domestic room heater in which the heat provided exceeds the electrical input, the difference being provided by cooling an airstream drawn from outside and subsequently discharged outside. The proposed device is shown schematically in Fig. 12-8. Air at atmospheric pressure (100 kPa) and temperature $(-5\,^\circ\text{C})$ is compressed adiabatically to a pressure of 200 kPa, thereby increasing its temperature. It then passes through a device that heats the room while cooling the air to $-18\,^\circ\text{C}$ before discharging it to atmosphere. No work interactions exist in this part of the unit.

Investigate the feasibility of the proposed design, assuming for the air that $R = 287 \, \text{J/kg K}$ and $\gamma = 1.4$.

SOLUTION The minimum possible amount of work needed to compress the air can be estimated by assuming that the process is reversible. In this case, the adiabatic process will occur isentropically and the temperature

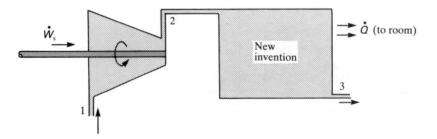

Figure 12-8 Proposed heating system.

increase will satisfy Eq. (12-45). Using $p_1 = 100\,\text{kPa}$, $T_1 = -5°C = 268\,\text{K}$ and $p_2 = 200\,\text{kPa}$, we obtain $T_2 = 326.7\,\text{K} = 53.7\,°C$. The work done by the adiabatic compressor on the air is shown by the steady-flow energy equation (11-19) to be $\dot{m}(h_2 - h_1)$ if changes in kinetic and potential energy are neglected. Since $h = c_p\Delta T$, and $c_p = \gamma R/(\gamma - 1) = 1005\,\text{J/kg K}$, the work done is $59.0\,\text{kJ/kg}$.

The heat emitted into the room by the air is determined by applying the steady-flow energy equation to the second process in which there is no work. Neglecting changes in kinetic and potential energy, the heat emitted is $\dot{m}c_p(T_2 - T_3)$, namely $72.1\,\text{J/kg}$. As predicted by the inventor, this exceeds the work input in the compressor.

To investigate the feasibility of the proposed design, we consider the change in entropy of the air as it passes through the device. Using Eq. (12-43) between the inlet and outlet sections where the pressures are equal, we find

$$s_3 - s_1 = c_p \ln\left(\frac{T_3}{T_1}\right) = -50.0\,\text{J/kg K} \qquad (12\text{-}49)$$

In principle, a decrease in energy is possible in a non-adiabatic process, but the amount of the decrease must satisfy Eq. (12-29). We must therefore consider whether this is possible in the hypothetical process.

The total amount of heat emitted is $72.1\,\text{J/kg}$. Common sense suggests that some of this will have been emitted while the air was close to its highest temperature, but we do not actually know how much was released at each temperature within the range $53.7\,°C$ to $-18\,°C$. Nevertheless, the value of $\int dQ/T$ must lie somewhere in the range $-72.1/326.7\,\text{J/kg K}$ to $-72.1/255\,\text{J/kg K}$. Since the actual reduction in entropy is well outside this range, we conclude that the inventor's design is impossible.

Comments on Example 12-4 Notwithstanding the impossibility of this particular design, it is indeed possible to design a room heater in which the heat provided is far in excess of the electrical input, the difference being provided by cooling the external environment. In such devices, known as heat pumps, a fluid circulates continuously around a closed loop changing from liquid to gas and back again once per cycle as indicated in Fig. 12-9.

A compressor receives the warm fluid in gaseous form and increases its pressure and temperature simultaneously. The fluid then passes through a heat exchanger where it cools and condenses, releasing its enthalpy of condensation (latent heat) to the desired location (e.g. the living room). Its pressure is then reduced to the original value by passing it through a throttle (valve). The cycle is completed when the fluid passes through a second heat exchanger in which it evaporates, the enthalpy of evaporation being provided by an external stream of air or water.

Example 12-5: Equilibrium Figure 12-10 depicts a rigid ball at rest on the bottom of a curved bowl containing a liquid. Investigate whether, in the absence of external heat and work, the contents of the bowl are in stable equilibrium. Assume that there are no field effects other than gravity.

SOLUTION It is possible to resolve this equation by considering buoyancy forces, but it is instructive to consider the entropy of the system instead.

Any small change in the position of the ball will involve an increase δz in its elevation and consequently an increase $\rho_B g V \, \delta z$ in its potential energy, where ρ_B and V denote its density and volume respectively. The net effect of the process is equivalent to a volume V of the fluid swapping places with the ball. Therefore the potential energy of the liquid of density

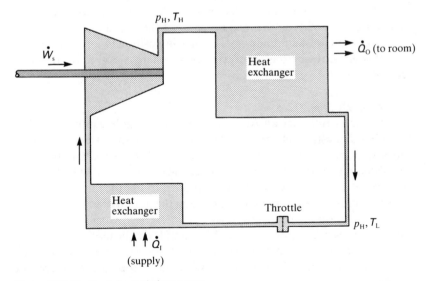

Figure 12-9 Idealized domestic heat pump.

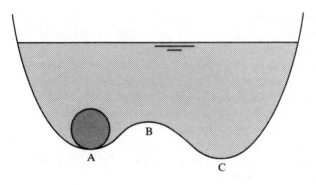

Figure 12-10 Stability of a ball in a bowl.

ρ decreases by $\rho g \mathcal{V} \, \delta z$. The net increase in potential energy of the system (defined as the contents of the bowl) is $(\rho_B - \rho) g \mathcal{V} \, \delta z$.

Since $Q = W = 0$, the first law equation (11-1) requires that the total energy of the system remains constant. There is no change in kinetic energy and so the increase in potential energy must be accompanied by an equal decrease in internal energy. By assuming that the ball and the liquid are both incompressible, we can therefore infer from (12-17) that the assumed increase in elevation would imply an entropy decrease of $(\rho_B - \rho) g \mathcal{V} \, \delta z / T$. The decrease is positive if $\rho_B > \rho$ and negative if $\rho_B < \rho$.

The second law equation (12-11) requires that the entropy of an adiabatic system cannot decrease, and so we deduce that the ball and liquid are in *stable* equilibrium if $\rho_B > \rho$. The conditions will be *unstable* if $\rho_B < \rho$ because an infinitesimal change can result in an increase in entropy. The third possibility is that the densities of the ball and the liquid are equal. In this case, an infinitesimal change in position involves no change in either entropy or (total) energy and the system is said to be in *neutral* equilibrium.

Comments on Example 12-5

1. It is a general rule that an *isolated* system is in stable equilibrium if all possible infinitesimal changes in its state would imply a decrease in its entropy. A more general rule is that an *adiabatic* system is in stable equilibrium if all possible infinitesimal changes at constant entropy imply an increase in energy. We would have needed to use the latter rule in the above solution if we had chosen to abandon the notion of potential energy and to regard the influence of gravity as an external work.
2. Analyses similar to the one used in this example can be used to investigate the stability of stratified flows. It is found, for example, that unstable conditions can exist in the atmosphere when the temperature decreases too rapidly with height.

3. When $\rho_B > \rho$, the ball is in stable equilibrium with respect to infinitesimal disturbances. However, a sufficiently large disturbance could cause it to move over the crest B to a new stable position. On a large scale, therefore, the ball is in *metastable* equilibrium. Metastable conditions can exist in some fluid flows such as transonic flow through a variable area duct with more than one throat. Some supersonic wind tunnels are deliberately designed in this manner.

Example 12-6: Stratified fluid Figure 12-11 depicts a stratified fluid in which a layer of warm fluid at a temperature T_1 overlies a layer of more dense fluid at a temperature T_2. Neglecting any convection effects and any heat transfer with the surroundings, show that any heat transfer between the two layers will be irreversible.

SOLUTION The upper layer cools by heating the lower layer. On average, the temperature of the upper layer is T_1 and so the entropy change resulting from an elemental heat transfer δQ will satisfy

$$\delta S_1 \geqslant \frac{\delta Q_1}{T_1} = -\frac{|\delta Q|}{T_1} \qquad (12\text{-}50)$$

On average, the entropy change in the lower layer will satisfy

$$\delta S_2 \geqslant \frac{\delta Q_2}{T_2} = \frac{|\delta Q|}{T_2} \qquad (12\text{-}51)$$

Hence the entropy change of the whole system satisfies

$$\delta S_1 + \delta S_2 \geqslant \left(\frac{1}{T_2} - \frac{1}{T_1}\right)|\delta Q| \qquad (12\text{-}52)$$

which is positive when $T_1 > T_2$. Since the combined system is adiabatic, a reversible process would involve no change in entropy. Hence the actual process is irreversible.

Comments on Example 12-6
1. It might be argued that the temperature at which the heat transfer actually takes place is the same for both layers (the interfacial temperature) and hence that $|\delta Q|/T$ is the same for both layers. Such an argument would be

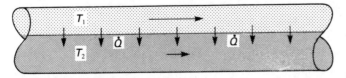

Figure 12-11 Stratified flow in a pipeline.

entirely valid and it would even be correct to deduce that the process of heat transfer in a layer of elemental thickness at the interface is reversible. Nevertheless, the overall result is irreversible because this line of reasoning cannot be sustained without hypothesizing non-equilibrium conditions within the layers themselves.

2. Stratified fluids occur commonly in practice, the atmosphere being an obvious example. Important cases of stratified flows include salt water wedges penetrating upstream in a river beneath a fresh water overlay and smoke movements in buildings. Thermal effects might be unimportant in the former case, but they have a pronounced influence on the latter.

Example 12-7: Shock waves At a certain distance from a large explosion, the ambient air pressure and temperature are 100 kPa and 10 °C. When the leading edge of the blast wave passes by, the pressure rises suddenly to 350 kPa. Assuming that the conditions are locally one-dimensional, estimate the speed of travel of the blast wave and show that the phenomenon is irreversible. For the air, use $R = 287$ J/kg K and $\gamma = 1.4$.

SOLUTION The blast wave is shown in Fig. 12-12a moving from right to left at a speed V_w. It is also shown in Fig. 12-12b in which all velocities are written relative to axes moving with the wavefront. Relative to these axes, we may assume steady-state conditions.

For a unit area of flow, the continuity, momentum and energy equations may be written as

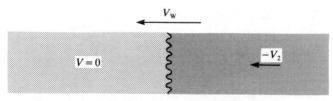

(a) Relative to stationary observer (V_2 is numerically negative)

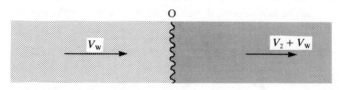

(b) Relative to moving observer

Figure 12-12 Flows close to a normal shock wave. (a) Relative to a stationary observer. (b) Relative to a moving observer.

$$\rho_1 V_w = \rho_2 (V_2 + V_w) \tag{12-53}$$

$$(p_1 - p_2) = \rho_1 V_w \{(V_2 + V_w) - V_w\} = \rho_1 V_w V_2 \tag{12-54}$$

and $$0 = c_p(T_2 - T_1) + \{\tfrac{1}{2}(V_2 + V_w)^2 - \tfrac{1}{2} V_w^2\} \tag{12-55}$$

Together with the equations of state on the two sides of the wavefront, namely $p_1 = \rho_1 R T_1$ and $p_2 = \rho_2 R T_2$, we have five equations in the five unknowns V_2, V_w, ρ_1, ρ_2 and T_2. With a little effort, these can be solved for the specified conditions to give:

$$V_2 = -339.5 \, \text{m/s} \qquad V_w = 598.0 \, \text{m/s} \qquad T_2 = 154.8 \, ^\circ\text{C}$$
$$\rho_1 = 1.231 \, \text{kg/m}^3 \qquad \rho_2 = 2.850 \, \text{kg/m}^3$$

The change in entropy between the states 1 and 2 can be obtained from Eq. (12-43). Using $c_p = \gamma R(\gamma - 1) = 1004.5 \, \text{J/kg K}$, we obtain $s_2 - s_1 = 55.3 \, \text{J/kg K}$. Since this is positive, the process is irreversible.

Comments on Example 12-7

1. Many unsteady-flow processes are irreversible because they disturb equilibrium conditions. The blast wave considered in this example is a strong normal *shock wave* in which the conditions change from state 1 to state 2 very quickly indeed—within a few molecular free paths, in fact. A strong discontinuity such as this involves a significant departure from equilibrium conditions.
2. If the above analysis is repeated assuming a reduced pressure, say 50 kPa, behind the shock, the continuity, momentum and energy equations will yield a solution, but the associated change in entropy will be negative. Since the conditions are adiabatic, this implies that the calculated solution is physically impossible. Negative shocks cannot exist in practice.
3. Weak shocks can approximate closely to reversible behaviour. In the limit, an infinitesimal shock is completely reversible. Sound waves may be regarded as an infinite succession of tiny shock waves, some positive and some negative.
4. The above analysis is based on an assumption that the flow is locally one-dimensional. In the open atmosphere, it would be more realistic to treat it as axi-symmetric or spherically symmetric. However, one-dimensional approximations are acceptable except close to the origin of the blast.

12-8 DIFFERENTIAL FORM OF THE UNIAXIAL ENTROPY EQUATION

The uniaxial entropy equation may be developed for control volumes of elemental length. Following closely the methods used for the continuity, momentum and energy equations, we write the entropy fluxes at the inlet and

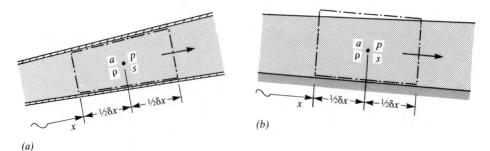

Figure 12-13 Elemental control volumes for uniaxial flows. (a) Pipe flow. (b) Channel flow.

outlet sections in Fig. 12-13 as

$$\dot{S}_1 = \dot{m}s - \tfrac{1}{2}\frac{\partial}{\partial x}(\dot{m}s)\,\delta x \qquad \dot{S}_0 = \dot{m}s + \tfrac{1}{2}\frac{\partial}{\partial x}(\dot{m}s)\,\delta x \qquad (12\text{-}56)$$

in which s should strictly be interpreted as a mean value defined by Eq. (12-23). By inspection the excess of the efflux over the influx is $(\partial/\partial x)(\dot{m}s)\,\delta x$.

The rate of change of entropy within the control volume is

$$\frac{\partial S_{cv}}{\partial t} = \frac{\partial}{\partial t}(s\rho a\,\delta x) \qquad (12\text{-}57)$$

Therefore, using $\dot{q}$ to denote the heat flux per unit length, the entropy equation (12-21) becomes

$$\frac{\dot{q}\,\delta x}{T} < \frac{\partial}{\partial t}(s\rho a\,\delta x) + \frac{\partial}{\partial x}(\dot{m}s)\,\delta x \qquad (12\text{-}58)$$

After dividing by δx and allowing it to approach zero, we obtain

$$\boxed{\frac{\dot{q}}{T} < \frac{\partial}{\partial t}(s\rho a) + \frac{\partial}{\partial x}(\dot{m}s)} \qquad (12\text{-}59)$$

Range of validity The comments given for the general entropy equation (12-21) also apply to its differential form. In addition, the flow must be uniaxial at all sections. The following simplified forms are valid for adiabatic and steady flows respectively.

Adiabatic

$$0 < \frac{\partial}{\partial t}(s\rho a) + \frac{\partial}{\partial x}(\dot{m}s) \qquad (12\text{-}60)$$

Steady

$$\frac{\dot{q}}{T} < \dot{m}\frac{\partial s}{\partial x} \qquad (12\text{-}61)$$

PROBLEMS

1 Air ($R=287\,\text{J/kg K}$, $c_\text{p}=1004\,\text{J/kg K}$) flows steadily through the machine shown in Fig. 12-14. If the pressure and temperature at the flow sections 1 and 2 are $p_1 = 1\,\text{MPa}$, $T_1 = 600\,\text{K}$ and $p_2 = 100\,\text{kPa}$, $T_2 = 300\,\text{K}$, determine the specific entropy difference $s_1 - s_2$ and hence decide whether the machine is a turbine or a compressor if the flow is adiabatic.

[35.1 J/kg K, compressor]

2 Show that the isothermal flow described in Problem 3 at the end of Chapter Eleven is irreversible.

3 A perfect gas is compressed in a polytropic process for which p/ρ^N is constant. Show that, if the process is adiabatic, $N > \gamma$.

4 A perfect gas is expanded polytropically and adiabatically. Show that $N < \gamma$.

5 In a certain power plant, the working fluid circulates continuously. Heat is supplied at only one stage in the process and the temperature of the source if 600 °C. Heat is rejected at only one stage and the temperature of the sink is 50 °C. Determine the maximum possible shaft power if the heat supplied to the steam is 750 kJ/kg. Also determine the maximum possible shaft power in a non-cyclic plant in which the enthalpies of the fluid at inlet and outlet are 3 MJ/kg and 2 MJ/kg respectively. The heat fluxes and the temperatures of the source and sink are the same as those for the first plant. Neglect changes in kinetic and potential energy.

[472 kJ/kg, 1472 kJ/kg]

6 Figure 12-15 depicts two streams of the same liquid meeting at a pipe junction and mixing at constant pressure. By considering the changes in entropy in the two streams, show that the process is irreversible.

7 Pretend that the process in Problem 6 can exist in reverse. Use it to devise a PMM2, that is a cyclic machine which delivers work and receives an equal amount of heat from a single source. Hence verify that the reversed process disobeys the second law of thermodynamics (12-1).

8 A 2 mm diameter water droplet ($\rho = 1000\,\text{kg/m}^3$, $\sigma = 0.074\,\text{J/m}^2$) falls freely under gravity. Show that it is not possible for the droplet to break up

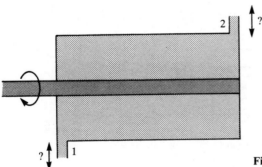

Figure 12-14 Turbine or compressor?

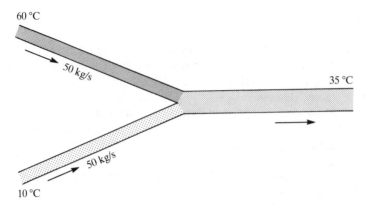

60 °C

50 kg/s

35 °C

50 kg/s

10 °C

Figure 12-15 Mixing streams with different temperatures.

adiabatically into eight 1 mm diameter droplets while falling through a height of 20 mm.

$$[h_{min} = 22.6 \text{ mm}]$$

9 Figure 12-16 depicts a surge wave in a rectangular section channel. Show that the conditions are physically possible only if $V_w > V_1$ and $V_1 > V_2$. (*Hint*: Consider flow relative to the wavefront and prove that this is from low depth to high depth.)

10 Repeat Example 12-1 (p. 337) assuming that the momentum and kinetic energy flux coefficients are $\beta_1 = 1.02$, $\beta_2 = 1.10$, $\alpha_1 = 1.06$, $\alpha_2 = 1.33$.

$$[14.1 \text{ m}^3/\text{s}]$$

11 Figure 12-17 shows a steady, adiabatic flow of a liquid induced by moving one flat plate parallel to another. Show that the temperature of the fluid increases downstream and hence prove that the flow is irreversible.

12 Sufficiently far from the source of the explosion described in Example 12-7, the induced pressure changes become extremely small. For this case, show that equation (12-54) may be written as $\delta p = -\rho c \, \delta V$ where c denotes the local speed of the wavefront. Hence show that the wavespeed is equal to $(\partial p/\partial \rho)^{1/2}$. Also use Eq. (12-55) to show that $\partial p/\partial T = \rho c_p$ and hence demonstrate that the wavefront propagates isentropically (and therefore reversibly).

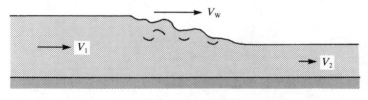

V_w

V_1

V_2

Figure 12-16 Flood surge in an open channel.

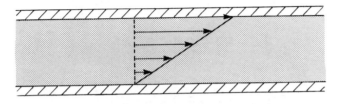

Figure 12-17 Flow between parallel plates.

FURTHER READING

Cravalho, E.G. and Smith, J.L. (1981) *Engineering Thermodynamics*, Pitman.
Haywood, R.W. (1980) *Equilibrium Thermodynamics*, Wiley (out of print).
Reay, D.A. and MacMichael, D.B.A. (1987) *Heat Pumps*, 2nd edn, Pergamon.
Turner, J.S. (1979) *Buoyancy Effects in Fluids*, Cambridge University Press.
Yih, C.-S. (1980) *Stratified Flows*, Academic Press.
Zemansky, M.W. and Dittman, R.H. (1981) *Heat and Thermodynamics*, 6th edn, McGraw-Hill.

A.1 PROPERTIES OF WATER AT A PRESSURE OF 1 bar

T	ρ	K	μ	ν	σ	u[†]	c_v	c_p	γ	s[†]	λ	Pr
°C	kg/m³	MPa	µPa s	mm²/s	N/m	kJ/kg	kJ/kg K	kJ/kg K	—	kJ/kg K	mW/m K	—
0	999.8	1963	1793	1.793	75.65	0.0	4.213	4.218	1.001	0.000	561.0	13.5
10	999.7	1986	1306	1.306	74.22	41.9	4.189	4.192	1.001	0.151	580.0	9.44
20	998.2	2192	1002	1.004	72.74	83.8	4.157	4.182	1.006	0.296	598.5	7.00
30	995.7	2267	797.6	0.798	71.20	125.7	4.117	4.178	1.015	0.436	615.6	5.41
40	992.2	2320	653.1	0.658	69.60	167.4	4.073	4.179	1.026	0.572	630.7	4.33
50	998.0	2352	546.9	0.554	67.95	209.2	4.028	4.181	1.038	0.703	643.6	3.55
60	983.2	2366	466.3	0.474	66.24	251.1	3.982	4.184	1.051	0.830	654.3	2.98
70	977.8	2366	403.8	0.413	64.49	272.0	3.931	4.189	1.066	0.954	663.0	2.55
80	971.8	2349	354.2	0.364	62.68	355.9	3.875	4.196	1.083	1.075	669.8	2.22
90	965.3	2320	314.2	0.325	60.82	376.9	3.805	4.205	1.105	1.192	675.0	1.96
99.63	958.6	2282	282.7	0.295	58.99	417.4	3.724	4.216	1.132	1.303	678.7	1.76
100[‡]	958.1	2281	281.5	0.294	58.92	419.0	3.721	4.216	1.133	1.307	678.8	1.75

[†] Arbitrary datum for u and s chosen at the triple point.

[‡] At $T = 100$°C, the pressure of saturated water is 101.325 kPa.

The assistance of ESDU Ltd in the compilation of this table is gratefully acknowledged.

A.2 PROPERTIES OF AIR AT A PRESSURE OF 1 bar

T °C	ρ kg/m^3	μ μPa s	ν mm^2/s	$u^\dagger$ kJ/kg	h kJ/kg	c_v kJ/kg K	c_p kJ/kg K	γ —	$s^\dagger$ kJ/kg K	λ mW/m K	Pr —
0	1.275	17.2	13.5	196.1	274.5	0.715	1.002	1.401	6.78	23.9	0.721
10	1.230	17.6	14.4	203.3	284.5	0.716	1.003	1.401	6.81	24.6	0.720
20	1.188	18.1	15.3	210.5	294.6	0.716	1.003	1.401	6.85	25.3	0.718
30	1.149	18.6	16.2	217.7	304.7	0.717	1.004	1.401	6.88	26.0	0.717
40	1.112	19.1	17.2	224.8	314.7	0.717	1.004	1.400	6.91	26.7	0.716
50	1.078	19.5	18.1	232.0	324.8	0.718	1.005	1.400	6.95	27.4	0.715
60	1.046	20.0	19.1	239.2	334.9	0.718	1.005	1.400	6.98	28.1	0.714
70	1.015	20.4	20.1	246.4	344.9	0.719	1.006	1.399	7.01	28.8	0.713
80	0.986	20.9	21.2	253.6	355.0	0.719	1.006	1.399	7.04	29.5	0.712
90	0.959	21.3	22.2	260.9	365.1	0.720	1.007	1.398	7.06	30.2	0.711
100	0.934	21.7	23.3	268.1	375.2	0.721	1.008	1.398	7.09	30.8	0.710

† Arbitrary datum for u and s chosen at absolute zero.

The assistance of ESDU Ltd in the compilation of this table is gratefully acknowledged.

INDEX

Absolute temperature, 23, 330
Absolute viscosity, 31, 84
Accelerating flow, 192, 194, 212, 247, 271, 313, 320
Acceleration, 10, 145, 149, 227, 295
Adiabatic process, 298, 308, 315, 326, 332, 336, 344, 348
Aero-engine, 50, 310
Aerofoil, 85, 257
Aerogenerator, 2, 137
Aircraft wing, 85
Airport runway, 117
Angular momentum, 216, 290
Angular momentum flux, 218, 225, 231
Angular velocity, 103, 117, 137, 216, 223, 277
Apparent stress, 95
Archimedes' principle, 60, 212
Axial flow pump, 233
Axial flow turbine, 233, 270

Barometer, 63
Bathtub vortex, 282
Bernoulli equation, 18, 122, 198, 237, 253, 259, 285, 305
  extended, 246, 264
Bernoulli sum, 239, 253, 264, 306
  variation across streamlines, 278, 283, 289
  variation along streamlines, 238, 242, 253

Bicycle pump, 46
Blasius, 106, 118
Blood flow, 2, 117, 166, 267
Body force, 50, 179, 205, 218
Boiler, 165
Boiling point, 24, 26, 330
Borda–Carnot loss, 264
Boundary layer, 4, 85, 110, 288
  growth, 88
  separation, 87, 243, 294
  thickness, 88, 99, 104
  transition, 89
Bourdon gauge, 66
Broad-crested weir, 269
Bubbles, 1, 26, 36, 45, 81, 149, 317
Buckingham (Π) groups, 123, 130, 148
Bulk modulus of elasticity, 28, 44, 47, 66
Buoyancy, 60, 74, 81, 212, 287, 343
Bursts of turbulence, 84

Canal, 79
Capillaries, 117
Capillarity, 35, 141, 179
Capillary wave, 144
Capsular flow, 176
Cavitation, 25, 140, 203, 267, 295, 317
Centre of pressure, 57, 58, 69
Centrifugal blower, 162
Centrifugal pump, 149, 158, 224, 235, 265, 283, 295, 315, 323